# INDUSTRIAL ELECTRON ACCELERATORS AND APPLICATIONS

# INDUSTRIAL ELECTRON ACCELERATORS AND APPLICATIONS

**E. A. Abramyan**
*Institute for High Temperatures*
*USSR Academy of Sciences*

**HEMISPHERE PUBLISHING CORPORATION**
A subsidiary of Harper & Row, Publishers, Inc.
Washington New York London

**DISTRIBUTION OUTSIDE NORTH AMERICA**
**SPRINGER-VERLAG**
Berlin Heidelberg New York London Paris Tokyo

**INDUSTRIAL ELECTRON ACCELERATORS AND APPLICATIONS**


Originally published as, Promyshlennye uskoriteli elektronov, by Energoatomizdat, Moscow, 1986.

1 2 3 4 5 6 7 8 9 0 EBEB 898

This book was set in Times Roman by Hemisphere Publishing Corporation. The editor was Victoria Danahy; the cover designer was Don Wittig; the production supervisor was Miriam Gonzalez; and the typesetter was Vera Kishinevsky.
Edwards Brothers, Inc. was printer and binder.

**Library of Congress Cataloging-in-Publication Data**

Abramian, E. A. (Evgeniĭ Aramovich)
Industrial electron accelerators and applications.

Translation of: Promyshlennye uskoriteli elektronov.
Includes bibliography and index.
1. Electron accelerators—Industrial applications.
2. Ionizing radiation—Industrial applications.
3. Electron beams—Industrial applications. I. Title.
QC787.E39A3713 1988 539.7′3712 87-19751
ISBN 0-89116-694-7 Hemisphere Publishing Corporation

**DISTRIBUTION OUTSIDE NORTH AMERICA:**
ISBN 3-540-18844-4 Springer-Verlag Berlin

# CONTENTS

# PREFACE

Advances in nuclear engineering and technology in the middle of this century greatly increased our knowledge of ionizing radiations, including their useful and detrimental effects on affected materials, and promoted the development of novel powerful radiation sources. Following laboratory studies of radiochemical processes and the realization of their efficiency in the 1950s, widescale industrial application of radiation technology began. Initially it was done on the basis of isotopic sources, but within 5 to 8 years, use was made of electron accelerators specially designed for these purposes. If, formerly, accelerated electrons found application only in limited equipment (for example, in x-ray machines), at present there are several virtually independent fields of application using accelerated particles with specific equipment and individual technique.

The beams of accelerated particles are particularly important in researching the structure of matter, for example, the structure of elementary particles and the nature of nuclear forces. The problems arising from this field of knowledge have fostered the advent and development of electron accelerators rated initially at minor energies of the order of 1 million electron volts (1 MeV) and then at medium and high energies up to $10^6$ MeV.

For applied purposes use, as a rule, is made of the electron accelerators featuring the energies within a few tens of MeV. Referred to as "industrial electron accelerators" are the electron-beam accelerators installed at industrial enterprises and used for irradiation of various materials, primarily in radiochemical and sterilization processes. Sometimes the term "industrial accelerators'' also applies to accelerators used for flaw detection and for medical purposes. Hereinafter we shall use the term in its latter, broader meaning, even though ''industrial'' may not be quite applicable to medical radiation installations. (Closely approximating the installations in question, both in performance characteristics and in some design features, are the installations used for electron-beam treatment of materials where electron energies do not exceed a few tens of keV.)

This book reviews the problems concerned both with generation of accelerated particles and with application of electron beams for exerting chemical, thermal, and other effects on matter. The book is designed for the experts specializing in designing and servicing electron accelerators or developing new production processes with the use of accelerated particles. As the number of people working with accelerators rapidly grows, some of them may have inadequate knowledge of the electron physics. This book provides some basic information on beams of accelerated particles, methods of electron acceleration, and processes of interaction of electrons with matter. The book describes the main types of industrial accelerators and technological installations. In addition to the electron accelerators in wide industrial use at present, descriptions of a number of experimental and pilot-scale accelerators are also given. The sections devoted to radiation technology cover both the well-known and commonly used technology, as well as processes that have only recently gone beyond the experimental stage or are still under development.

*E. A. Abramyan*

CHAPTER

# ONE

# INTRODUCTION

## 1.1 BRIEF REVIEW OF ELECTRON ACCELERATOR DEVELOPMENT AND APPLICATION OF RADIATION EFFECT

Investigations on the effect of radiation on matter began at the end of the last century. After the discovery of bremsstrahlung radiation by Roentgen in 1895 and of radioactivity by Becquerel in 1896, experiments were conducted with irradiation of glass, porcelain, water, and other materials. Then radiation began to be used for roentgenoscopy and flaw detection. At the beginning of the 20th century, radiation effects were studied on the basis of radiation of radon.

The next stage in application of radiation effects began with the advent of artificial radiation sources—accelerators of particles with an energy of a few hundred kiloelectronvolts. In 1925, Coolidge made use of a voltage source of up to 250 kV for acceleration of electrons and subjected a number of substances to experimental irradiation. In 1927, using a three-stage accelerating tube, he was the first to succeed in obtaining an electron beam of 900 keV extracted into the atmosphere. During 1930 to 1931, accelerating equipment based on the Tesla transformer provided acceleration of electrons to 2 MeV and protons to an energy of approximately 1

MeV. In the early 1930s, Van de Graaff designed an electrostatic generator rated at a voltage of 1.5 MV and a current of 25 uA, Cockroft and Walton developed a cascade generator of 0.7 MeV and a current of 10 mA, Lawrence invented a cyclotron for accelerating hydrogen ions to an energy of 1.25 MeV, and finally a linear resonance accelerator was developed. In the 1920s, the principles of inductive acceleration of particles were formulated. (However, the first actually operative betatron was designed by Kerst in 1941.) In fact, within a few years a number of fundamental schemes of electron acceleration were proposed that served as a basis for further development of the electron accelerating technology in the subsequent decades.

This period of time was also characterized by a rapid development of electron-wave optics, experimental treatment of metals by means of focused electron beams, advancement of vacuum technique, and other allied fields of science and technology without which further improvement of the electron accelerators could hardly have been possible.

The most radical achievement in particle acceleration physics in the following decades was the discovery of the principle of phase stability by Soviet physicist Veksler in 1944 (and, independently in the United States, by MacMillan in 1945). This principle served as a basis for the development of a number of new types of accelerators, such as the synchrotron, linear accelerator, and microtron. At about the same time, a high-voltage pulsed x-ray technique with field-emission cathodes originated. The first pulsed installations for roentgenography of fast processes were invented in the early 1940s.

It should be noted that even the comparatively low-current early accelerators ensured qualitatively new capabilities as compared with the radioactive sources. For example, while a flux of alpha particles from 1 g of radium amounts to $3.7 \times 10^{10}$ particles per second in a round solid angle of $4\pi$, the quantity of the alpha particles obtained by the use of the accelerator at a current of 10 $\mu$A is $3.125 \times 10^{13}$ particles per second and they travel in one direction and form a flux of a limited cross section. The average current of an electron beam in modern industrial electron accelerators is 0.1 to 0.5 A which corresponds to $10^{17}$ particles per second.

In the late 1920s and early 1930s, a number of experiments were performed to study the effects of irradiation on the genetic mechanism of an organic cell with a view to obtaining new varieties of plants. In the 1930s, a very active research was started to study the germicidal effect of radiation. A proposal was made to carry out sterilization of food. This process was subjected to intense investigations after the results of successful experiments on radiation treatment of food had been published in 1947. The late 1940s were noted for the discovery of the processes of cross-linking polyethylene and rubber. These processes began to be actively investigated in 1950s. In the middle of the 1940s, research of atomic problems resulted in organization of scientific bodies carrying on wide-scale investigations on radiochemistry.

The third stage in the history of research and application of radiation dates back to the 1950s and includes wide-scale industrial use of radiation for technological and other applied purposes. By that time, powerful radiation sources and new radiation meters had been developed, and the need arose to study the effect of intense radiation on construction materials. Some new processes were discovered and used as a basis for the industrial radiation technology.

Irradiation sterilization of medicinal facilities and modification of polyethylene should be considered as the beginning of use of radiation as a technological instrument. For instance, in the second half of the 1950s, industrial irradiation plants were devised on the basis of isotopic sources in the United States, Denmark, and some other countries. Since the 1960s, electron accelerators have been used for this purpose (e.g., by the firm Research Establishment DAE in Denmark and SRTI/CARI in France). Starting from 1956, articles of irradiated polyethylene have been manufactured by General Electric Company in the United States. In the 1950s, electron beams found application in therapy. During the same period, isotopic sources of radiation found wide application for sterilization of insects to reduce their population over a vast area in the state of Florida (U. S.). At the same time, the first pilot-scale and then commercial installations based on the electron accelerators and intended for radiation hardening of varnish and paint coatings and for some other industrial processes had been developed.

In the mid-1950s, electron beams with an energy of 0.03 to 0.05 MeV found wide application for heat treatment of metals, primarily in atomic, aviation, and aerospace engineering. Electron accelerators are being developed for applied purposes in two ways: (1) for radiation technological installations with an energy of 0.15 to 10 MeV and a power of the order of 10 kW, and (2) for flaw-detection and radiation-therapy apparatuses rated at an energy of 35 to 45 MeV and a materially lower power. Industrial radiation installations should meet the requirements of high reliability, efficiency, ease of servicing, and low cost. Development of electron accelerators with energies of about 1 MeV is traced in Fig 1.1. Though the curves depicted indicate one of the existing types of the industrial accelerators—dynamitron (Radiation Dynamics, U.S.)—they also present an approximate illustration of rates of improvement of the parameters of the other types of electron accelerators. At that time, some other companies in the United States were engaged in developing industrial accelerators: an insulating-core transformer was developed by the High Voltage Engineering Corp., a shock-excited tuned transformer by General Electric, and accelerators for hardening of coatings by the Ford Motor Co. Ion Physics Corp. (a branch company of High Voltage Engineering Corp.) was one of the pioneers that started commercial production of the sources of high-power pulsed bremsstrahlung radiation. In the late 1960s, a new company, Energy Science, Inc., specialized in manufacturing electron accelerators rated at energies of 150 to 300 keV and in fabrication of production flow lines with the use of such accelerators. In France, the Sames Company (which later became the Tunzini-Sames Company) developed rotor generators. In England, the Vicers Company Ltd. started production of linear accelerators for industrial application. Betatrons and other accelerators were produced by the Brown Boveri Company in Switzerland. The total power of generated radiation in the industrial plants based on electron accelerators amounted to 300 kW in 1967, 800 to 1000 kW in 1970, and now exceeds 20,000 kW. The annual increase of the radiation power of the accelerators is 15 to 20 percent. Figure 1.2 demonstrates the progress of industrial radiation facilities in Japan.

In the Soviet Union, accelerators for industrial use were initially developed in the Nuclear Physics Institute in Novosibirsk

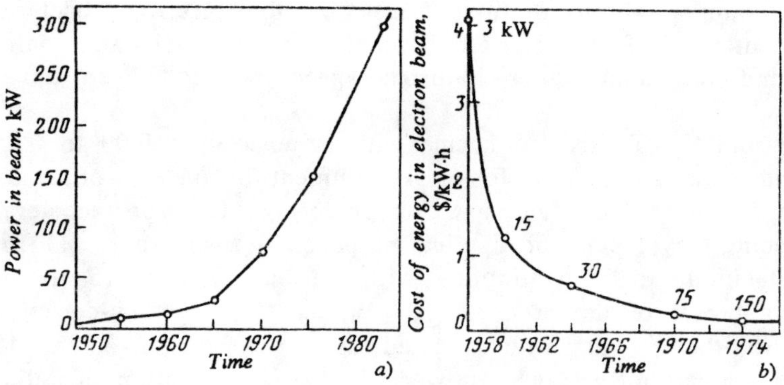


**Figure 1.1** Development of industrial accelerators with energy of approximately 1 MeV. *a*) Progress of power of dynamitron industrial accelerators. *b*) Decrease of cost of energy of electron beam.

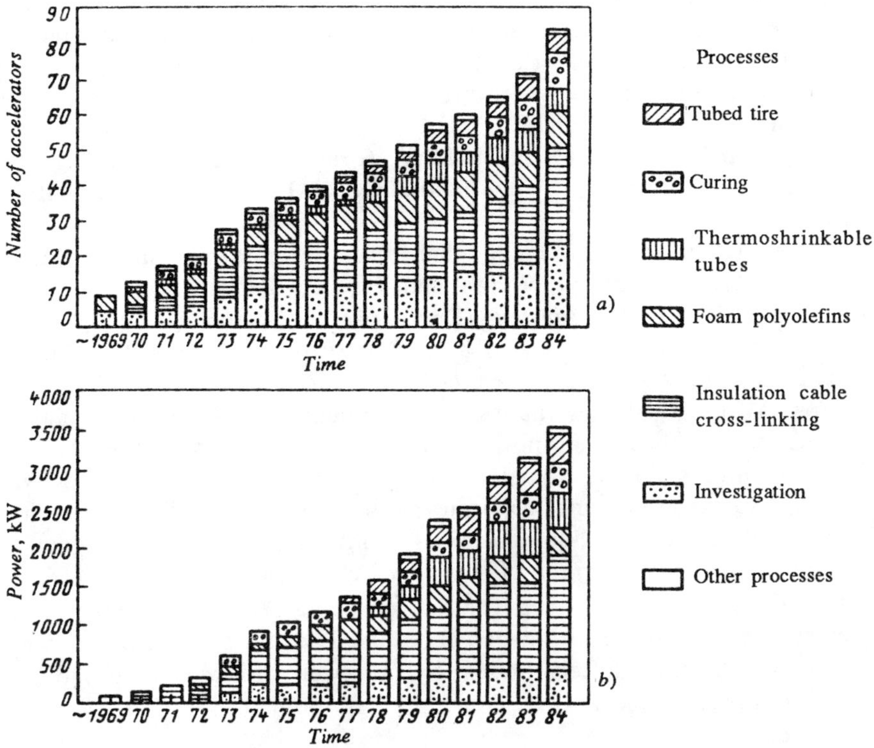


**Figure 1.2** Development of industrial accelerators in Japan. *a*) Number of accelerators. *b*) Total power of accelerators.

(type ELT-1.5, 1964), in the Electrophysical Apparatus Institute (NIIEFA) in Leningrad (type Electron-3, 1970), in the Electronics Institute in Moscow (type ELU 5-1.5, 1965), in the Polytechnical Institute in Tomsk, and at the Moscow transformer plant (betatron, 1950s). Industrial use of electron accelerators in the USSR for radiation technology began in 1968 when a facility for irradiation of cable insulation was put into service. Production of thermoplen (thermally shrinkable film) began in 1975. Radiochemical treatment of fabrics has been performed since 1967.

Industrial electron accelerators are presently also manufactured in Japan (Nissin Company-HV and others), in France (CGR-MeV), in F.R. Germany (Polymer Physik GMBH), and in Switzerland (E. Haefaly). Industrial-purpose electron accelerators are also being developed in China, Hungary, Czechoslovakia, and some other countries.

In addition to cross-linking of polyethylene and sterilization of medicines, irradiation facilities are finding ever increasing use for such processes as hardening of varnish and paint coatings and graft polymerization. The total number of electron and ion accelerator units used for medical therapy exceeds 1500. These accelerators are manufactured by many firms, including Varian (U.S.), Radiation Dynamics (U.S.), Philips/MEL (U.K.), SHIN/EMI (U.K.), Mitsubishi (Japan), Toshiba (Japan), CGR-MeV (France), Siemens (F.R. Germany), Brown Boveri (Switzerland), CGR/AECL (Canada), NIIEFA (USSR).

Flaw-detection equipment that uses bremsstrahlung radiation is under development by Varian and Hewlett Packard (U.S.); Brown Boveri (Switzerland); Andrex (Denmark); Gilardoni (Italy); Müller (F.R. Germany); Pantak (U.K.); NIIEFA, Burevestnik, Spektr, Aktyubrentgen, and Mosrentgen (USSR), and others.

As a result of continuous advancement of high-voltage pulsed installations, it has been possible to develop some portable (1 to 2 kg) apparatuses for flaw detection, for example, types IRA, RINA, MIRA (Burevestnik) and superpower bremsstrahlung generators, the largest of which is installation Aurora (U.S.) with an electron energy of 15 MeV, current of 1.6 MA, and a pulse duration of ~120 ns (with the mass of the installation being 7.500 t). The Aurora generator and other high-power pulsed radiation sources were developed in the United States by Maxwell, Physics International, and Ion Physics Corp.

## 1.2 TYPES OF RADIATION AND UNITS OF MEASUREMENT

Various kinds of radiation find industrial application:

- fluxes of particles (electrons, protons, neutrons, and heavier particles);
- shortwave electromagnetic and microwave radiation;
- luminous and thermal radiation;
- coherent electromagnetic radiation (laser);
- synchrotron radiation (under development).

This book reviews, basically, radiations that are generated by electron accelerators and that find industrial use, i.e., electron fluxes and shortwave electromagnetic radiation (bremsstrahlung).

To measure ionization, which is one of the basic radiation properties, it is necessary to determine the amount of energy released when the radiation passes across the affected matter. For measuring the absorbed radiation dose (in the SI system), use is made of the unit gray (Gy), which is the absorption of energy of 1 J in 1 kg of matter:

$$1 \text{ Gy} = 1 \text{ J/kg}$$

Correspondingly, the capacity of technological radiation installations in kilograms per hour (kg/h) will be equal to 3.6 $PD^{-1}\eta$, where $P$ is the power of the radiation source in kilowatts (kW), $D$ is the absorbed radiation dose in megagrays (MGy), and $\eta$ is the radiation utilization factor (in fractions). Until very recently, the absorbed dose was measured in rads (rd) where 1 rd corresponds to absorption of energy of 100 erg in 1 g of matter:

$$1 \text{ Gy} = 100 \text{ rd}$$

The rate of absorbed dose buildup is referred to as the dose rate. The measurement unit of the latter is 1 Gy/s. For determination of ionization effect of bremsstrahlung or flux of gamma quanta in air, the SI system contemplates an exposure dose expressed in coulombs (C) per kilogram: 1 C/kg equals an exposure dose of bremsstrahlung or gamma radiation at which, in dry atmospheric air of 1 kg mass, ionization takes place where ions carry opposite electric charges equal to 1 C. The exposure dose unit roentgen (R), which was accepted earlier, relates to coulombs per kilogram as:

$$1 \text{ R} = 2.58 \times 10^{-4} \text{ C/kg}$$

With the radiation energy lying within 3 MeV, 1 R = 0.88 rd.

For determining the aftereffects of irradiating living organisms, use is made of the unit *sievert* (Sv) which is identified as an equivalent dose and corresponds to the absorption of one joule per kilogram in living tissue:

$$1 \text{ Sv} = \frac{1 \text{ J/kg}}{Q}$$

The factor $Q$ determines the level of aftereffects of irradiation of living organisms: $Q \geq 1$. The unit used before was rem (medical roentgen-equivalent:

$$1 \text{ rem} = 10^{-2} \text{ Sv}$$

For measuring the radioactivity of isotopic sources in the SI, the *becquerel* (Bq) is adopted, which corresponds to activity of matter with one radioactive decay per second: 1 Bq = 1 decay/s. The ratio between this unit and the *curie* (Ci) is:

$$1 \text{ Ci} = 3.7 \times 10^{10} \text{ Bq}$$

1 curie corresponds to $3.7 \times 10^{10}$ decays per second.

The most widespread unit of measurement of energy of accelerated elementary particles is the *electronvolt* (eV). One electronvolt is defined as the energy acquired by an electron or a particle of any mass with unit charge $z = 1$ passing through an electric field potential difference of 1 V. The relationship to other units of measurement is

$$1 \text{ eV} = 1.6 \times 10^{-19} \text{ J} = 1.6 \times 10^{-12} \text{ erg}$$

## 1.3 BEAMS OF ACCELERATED CHARGED PARTICLES: GENERAL CONCEPTS

A beam of accelerated charged particles is commonly understood as a flux of particles traveling in one direction in approximately parallel paths. The energy spread and the ratio between the transverse and longitudinal velocities do not exceed a few percent. The most typical medium of propagation of electron beams is a vacuum of $10^{-5}$ to $10^{-3}$ Pa. The beam of particles of one kind may contain different kinds of interacting particles. If this interaction becomes so pronounced that the beam particles acquire great transverse velocities, such a beam actually loses the main distingushing feature mentioned above. This is particularly characteristic of beams in high-pressure media, although a limited distance can be covered by the beam through dense media.

For considering the processes of acceleration of elementary particles and their interaction with matter, it is essential that at least the following characteristics of the particles be known: rest mass $m_0$, charge $ze$, and kinetic energy $W$ (or total energy $\varepsilon$). The particles moving in electromagnetic fields are affected by the following forces:

1. $F_E = zeE$ in an electric field with intensity $E$
2. $F_B = zec^{-1}vB$ (Lorentz force) in a magnetic field with induction $B$, where $v$ and $c$ stand for the particle charge in terms of units $e = 1.6 \times 10^{-19}$ C $= 4.8 \times 10^{-10}$ electrostatic units.

In the course of movement through an electric field, the particle energy tends to vary. Increments or decrements of energy in a potential field are directly proportional to the difference of potentials at the beginning and at the end of the particle trajectory: $\Delta W = ze\Delta U$.

The velocity of an electron with an energy of ~1 MeV or more very much approximates the velocity of light. In this case, the movement of the particle can be described with formulas of relativistic mechanics. The total energy and mass of the particle are interrelated by the ratio $\varepsilon = mc^2$. The resting energy of an electron $\varepsilon_0 = 0.511$ MeV, and that of a proton is 938 MeV. (It will be recalled that the zero rest mass of an electron $m_0 = 9.11 \times 10^{-28}$ g and the velocity of light, $c = 3 \times 10^{10}$ cm/s.

The following is the universally adopted notation:

relative particle velocity $\beta = v/c$
total energy $\varepsilon = \varepsilon_0 + W$
relativistic factor $\gamma = \varepsilon/\varepsilon_0 = (\varepsilon_0 + W)\,\varepsilon_0^{-1}$

$\beta$ is related to $\gamma$ by the ratio

$$\beta = \sqrt{1 - \gamma^{-2}} = \sqrt{1 - \varepsilon_0^2\,(\varepsilon_0 + W)^{-2}}$$

The relationship between the relative velocity $\beta$ of the electrons and kinetic energy $W$ is shown in Fig. 1.3. Kinetic energy $W$ of the relativistic particle is related to its pulse $p = mv$ by the formula

$$W = \sqrt{c^2p^2 + \varepsilon_0^2} - \varepsilon_0$$

When an electron moves in a magnetic field, the centrifugal force $mv^2R^{-1}$ at every point of the trajectory is equal to the Lorentz force: $mv^2R^{-1} = ec^{-1}vB$. Consequently, $p = ec^{-1}BR$ where $R$ is the radius of curvature of the electron trajectory. Thus, the above-stated relationship for $W$ will take the form

$$W = \sqrt{9\;B^2R^2 + 0.511^2} - 0.511$$

where $W$ is expressed in MeV, $B$ in T, and $R$ in cm (see also Fig. 1.3). At $W \gg 1$ MeV, $W = 3\,BR$. The total energy, kinetic energy, and particle momentum may be written as $\gamma m_0c^2$, $(\gamma - 1)\,m_0c^2$, and $\beta\gamma m_0c$.

Returning to the discussion of the forces acting on elementary particles in electric and magnetic fields, it might be well to point out that for curving the trajectories of relativistic particles, use is usually made of a magnetic field. This is done because, in fields realized approximately with a similar degree of technological difficulties, the Lorentz force is far in excess of the force of an electric field. Indeed, $F_B/F_E = 300\,\beta\,B/E$ ($B$ is expressed in terms of T, and $E$ in MV/m). For example, for electrons with $W = 1$ MeV and for fields with $B = 1.5$ T and $E = 15$ MV/m, $F_B/F_E \cong 28$.

In addition to the forces of the external electromagnetic field, the beam particles are affected by the forces of the intrinsic electric and magnetic fields of the beam ($F_e$ and $F_B$). Beginning with a certain density of particle in the beam $N$ (particles/cm$^3$), the said forces may exert an essential influence on movement of the beam particles. Coulomb force $F_e$, which tends to separate like-sign charges, may bring about an increase of the cross section of the particle flux. The magnetic construction force $F_B$ acts in the opposite direction. Thus, in a beam of bunched particles moving at relative velocity $\beta$ and possessing uniform density $N$ and a round section with radius $a$, the intrinsic field will feature the maximum characteristics on the beam surface; the intrinsic electric field intensity $E_e = 2\pi\, eNa$ and magnetic field induction $B = 2\pi\, eN\beta a$ (Fig. 1.4*a*). The total force applied to the particle: $F_e - F_B = eE_e - e\,\beta B = 2\pi\, e^2Na\,(1 - \beta^2) = F_e\gamma^2$. For example, for electrons with an energy of $W$ = 1 MeV, $\gamma \cong 3$, and the Coulomb repulsive force $F_e$ is attenuated approximately by one order.

The inherent space charge has a pronounced effect on the longitudinal motion of the beam. In the case of acceleration of electrons in the external electric field, the intrinsic charge attenuates the accelerating field near the cathode and imposes the restriction on the beam current. This restriction for the current between the cathode and anode, which are essentially two infinite planes, is determined by the Langmuir law, i.e., $j = 2.3 \times 10^{-6}\, U^{3/2}$ where $j$ is the limiting density of the current in A/cm$^2$ and $U$ is the potential difference between the cathode and anode (in volts). As far as nonrelativistic particles are concerned, the limiting current that can flow in the given electron-optical system $I = \mathcal{P}U^{3/2}$ where $\mathcal{P}$ is the perveance whose value is dependent upon the geometry of the system. In the case of relativistic energies the nature of the current-voltage relationship changes: at $\gamma \gg 1$, $I \cong \mathcal{P}U$.

The potential inside the electron beam depends on the density of the space charge. In the case of similar current and energy of the beam particles, the difference of potentials in the center on the edge of the beam will also be dictated by its dimensions and geometry (Fig. 1.4*b*). This potential difference is much less in sheet and hollow-ring beams. Correspondingly, sheet and hollow-ring beams allow a greater current to be passed, as compared to a cylindrical beam.

Appreciable effects on the motion of the particles may be produced by ion compensation of the beam space charge. Ionization of residual gas by the beam electrons gives rise to positive ions and slow electrons. Due to a prompter escape of the generated electrons from the beam, negative charge of the beam may be partially or entirely balanced out by the ions. This results in attenuation of the electrostatic repulsive force. The compensation time $\tau$ of the beam charge in a medium with pressure $P$ can be approximately determined by the ratio $\tau = 0.04\, P^{-1}$ where $\tau$ is expressed in microseconds ($\mu$s) and $P$ in pascals (Pa). Thus, in a technological vacuum of $10^{-3}$ to

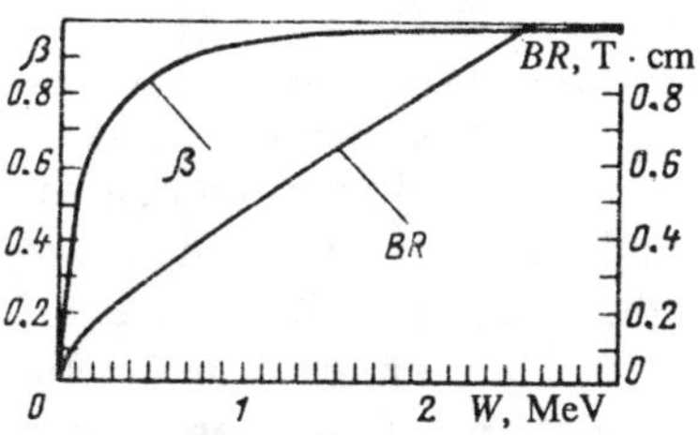


**Figure 1.3** Relative electron velocity β and parameter $BR$ versus kinetic energy $W$.

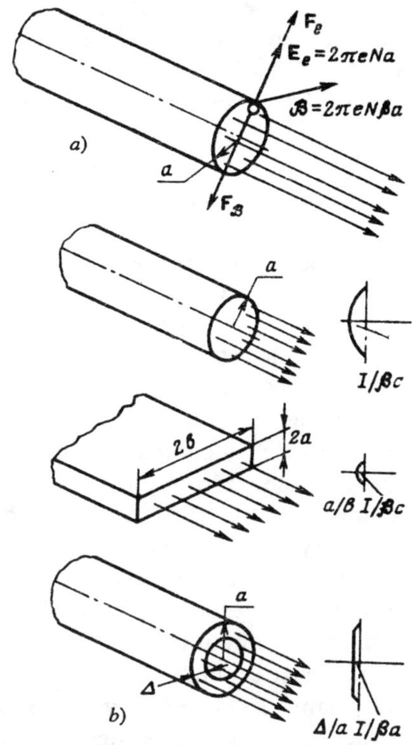


**Figure 1.4** Intrinsic fields of beam. *a*) Intrinsic electric field intensity $E_e$ and intrinsic magnetic field induction $\mathcal{B}$ of a round electron beam with uniform density $N$ (particles/cm³). $r$ = distance to beam axis; $F_e$ and $F_{\mathcal{B}}$ = forces produced by intrinsic fields of electron beam. *b*) Distribution of intrinsic electric field of cylindrical, sheet and hollow-ring beams. Vertical dimension of beam is plotted on ordinate axis and potential of inherent charge of electrons is plotted on abscissa. Potential at beam edge is taken to be zero. $I$ = total beam current; β = relative velocity of electrons.

$10^{-4}$ Pa in a lapse of 10 to 100 μs after emission of the beam, the ion density in the beam area reaches the same order of magnitude as the density of the electrons.

Distribution of the electrons over the beam cross section depends both on the performance characteristics of the electron injector and on the transport conditions. Considered to be most typical is the gaussian (or normal) distribution. In some cases, the beam features a tubular structure or is broken into individual jets. As a rule, electron fluxes have no well-defined boundaries. If it becomes necessary to cut out the external layers of the beam, where a minor amount of current is generally contained, use is made of a special orifice plate (diaphragm). On passing the apertures of the diaphragm, the beam acquires a more distinct contour. However, in the course of subsequent transport, the boundaries of the beam tend to diverge under the action of electrons scattered at the edges of the diaphragm, as well as due to interaction of electrons with the media and because of other effects.

Each electron of the beam at any moment is characterized in the cartesian coordinate system by its position coordinates $x$, $y$, and $z$ and by the projection of the momentum on each of axes $P_x$, $P_y$, and $P_z$. The volume occupied by the set of electrons in a six-dimensional phase space is referred to as a phase volume of the electron beam. According to the Liuwill theorem, it is invariant along the trajectory of the beam particles. The area of projection of the phase volume on the shift-gradient plane ($R$, $R'$) is termed the emittance $\varepsilon$ of the beam (here $R' = dR/dz$, where $z$ is the coordinate coincident with the direction of the beam). The emittance is a quantitative characteristic enabling determination of the angular divergence of the beam, its capability of being focused by the external magnetic field, etc. The equation for accurate calculation of the emittance takes the form

$$\varepsilon = \int dR dR'$$

The approximate value of the emittance can be derived by multiplying the beam diameter $d$ by its angular spread $\psi$: $\varepsilon \cong d\psi$. In the case of paraxiality of the trajectories and absence of acceleration, the emittance of the beam remains unchanged along the transport channel. Therefore, the divergence of the beam is maximal where the cross section of the beam is of minimum diameter. There is also a concept of "brightness" of the beam which determines the current flowing across a unit area of the beam to a unit space angle: $4I/\pi^2\varepsilon^2$, where $I$ is expressed in amperes and $\varepsilon$ in rd/cm.

The physical phenomena determining the motion of beams of charged particles in vacuum will be mentioned only briefly, since their description can be found in a number of monographs, for example, [1–3]. Physical and chemical processes resulting from interactions of electron beams with matter form the basis for industrial use of accelerated electrons and are considered in subsequent chapters.

## 1.4 ISOTOPIC SOURCES

Up-to-date irradiating installations use two types of radiation sources: radioactive isotopes and accelerators. The former had found application for wide-scale processing of materials before accelerators were used. At present, isotopic sources are used approximately in one-third of all the effective industrial radiation installations, and preference is given to $^{60}Co$. As a result of a decay, the $^{60}Co$ nucleus emits two $\gamma$ quanta with energies of 1.17 and 1.33 MeV. The half-life of $^{60}Co$ is 5.27 years (the activity drops every month approximately by 1 percent). $^{60}Co$ is prepared for use in a nuclear reactor by neutron irradiation of special vials of $^{59}Co$. To a lesser extent, $^{137}Cs$ (with $\gamma$ quanta energy of 0.66 MeV and a half-life of 30 years) is used as an isotope source. In the isotopic sources, use is mainly made of gamma radiation, since the electrons produced in the process of decay are absorbed by the mass of the substance of the source proper. An exception is beta-active films based on $^{32}P$ and other radioactive isotopes, which are sources of fast electrons.

The fuel elements of nuclear reactors also contain radioactive isotopes and may find application for commercial radiation. However, the use of the fuel elements presents difficulties due to the comparatively quick decay of their radioactivity. Use is also made of closed loops for circulation of matter capable of being activated under the action of the reactor radiation. The articles subject to irradiation are positioned at loop portion located beyond the reactor. The purpose of the working medium of the loop may be served by fluid, gas, and solid (spherical fuel elements). It is conceivable that with further progress and wide-scale application of atomic power plants, some novel complex installation will be invented in which radiation from nuclear fission products in nuclear reactors will be used for performing radiation technological processes.

The power released in $^{60}Co$ of 1 Ci is $(3.7 \times 10^{10}) \times (1.17 + 1.33) \times 10^6 = 9.25 \times 10^{16}$ eV/s = 14.8 mW. The radiation of a cobalt source with an activity of 1 MCi = $3.7 \times 10^{16}$ Bq is equivalent in its ionization effect to that of an electron beam with a power of 14.8 kW. This comparison holds true only from the viewpoint of energy, since electrons and gamma radiation are substantially different in penetrating power. Unlike gamma radiation of a powerful isotopic source, electron beams can be localized. Compared to the electron accelerators, isotopic sources are simple to use and less troublesome. However, the electron accelerators ensure higher radiation capacity which makes them very useful in production assembly lines. Additionally, accelerators excel in high efficiency and low cost of radiation. When de-energized, electron accelerators feature radiation safety. The maximum dose rate of a cobalt source is $\sim 10^4$ Gy/h, while that of modern industrial accelerators exceeds $10^8$ Gy/h. As a rule, application of isotopic sources is found more reasonable for long-term irradiation of objects by minor doses and for irradiation of very large articles.

## 1.5 APPLICATION OF BEAMS OF ACCELERATED CHARGED PARTICLES AND OTHER KINDS OF RADIATION

Fluxes of fast-moving, charged particles feature various and sometimes unique properties enabling their application in many devices and installations. In electron tubes, for example, an electron beam provides closure of the circuit and is capable of changing the resistance of the circuit in compliance with the respective law. When used in microwave devices, the electron beam transfers energy by accepting it in some sections of the electron transit space and supplying it to other sections. In electronic welding machines, the electrons at first are accelerated and then release energy in the form of heat, thereby melting and welding metals. Operation of cathode-ray tubes (in television sets, oscilloscopes, and other apparatuses) is based on the capability of electrons to cause luminophor (phosphor) to glow.

Other properties of fast electrons are found useful in radiochemical facilities and x-ray apparatuses. In radiochemical installations, electrons serve to ionize the matter under processing and disturb the molecular bonds, thereby changing the chemical structure and properties of the matter. In x-ray apparatuses, electrons generate high-energy electromagnetic radiation with a great penetrating capability which allows examination of the internal structure of a material. Accelerated electrons and bremsstrahlung are used for therapy and for sterilization of medicines and food.

When curving the trajectories of electrons by the use of a magnetic field, synchrotron radiation is generated which may find use for research purposes and, potentially, for technological applications. The problems presently being studied include, for example, applicability of electron beams for transporting energy over long distances and utilizing electron and ion beams for heating plasma in future thermonuclear power plants. Ion beams are used mainly for implantation and alloying of metals, for study of radiation-damaged materials, and for therapy of malignant tumors.

Table 1.1 indicates typical applications of fluxes of charged elementary particles and radiations generated by electron beams and isotopic sources. Table 1.2 lists radiochemical and radiobiological processes in which modern industrial electron accelerators generally find wide application.

**Table 1.1 Applications of fluxes of elementary particles and radiations generated by electron beams and isotopic sources**

| Kind of radiation | Characteristic energy, MeV* | Process or device |
|---|---|---|
| Electrons | 0.0001–0.2 | Electron tubes |
| | 0.03–0.2 (1–2) | Microwave devices |
| | 0.010–0.025 | Cathode-ray tubes |
| | 0.02 | Lithography |
| | 0.1–0.3 | Ionization laser |
| | 0.05–0.1 (3) | Electron microscope |
| | 0.02–0.150 | Beam radiation treatment of materials |
| | 0.03–0.3 | Treatment of metal surfaces |
| | 0.15–10 (15) | Radiochemical and radiobiological technology (see Table 1.2) |
| | 4–40 | Radiotherapy |
| | 1–30 | Flaw detection |
| | (5–20) | Distant transfer of energy |
| | (0.5–5) | Research of controllable thermonuclear reaction |
| | Up to $3 \times 10^4$ | Physical investigations |
| Secondary radiations generated by electron beam | | |
| Bremsstrahlung | 0.03–0.15 | Roentgenoscopy |
| | 0.05–25 (35) | Flaw detection |
| | 1–10 (15) | Sterilization of medical facilities |
| | 1–2 (10) | Preservation treatment of foodstuffs |
| | 0.1–40 | Radiotherapy |
| Synchrotron radiation | Energy of electrons of up to $10^4$ MeV; energy of photons of up to 75 keV | Examination of surfaces, spectroscopy, etc. |
| Gamma radiation of isotopic sources $^{60}Co$ and $^{137}Cs$ | 0.66–1.33 | Radiochemical processes |
| | | Gamma defectoscopy |
| | | Gamma therapy |
| | | Sterilization of medical facilities |
| | | Preservation treatment of foodstuffs for storage |
| | | Radiation treatment of seeds |
| | | Radioactive indicators |

(*See footnote on page 16.*)

**Table 1.1 Applications of fluxes of elementary particles and radiations generated by electron beams and isotopic sources** *(Continued)*

| Kind of radiation | Characteristic energy, MeV* | Process or device |
|---|---|---|
| Ions | 0.03–0.5 (5) | Ionic implantation |
| | 0.05 | Lithography |
| | 3.25 | Activation analysis |
| | 8–45 | Production of radioactive isotopes |
| | 50–400 per nucleon | Radiotherapy |
| | (0.5–5) | Research of controllable thermonuclear reaction |
| | Up to $4.10^5$ | Physical investigations |
| Neutrons | 14** | Neutron logging |
| | 14 | Neutron activation analysis |
| | 14 | Neutron therapy |

*Numbers in parentheses refer to data relating to devices that have not yet found wide use or are in experimental stages.

**Neutrons with energies of up to 14 MeV are obtained by bombarding a tritium target by deuterium ions with an energy of 150 to 500 keV: reaction $T\,(d,\,n)\,^4He$. For producing neutron fluxes of higher energies, use is made of cyclotrons and other accelerators.

**Table 1.2 Application of industrial accelerators in radiochemical and radiobiological technology**

| Process | Energy, MeV | Dose, kGy |
|---|---|---|
| Cross-linking of polyethylene | 0.3–4 | 50–300 |
| Production of thermoshrinkable articles | 0.3–4 | 100–250 |
| Hardening of coatings | 0.15–0.5 | 20–500 |
| Graft polymerization | 0.3–2.5 | 10–300 |
| Modification of textiles | 0.3–1.5 | 5–50 |
| Fabrication of fiber glass plastics | 0.5–1 | 10–400 |
| Vulcanization of rubbers | 0.5–15 | 50–300 |
| Sterilization of medicinal facilities | 1–10 (15)* | 20–50 |
| Deinsectization of grain | 1 | 0.5–1 |
| Sewage treatment | 0.54 | 0.5–1 |
| Purification of exhaust gases | 0.3–1.5 | 10–15 |

*See footnote for Table 1.1.

CHAPTER

# TWO

# INTERACTION OF RADIATION WITH MATTER AND APPLICATIONS OF ELECTRON ACCELERATORS

## 2.1 IONIZING EFFECT OF ELECTRON BEAM

After penetrating matter, the accelerated electrons interact electrostatically, with electrons and atomic nuclei either being repulsed from or attracted to them. Most of the electron beam energy is consumed in the interaction with orbital electrons, i.e., by exciting and ionizing atoms. Fast-moving electrons that travel close to a nucleus may result in elastic or inelastic scattering with emission (in the case of inelastic scattering) of a photon of electromagnetic radiation. Interacting with nuclei of some substances, photons may invite nuclear reactions which, in turn, cause induced radioactivity and neutron radiation. The main part of the energy absorbed by the affected matter is converted, finally, into heat which brings about heating up of the material under irradiation.

When interacting with orbital electrons, the primary particle imparts a part of its own energy to these electrons, thereby enabling them to leave their orbits for new ones of greater diameters (excitation) or to be detached from atoms (ionization). The average energy used for forming a pair of ions is virtually independent of the energy of incident electrons and for many substances comes to ~30 eV (or 34 eV for air). This energy is just negligibly dependent on the type of ionizing radiation (electrons, photons, or ions). The ionization potential of most ordinary molecules approximates 10 eV. The remaining two-thirds of the energy of the primary particle is expended in excitation and other processes.

As the energy of the electrons generated by industrial accelerators greatly exceeds the expenditures of energy for production of a pair of ions, the full loss of the energy occurs at the expense of numerous collisions. With each interaction, the direction of the bombarding electron varies, as a result of which the trajectories of the primary particles are in the form of a broken line (Fig. 2.1). The transit time of the electron between the ionization events is $\sim 10^{-15}$ s. A major portion of the electrons detached from atoms (secondary electrons or δ electrons) feature an energy of tens of electron-volts and also take part in the processes of excitation and ionization. The secondary electrons are capable of ionizing an average of two or three molecules. The secondary electrons advance over a range of $10^{-8}$ to $10^{-7}$ cm and bunch up ions near the trajectory of the primary particle.

Some secondary electrons escape the surface of the material at different angles toward the primary beam (Fig. 2.2). In industrial accelerators, these secondary particles carry away some energy, thereby decreasing the radiation dose. The essential portion of the secondary particles gets on the parts of the beam extractor and

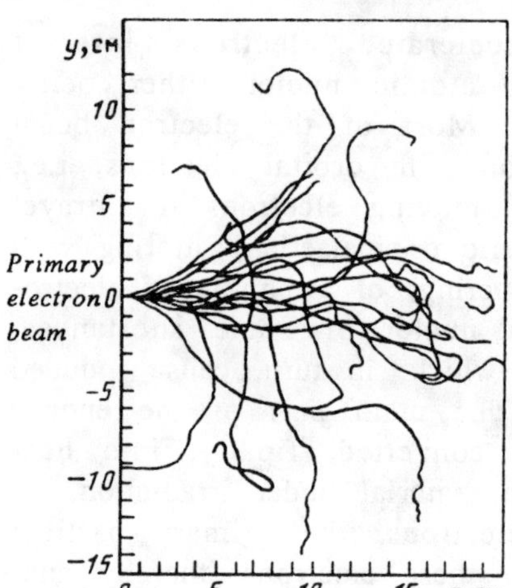


**Figure 2.1** Typical trajectories of electrons with initial energy of 0.15 MeV extracted into atmosphere (calculated by Monte-Carlo method for room air temperatures).

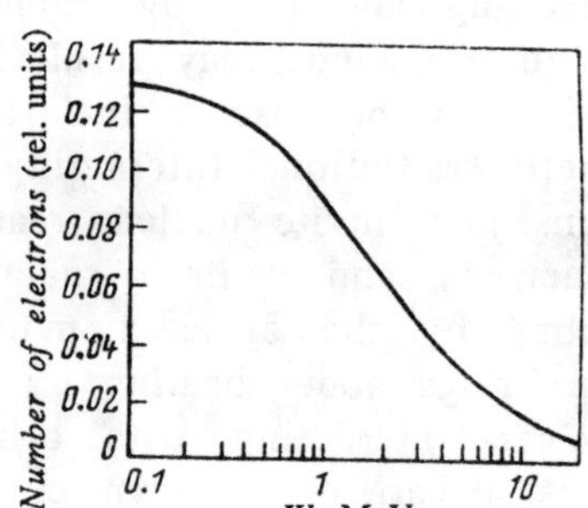


**Figure 2.2** Ratio between flux of secondary electrons and flux of primary electrons incident at right angle to aluminum surface versus energy of primary electrons (thick target).

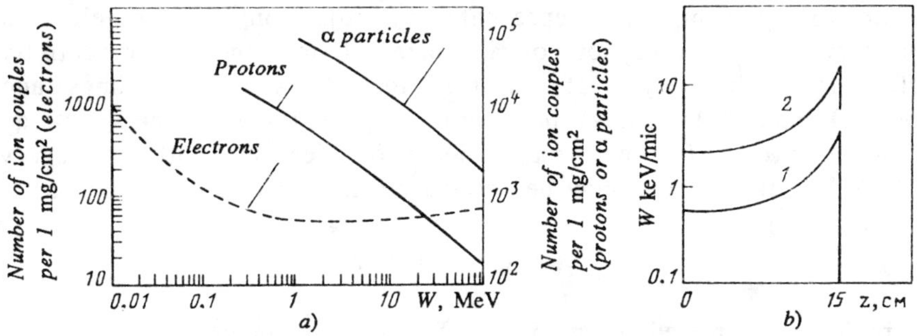


**Figure 2.3** Energy loss by different particles in matter. *a*) Specific ionization of air by electrons, protons and α particles. *b*) Linear losses of energy along tracks of heavy particles [4]. 1—protons with energy of 150 MeV; 2—α particles with energy of 600 MeV.

other assemblies of the accelerators, thereby presenting difficulties for normal operation of the equipment.

Ionization losses of the electrons per unit of the trajectory length are constant, beginning with energy levels of 0.5 MeV and higher. The curve of the specific ionization depicted in Figure 2.3 indicates the rate of energy loss by primary particles, since the energy expended in composing a pair of ions almost remains unchanged. The losses in question tend to increase appreciably at the end of the path of the electrons, i.e., at energies below 100 keV. The length of the electron trajectory in matter is approximately proportional to the initial velocity of the electron. The electron

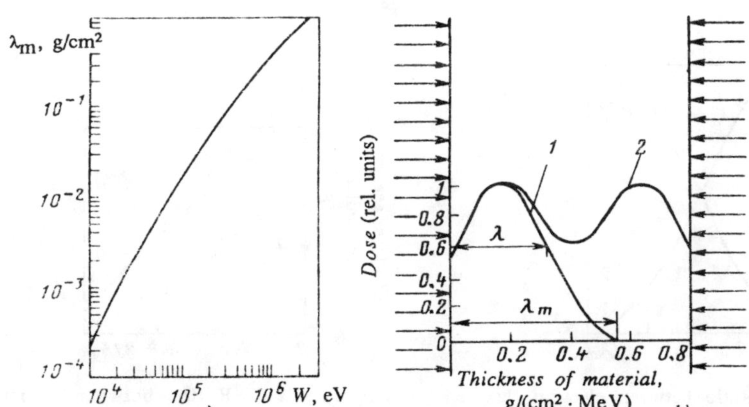


**Figure 2.4** Absorption of electrons by matter. *a*) Electron penetration depth versus energy. *b*) ionization density distribution (in relative units) for *1*—one-sided and *2*—two-sided irradiation.

penetration depth $\lambda_m$* represents a projection of the electron trajectory in the direction of its original motion and essentially differs from the length of the trajectory (less by 1.5 times and more). Figure 2.4*a* shows penetration $\lambda_m$ versus the kinetic energy of the electrons. The penetration depth of electrons with an energy of 10 keV < $W$ <1 MeV may be derived by the use of the formula [5]

$$\lambda_m = 6.7 \times 10^{-11} \frac{W^{5/3}}{\rho}$$

where $\lambda_m$ is expressed in cm, $W$ in eV, and $\rho$ in g/cm³.

The experimentally revealed distribution of matter-absorbed energy (ionization density, radiation dose, or absorbed dose) in the range of energies of primary electrons within 1 and 10 MeV is shown in Fig. 2.4*b*. The layer thickness $\lambda$ of the substance with density $\rho$ and a radiation dose nonuniformity of ± 20 percent is determined by the equation

$$\lambda \cong 0.33 \frac{W}{\rho}$$

where $\lambda$ is expressed in cm, $W$ in meV, and $\rho$ in g/cm³.

In the case of double-sided irradiation

$$\lambda \cong 0.8 \frac{W}{\rho}$$

*The most commonly used units of measurement of $\lambda_m$ are cm, g/cm², and g/cm² × MeV).

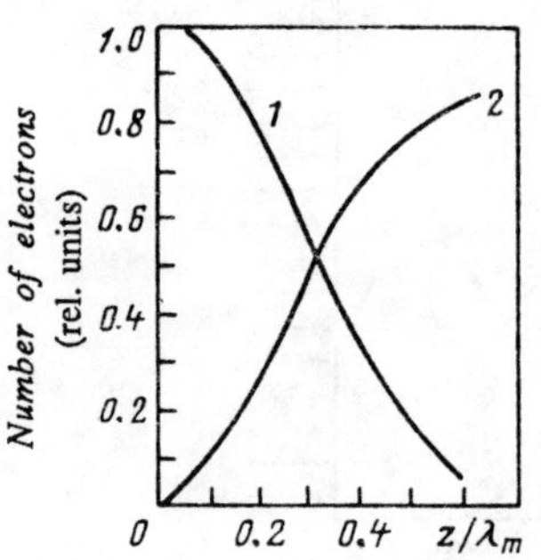


**Figure 2.5** Ratio between electron flux after penetration and inital flux (*1*) and matter-absorbed quota of initial flux (2) versus thickness of irradiated matter *z*. $\lambda_m$ = electron penetration depth; electron energy = 0.1 to 1 MeV; material = titanium.

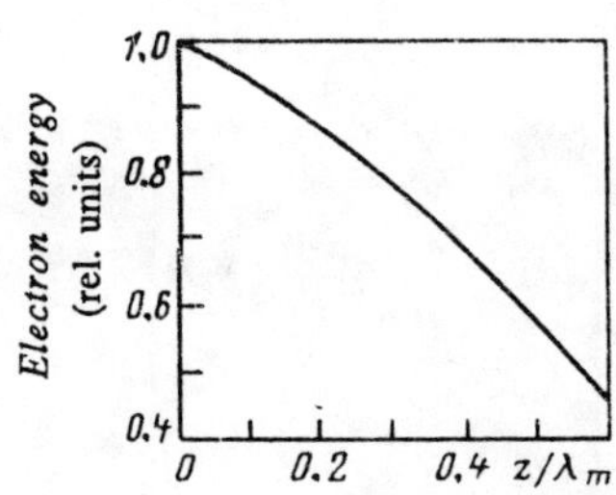


**Figure 2.6** Ratio between most probable energy of electrons after penetration of matter and initial energy versus material thickness *z*. $\lambda_m$ = penetration depth of electrons; initial electron energy = 0.1 to 1 MeV; material = titanium.

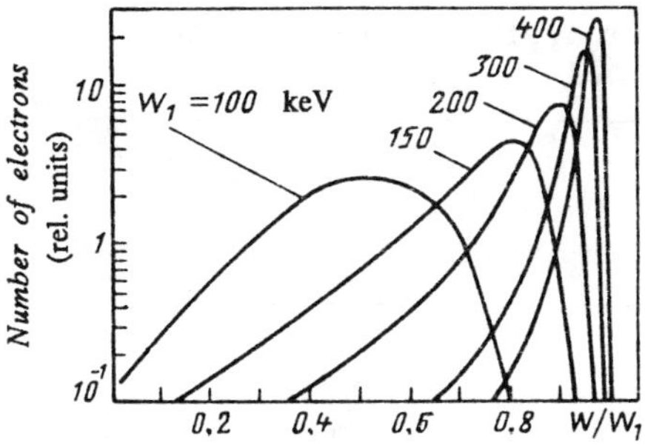


**Figure 2.7** Energy spectra of electrons passing through foil (titanium, 11.43 mg/cm$^2$). $W_1$ = energy of primary electrons; $W$ = energy of electrons after passing across foil.

The material may be irradiated more uniformly, in case of thin specimens, by the use of additional filters as well as by irradiating the material with the aid of a non-monoenergetic beam with a specially selected energy spectrum. The characteristic curves shown in Fig. 2.5 display the absorption of electrons in matter. Figures 2.6 and 2.7 illustrate variations of the parameters of the electron beam after passing across thin layers complying with thickness of the foils used for extraction of the electrons from vacuum into the atmosphere. The angular distribution of the electrons passing through the extraction window is shown in Table 2.1 [6].

It should be noted that the heavy particles (protons and ions) also lose their energy, basically, by ionization of atoms. However,

**Table 2.1 Angular distribution of electrons passed through aluminum and titanium foils (with quota of electrons in cone around incident direction)**

| Material | Initial energy, keV | Thickness of foil, mg/cm$^2$ | Cone angle 5° | 15° | 30° | 60° |
|---|---|---|---|---|---|---|
| Aluminum | 150 | 6.86 | 0.013 | 0.112 | 0.374 | 0.849 |
| | 150 | 13.72 | 0.011 | 0.090 | 0.320 | 0.821 |
| | 300 | 6.86 | 0.041 | 0.302 | 0.719 | 0.973 |
| | 300 | 13.72 | 0.020 | 0.164 | 0.487 | 0.907 |
| | 400 | 6.86 | 0.050 | 0.381 | 0.826 | 0.988 |
| | 400 | 13.72 | 0.025 | 0.209 | 0.611 | 0.953 |
| Titanium | 150 | 11.43 | 0.011 | 0.084 | 0.304 | 0.808 |
| | 150 | 22.86 | 0.009 | 0.077 | 0.309 | 0.802 |
| | 300 | 11.43 | 0.014 | 0.125 | 0.414 | 0.881 |
| | 300 | 22.86 | 0.011 | 0.090 | 0.322 | 0.810 |
| | 400 | 11.43 | 0.022 | 0.188 | 0.539 | 0.929 |
| | 400 | 22.86 | 0.012 | 0.0102 | 0.364 | 0.841 |

**Table 2.2 Comparison of penetration depth $\lambda$ of electrons and protons in aluminum**

| | $\lambda_m$, mg/cm$^2$ | | | $\lambda_m$, mg/cm$^2$ | |
|---|---|---|---|---|---|
| Energy, MeV | Electrons | Protons | Energy, MeV | Electrons | Protons |
| 0.1 | 13 | 0.2 | 1 | 400 | 2.8 |
| 0.2 | 40 | 0.36 | 2 | 900 | 8.6 |
| 0.5 | 170 | 1 | 5 | 2600 | 42 |

due to their great mass, ions coming into collision with orbital electrons are deflected only slightly and, therefore, their trajectory approximates a straight line. In the case of both electrons and ions, secondary electrons provide a dense ionization close to the trajectory of the primary particle. The rate and depth of penetration of ions into matter are substantially less than those of electrons with the same energy (see Table 2.2). Heavy particles lose more energy per unit track length and correspondingly produce a greater amount of ions. Figure 2.3*b* indicates linear losses of energy (losses per unit length of the trajectory) for protons and alpha particles. The losses tend to increase with decrease of energy. The curved portion corresponding to a sharp rise of losses at the end of the track is referred to as the Bragg peak.

A small portion of the secondary electrons arising from the effects of an ion beam features high energy. As a consequence, in addition to dense ionization along the tracks of the primary ions (under the action of the ions proper and secondary electrons with a low energy), some additional ionization areas (spurs) arise along traveling fast-moving secondary electrons.

## 2.2 GENERATION OF BREMSSTRAHLUNG

Accelerated electrons moving near an atomic nucleus change their initial direction of motion and impart some of their energy to the nucleus. However, because of the pronounced difference in mass between the electrons and nucleus, the imparted energy is too small to shift the nucleus to effect ionization. Apart from such an elastic scattering, some inelastic scatterings of the electron over the nucleus take place with a certain degree of probability. As a result, some of the electron's energy is transformed into a shortwave braking radiation (bremsstrahlung). The bremsstrahlung features a continuum, for the energy of radiated photons ranges from zero to the kinetic energy of the primary electrons. The maximum intensity of radiation falls to the share of the photon energy which is half as much as the energy of the primary particles. The ratio of the energy lost by the electrons for creating bremsstrahlung to the losses for ionization and excitation equals $Wz/800$, where $W$ is the kinetic energy of the electron in megaelectronvolts and $z$ is the nuclear charge. Thus, in the case of irradiation of polymers (generally with $z$ = 6 to 8) by electrons with energy $W$ = 1 MeV, about 1 percent of the energy of the primary beam is transferred into the bremsstrahlung. The intensity of radiation depends on the beam current, energy of electrons, and target material and its thickness. Figure 2.8 illustrates exposure dose rate in the beam direction versus the energy of the electrons and the target material. The intensity of bremsstrahlung tends to grow almost quadratically with increase of nuclear charge of the target material. The target is generally fabricated of tungsten with $z$ = 74, tantalum with $z$ = 73, and other heavy metals. These materials are selected for their high melting points, which makes it possible to increase the beam density and, correspondingly, to decrease the diameter of the focal spot. This is particularly important for radiation defectoscopy of articles (Section 2.6). The melting temperature of tungsten is 3380 K. In addition, tungsten features high heat conduction, which also aids in development of more perfect fine-focused systems. Increase of thickness of electron targets results, on the one hand, in growth of

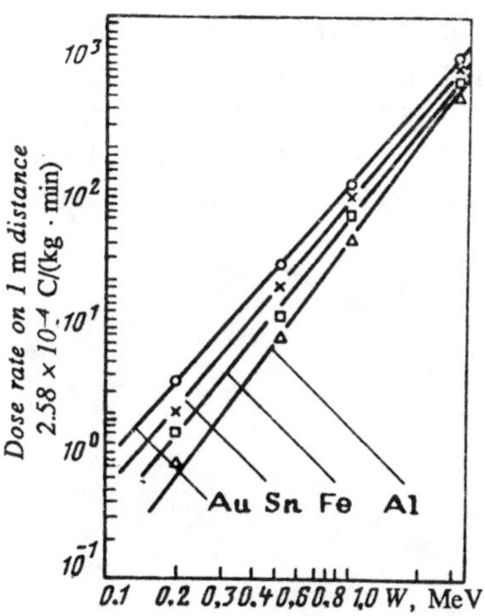


**Figure 2.8** Power of bremsstrahlung dose rate versus electron energy and target material.

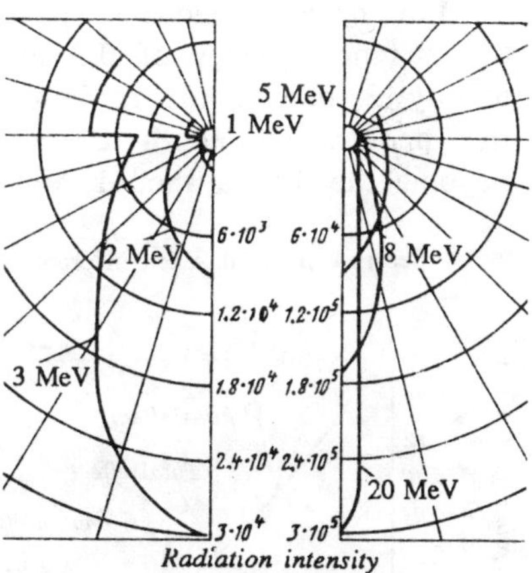


**Figure 2.9** Angular distribution of bremsstrahlung intensity at 1 m distance from tungsten target.

quota of electron beam energy converted into bremsstrahlung, and, on the other hand, in higher absorption of the bremsstrahlung. The optimal thickness of a tungsten target for energies ranging from 3 to 10 MeV is 1 mm. Figure 2.9 shows the angular distribution of bremsstrahlung intensity referred to the power of the electron beam. With the increase of energy of the primary electrons, the radiation propagates forward within a narrower angle. For the case of relativistic particles this angle is $I/\gamma$.*

## 2.3 PENETRATION OF BREMSSTRAHLUNG THROUGH MATTER

Interaction cross section of bremsstrahlung photons with matter under effect is materially less than that of electrons of the same energy. Due to this fact, the penetration depth of the bremsstrahlung is much greater (sometimes almost by two orders). This enables the bremsstrahlung to be applied for radiation treatment of large articles.

The principal mechanisms of energy transfer by means of bremsstrahlung are the photoelectric effect, the Compton effect, at a photon energy exceeding 1.022 MeV, formation of electron-positron pairs. As illustrated in Fig. 2.10, the photoelectric effect takes place at low energy levels (approximately ≤ 200 keV). Then the scattering is dictated by the Compton effect. With energy levels of 2 to 3 MeV and more, the contribution of the pair-production process tends to intensify.

For narrow electron beams, absorption occurs exponentially. At the quantum energy of 0.3 to 1 MeV per g/cm$^2$ of matter, approximately 3 percent of radiation may be absorbed. When calculating the penetration depth of a narrow beam, the scattered photons beyond the beam should not be taken into account. But in actual

* See new material added in proof on page 31.

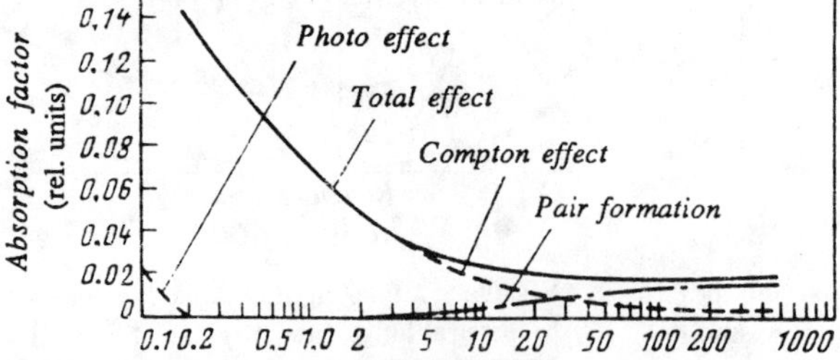


**Figure 2.10** Bremsstrahlung absorption coefficient versus photon energy.

cases, for instance, in industrial flaw detection, in calculations of radiation shielding of the accelerators, such an assumption may lead to material errors, because scattered photons retard the process of decrease of intensity of the bremsstrahlung. Table 2.3 gives the thicknesses of concrete, iron, and lead for different multiplicity factors of attenuation of a broad beam of bremsstrahlung (or gamma radiation) [8]. Attenuation of broad radiation fluxes in particular versions of the accelerators may be materially dependent on the geometry of the system. When calculating and fabricating the radiation shielding, it is good practice to measure the dose of the exposure field and, if necessary, to introduce the needed design corrections.

## 2.4 PHOTONUCLEAR REACTIONS

Beginning with a certain energy of photons of bremsstrahlung, the photonuclear reactions $\gamma, n$ are initiated in matter, which results in production of radioactive isotopes. Reactions $\gamma, p$; $\gamma, d$; $\gamma, \alpha$ are usually ignored, for their cross section is materially less, and therefore predominantly stable isotopes are produced. Threshold energies of photonuclear reactions are not similar for different elements. The minimum energy of the photons and, respectively, the minimum energy of the electrons generating bremsstrahlung (i.e., the energy providing reaction $\gamma, n$) is ~1.67 MeV. In this case, reaction $^9$Be $(\gamma, n)$ $^8$Be occurs.

| Reaction: | $^9Be(\gamma,n)^8Be$ | $^2H(\gamma,n)^1H$ | $^6Li(\gamma,n)^5Li$ | $^{127}I(\gamma,n)^{126}I$ | $^{19}F(\gamma,n)^{18}F$ |
|---|---|---|---|---|---|
| Threshold energy, MeV: | 1.67 | 2.23 | 5.35 | 9.13 | 10.4 |
| Reaction: | $^{14}N(\gamma,n)^{13}N$ | $^{79}Br(\gamma,n)^{78}Br$ | $^{35}Cl(\gamma,n)^{34}Cl$ | $^{16}O(\gamma,n)^{15}O$ | $^{12}C(\gamma,n)^{11}C$ |
| Threshold energy, MeV: | 10.5 | 10.65 | 13.0 | 15.6 | 18.6 |

Radioactive materials are produced basically at electron energies exceeding 9 to 10 MeV. Cross sections of most reactions lie within 0.01 to 1 barn ($10^{-26}$ to $10^{-24}$ cm$^2$); 1 barn = $10^{-24}$ cm$^2$). The overwhelming majority of electron accelerators used for applied purposes feature energies not in excess of 1.5 MeV and their beams do not provide induced activity. Accelerated electrons also find application for generating neutrons. For instance, by means of a linear accelerator with an energy of 4.2 MeV and an average current of the electron beam of 50 $\mu$A, it is possible to obtain a thermal

**Table 2.3 Thickness of matter for different multiplicity factors of attenuation of broad-beam bremsstrahlung of various energy levels**

| Material and attenuation multiplicity factor | Thickness (cm) by various energy levels, MeV | | | | | | | | | |
|---|---|---|---|---|---|---|---|---|---|---|
| | 0.1 | 0.3 | 0.5 | 0.7 | 1.0 | 1.5 | 2.0 | 3.0 | 4.0 | 6.0 |
| Concrete, $\rho = 2.3$ g/cm$^3$ | | | | | | | | | | |
| 10 | 8.2 | 19.7 | 25.8 | 27.6 | 29.9 | 34.0 | 37.6 | 43.4 | 47.5 | 51.6 |
| $10^2$ | 11.5 | 28.9 | 39.9 | 45.3 | 64.6 | 74.8 | 84.5 | 101.0 | 110.4 | 124.4 |
| $10^3$ | 15.5 | 39.2 | 55.2 | 62.5 | 82.8 | 97.0 | 110.9 | 132.7 | 146.8 | 166.7 |
| $10^4$ | 18.8 | 48.5 | 69.3 | 79.1 | 102.0 | 120.4 | 136.2 | 164.9 | 181.4 | 206.6 |
| $10^5$ | 30.5 | 64.6 | 82.8 | 93.5 | 119.7 | 142.5 | 162.5 | 196.0 | 214.8 | 247.1 |
| $10^6$ | 49.3 | 79.8 | 97.0 | 109.1 | 137.0 | 165.8 | 189.0 | 227.0 | 250.1 | 286.6 |
| $10^7$ | 64.0 | 95.7 | 111.0 | 124.6 | 142.0 | 173.0 | 197.0 | 236.5 | 259.4 | 299.4 |
| Iron | | | | | | | | | | |
| 10 | 1.9 | 4.6 | 6.3 | 7.3 | 8.5 | 10.0 | 11.0 | 12.2 | 12.6 | 13.2 |
| $10^2$ | 3.4 | 8.1 | 10.8 | 12.5 | 14.5 | 17.3 | 19.5 | 22.1 | 23.3 | 25.0 |
| $10^3$ | 4.5 | 11.0 | 15.0 | 17.5 | 20.5 | 24.4 | 27.5 | 31.7 | 33.7 | 35.0 |
| $10^4$ | 6.8 | 14.7 | 19.0 | 22.3 | 26.0 | 31.3 | 35.5 | 40.9 | 43.7 | 46.5 |
| $10^5$ | 10.0 | 18.2 | 23.0 | 26.7 | 31.5 | 38.0 | 43.2 | 50.0 | 53.4 | 58.3 |
| $10^6$ | 12.8 | 21.4 | 26.7 | 31.2 | 37.0 | 44.7 | 50.6 | 58.8 | 63.3 | 69.0 |
| $10^7$ | 15.0 | 24.3 | 30.5 | 35.8 | 42.4 | 51.3 | 57.9 | 67.5 | 73.1 | 79.4 |
| Lead | | | | | | | | | | |
| 10 | 0.3 | 0.9 | 1.6 | 2.6 | 3.8 | 5.1 | 5.9 | 6.5 | 6.4 | 5.5 |
| $10^2$ | 0.5 | 1.6 | 3.0 | 4.7 | 7.0 | 9.65 | 11.3 | 12.2 | 12.1 | 10.9 |
| $10^3$ | 0.7 | 2.4 | 4.4 | 6.95 | 10.2 | 14.1 | 16.5 | 18.0 | 17.8 | 16.5 |
| $10^4$ | 1.05 | 3.3 | 5.9 | 9.1 | 13.3 | 18.3 | 21.5 | 23.5 | 23.4 | 22.0 |
| $10^5$ | 1.15 | 3.8 | 7.2 | 11.1 | 16.5 | 22.7 | 26.2 | 28.9 | 28.9 | 27.5 |
| $10^6$ | 1.45 | 4.7 | 8.7 | 13.1 | 19.5 | 26.8 | 31.0 | 34.3 | 34.4 | 33.0 |
| $10^7$ | 1.7 | 5.4 | 10.1 | 15.2 | 22.5 | 31.2 | 35.8 | 39.7 | 39.9 | 38.4 |

neutron flux of $\sim 10^7$ particles/$cm^2$ s, provided that a beryllium target and a paraffin block for moderating neutrons are used. At an energy level of 30 MeV and a current of 145 $\mu$A, with the use of a tantalum or tungsten target, a flux of fast neutrons of $\sim 10^{10}$ particles/$cm^2$ s can be obtained.

## 2.5 ELECTRIC CHARGE IN MATERIALS UNDER IRRADIATION

As noted above, one of the principal mechanisms of absorption of energy of electron beam and bremsstrahlung in matter is ionization of atoms. Along the trajectories of primary and secondary electrons, some ionized articles are bunched into aggregates. Free electrons as well as ionized atoms and molecules are preserved for a long time in materials featuring low electric conductance. Charge relaxation time $\tau = \varepsilon \varepsilon_0/\delta$, where $\tau$ is the time expressed in seconds, $\varepsilon$ is the dielectric permeability of the medium, $\varepsilon_0$ is the electric constant $\Phi$/M and $\delta$ is the electrical conductivity in (S/m) (S/m corresponds to 1 $Ohm^{-1} \cdot m^{-1}$). For example, for $\varepsilon = 5$ and $\delta = 10^{-12}$ S/m,

$$\tau = \frac{5 \times 10^{-12}}{10^{-12}} = 5 \text{ s}$$

For $\varepsilon = 5$ and $\delta = 10^{-17}$ S/m,

$$\tau \cong 10^5 \text{ s}$$

For metals, $\tau$ is negligible. Thus, for $\delta = 1$ S/m

$$\tau \cong 10^{-10} \text{ s}$$

Therefore, the surplus charge in a metal will quickly emerge on metal surface and its leakage time depends on the properties of the surrounding medium (air, vacuum, etc.). A surplus negative charge in an insulator may cause internal breakdown of materials, sometimes in the form of free branches (dendrites) which can be easily viewed in transparent dielectrics. Deceleration of electrons of the primary beam by the action of the electric field of a surplus charge may cause most of the absorbed dose to shift to the surface of the dielectric under irradiation [9]. The yields of the radio-chemical reactions will correspondingly vary over the thickness of the specimen. In the process of irradiation, the conductance of the dielectric may materially increase due to emergence of additional electric current carrier. Charged microareas (of the order of $10^{-8}$ to $10^{-7}$ cm) are produced in the irradiated dielectrics and preserved for a period of $\tau$. In this case, negative and positive charges are not distributed in the material uniformly. Internal electric fields may deteriorate mechanical strength of materials and change solubility and absorptive property of materials. The relaxation time of charges tends to sharply decrease with rise of temperature of materials.

## 2.6 RADIATION DEFECTOSCOPY

Attenuation of bremsstrahlung depends on the density and thickness of materials involved. Therefore, the shadow picture obtained in the substance after beam penetration and displayed by the use of an x-ray film or on a fluorescent screen will show the existence and dimensions of structural heterogeneities in the material under examination. This technique allows nondestructive testing for detection of various flaws such as, cracks, caverns, and inclusions of extraneous materials.

The quality of the image is improved by decreasing the area of the radiation source, decreasing the distance between the source and the object to be scrutinized, and lastly, with bringing this place nearer to the object under examination. One of the main restrictions of radiography is the minimal dimension of the revealed defects. Bremsstrahlung penetration into the matter under inspection causes a scattered radiation (scattered photons, secondary bremsstrahlung caused by electrons produced in the material under examination). As a result, the contrast of the image is inadequate. Therefore, when thicker material is to be x-rayed, use is made of harder radiation. For example, for defectoscopy of plastics and light metals as well as steel of 120 to 150 mm thickness, use is made of radiation with an energy of a few hundred kiloelectronvolts; for steel of 500 to 600 mm in thickness, the radiation of some 10 MeV is required. Usually, the typical dimension of the detectable defects is not less than 1 percent of the thickness penetrated by the radiation beam.

In addition, application of bremsstrahlung can be performed by the use of accelerated electrons. Along with the other methods of radiography for defectoscopy by fast electrons, use is made of the dependence of the intensity of the electron flux passed through the matter on the thickness and density of the object under examination. As a rule, for this purpose, use is made of electrons with an energy of 1 to 30 MeV.

Surface-layer flaws of materials may be detected by measuring the current of secondary backscattered electrons. This method is found particularly efficient at small angles of incident of electrons on the material surface, since in this case the current of secondary particles tends to rise materially.

## 2.7 ACTIVATION ANALYSIS

Activation analysis based on the principle of application of photonuclear reactions enables detection of minor admixtures of various substances in ores, biological objects, as well as in production of semiconductors, in research of contamination of environments, and so on. Irradiation by bremsstrahlung photons starting from threshold energies of reactions $\gamma, n$ results in activation of matter. When studying the relationship between the production of radioactive isotopes and radiation energy, the qualitative composition of the object under study is determined. For qualitative analysis, it is generally necessary to resort to comparing the activity induced under similar conditions in a reference specimen containing a known amount of certain elements. This method allows rapid analyses of materials and determination of impurities with a concentration of up to $10^{-5}$ to $10^{-3}$ percent.

Activation of matter may also be effected by the use of neutrons (reaction $n, n$). Use is made of both thermal neutrons of reactors and fast neutrons produced with the aid of neutron generators on the basis of accelerators of charged particles. Each of these methods is intended for particular application. The gamma-activation analysis, as compared with the neutron-activation analysis, ensures a higher sensitivity in determining content of light elements. The threshold nature of photonuclear reactions permits accomplishment of sampling analyses. For investigating the composition of rocks at different depths, use is made of portable electron accelerators and neutron generators which are sunk into drill holes.

## 2.8 MAINTENANCE OF NON-SELF-MAINTAINED DISCHARGE IN ELECTROIONIZATION LASER

In lasers operating on high-pressure gas, the electron beam is used for ionization of large volumes of gas (Fig. 2.11 shows a diagrammatic illustration of a laser). Voltage from an external power source (6) is applied between the cathode (3) and anode (7). The space between the cathode and anode is essentially gas-discharge chamber (4). An accelerated electron beam (1) injected into the chamber invites ionization, thereby initiating production of low-energy secondary electrons. The latter move across the electric

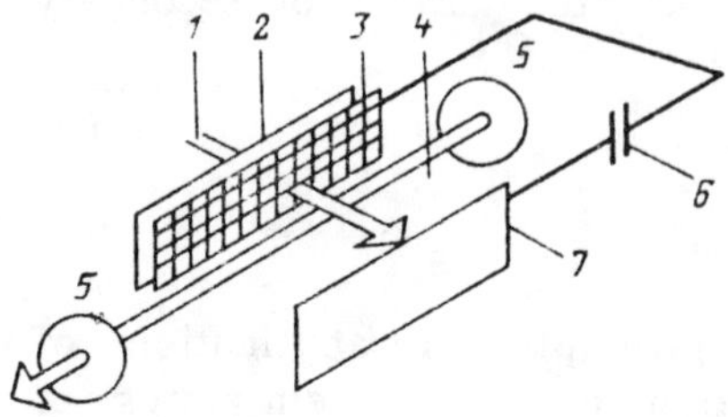


**Figure 2.11** Diagram of electroionization laser. *1*—Beam of accelerated electrons; *2*—foil separating accelerator vacuum chamber from gas chamber; *3*—reticular cathode of gas-discharge chamber; *4*—gas-discharge chamber; *5*—mirrors; *6*—discharge power supply; *7*—anode of gas-discharge chamber; *8*—coherent radiation.

field and induce the quantized levels. Concentration of these electrons is governed by the intensity of the primary electron beam. Optimum pumping conditions can be obtained by adjusting the ratio between the intensity of the electric field and the pressure (i.e., $E/P$).

The typical parameters of an electroionization laser on vibrational-rotational transitions of $CO_2$ molecules at atmospheric pressure are as follows: electron beam energy, 100 to 300 keV; average density of electron beam [determined generally by strength of foil (2)], from 2 to 5 A/m$^2$; voltage between discharge gap electrodes (3 and 7) ranges from 150 to 250 kV; $E/P$ = 10 V/mPa; specific pumping energy, ~5 J/m$^3$Pa. The length of pumping pulses in compliance with the relaxation time for mixtures of $Co_2$, $N_2$, He is ≤ 10 μs.

## 2.9 SYNCHROTRON RADIATION

When an electron beam moves across a magnetic field, a magnetic braking radiation or a synchrotron radiation arises which is tangent to the trajectory in the direction of travel of electrons [10]. Synchrotron radiation energy losses per orbital revolution amount to $\Delta W = 10^{-5}\ W^4/R$, where $\Delta W$ is energy loss (in eV), $W$ is the energy (in MeV), and $R$ is the orbit radius (in cm). Synchrotron radiation intensity becomes material starting from electron energies of the order of a 100 MeV. The unique potentialities of synchrotron radiation make it applicable for a number of applied and technological purposes. Synchrotron radiation average intensity is some three to five orders of magnitude higher than that of powerful gas-discharge tubes and x-ray tubes. Besides, synchrotron radiation is characterized by a negligible angular divergence ($10^{-3}$ to $10^{-4}$) and a continuous spectrum from the infrared region to the x-ray region ($10^{-4}$ to $10^{-12}$ cm). Use of synchrotron radiation in x-ray microscopy makes it possible to reduce by many times the exposure time and to improve the resolution up to 100 Å. This allows for microfilming of dynamics of processes in living objects. In medicine, this opens the way for studying distribution of medium and heavy chemical elements in human organs. Qualitatively new possibilities are offered by the use of synchrotron radiation in lithography, molecular microsurgery, etc. For technological applications of synchrotron radiation, it is essential that the capacity and efficiency of the relevant generators should be enhanced materially. It is not inconceivable that in the future the efficiency of synchrotron radiation generators with superconducting magnets and resonant cavities will be near to 100 percent [11].

## MATERIAL ADDED IN PROOF

| | Ta | W | Au |
|---|---|---|---|
| Photon transmission % | 7.60 | 7.61 | 7.57 |
| Photon reflection % | 3.54 | 3.57 | 3.81 |
| Electron reflection % | 12.51 | 12.89 | 14.11 |
| Energy deposition % | 75.99 | 74.96 | 73.81 |
| Specific gravity (g/cm³) | 17.1 | 19.3 | 19.3 |
| Specific heat (kcal/kg°C) | 0.033 | 0.032 | 0.031 |
| Thermal conductivity (kcal/m h °C) | 52.8 (20°C)<br>63.6 (500°C) | 170 (20°C) | 254 (20°C) |
| Thermal expansion ($10^{-6}$/°C) | 6.5 (20°C) | 4.3 (20°C) | 14.2 (20°C) |
| Melting point °C | 2850 | 3410 | 1063 |
| Readiness for machining | good | no | good |
| Cost | high | low | expensive |

CHAPTER

# THREE

# CHEMICAL PROCESSES OCCURRING UNDER THE EFFECT OF IONIZING RADIATION

## 3.1 PECULIARITIES OF RADIOCHEMICAL PROCESSES

As was pointed out in the preceding chapter, the energy of particles of primary radiation of industrial accelerators exceeds by tens of thousands times the energy required for an ionization event, i.e., for disturbing one chemical bond. When passing across any matter, the primary particles apply energy to numerous molecules, thereby causing excitation and ionization.* Charged and uncharged fragments of molecules resulting from bond breaking feature high chemical activity and tend to quickly react with one another and with other molecules. New active particles (free radicals and secondary ions) arising from this activity also start reacting. As a consequence, the molecular structure of the material undergoes transformation and a new substance with new properties is obtained. High chemical activity of the reacting particles enables radiochemical reactions to be carried on at very low temperatures (up to the temperature of liquid helium). It opens up possibilities for chemically combining substances incompatible by other

* Regardless of the kind of primary radiation, most of its energy is distributed between secondary electrons which, in the final analysis, provide excitation and ionization. With the radiation effect of primary electrons, secondary electrons ensure about half of all ionization events.

methods and for obtaining strong multilayer and combination materials. Exclusion of any chemical initiating agents and catalysts makes it practical to produce materials featuring high purity.

The other distinctive feature of radiochemical reactions is that they cover thick layers. Reaction zones are positioned along the tracks of primary and secondary particles. The spacing of the tracks is dependent on the kinds of ionizing particles, their ionization capacity and intensity, and duration of the irradiation event. The rate of some reactions are comparatively easy to regulate by varying the radiation intensity. Reactions can be successfully performed in solids; therefore, the relevant processes in preliminarily molded articles are feasible. Thus, the radiochemical process follows the molding of the articles, which makes the process more convenient than existing traditional technology. In some cases, the use of ionizing radiation makes it possible to impart to the materials some new properties not attainable by other methods.

The energy capacity of a radiochemical reaction is determined by radiochemical yield $G$ which is equal to the number of variations of the given kind (decayed or generated molecules, ions, radicals, etc.) for every 100 eV of the absorbed energy [12]. For example, the air ionization yield $G = (100\ \text{eV})/(34\ \text{eV}) = 2.9$. This determination method does not take into account the mechanism responsible for this change, nor is it based on the belief that all the absorbed energy is consumed just for this change. Depending on the value of the radiochemical yield $G$, the processes are subdivided into three groups:

1. $1 < G < 10$. Energy-consuming processes with high activation energies, including endothermic reactions.
2. $10 < G < 20$. Reactions with low radiation energy and formation of short chains, including, sometimes, exothermic reactions. The potential barrier of reacting components is low. Reactions between molecules may proceed with a low degree of excitation, which ensures more efficient use of radiation energy.
3. $G > 20$ (up to $10^6$ or more). These are generally chain reactions. Initial radiation-chemical event initiates multiple transformation on the molecular structure of matter.

By using ionizing radiation, it becomes possible to effectively perform different chemical processes: polymerization, cross-linking, destruction of polymers, etc. In some processes, use is made of ultraviolet radiation instead of high-energy radiation. Ultraviolet radiation may provide excitation and, in some cases, ionization of atoms. The principal difference between interaction of photons of some 10 eV with matter and interaction of high-energy particles and photons is that each photon is capable of exciting only one molecule. The excited products are uniformly distributed over the planes perpendicular to the direction of incident of light. The amount of the reaction products is little, and the latter occupy the position near the surface of materials under irradiation. Complex reactions arising along tracks of fast particles due to high concentration of excited and ionized molecules are practically excluded. In treating very thin surface layers of materials, the use of ultraviolet radiation may sometimes prove to be more efficient. Sometimes use is also made of photoradiation combination treatment of materials.

## 3.2 RADIATION POLYMERIZATION

In the process of polymerization, the unsaturated molecules of a low-molecular compound (monomer) combine with one another, thereby producing a polymer—a compound of high-molecular weight (mass). The molecular chain of a polymer may be composed of similar structural links (homopolymer) or comprise links of different kinds (copolymer) (see Fig. 3.1). In formation of linear or branched polymers, the structural links gradually join into union with one another. Macromolecules represent long chains (in some cases with side branches). Molecules of two-dimensional polymers form flat structures, while in spatial polymers molecules grow in three directions.

Ionizing radiation is capable of stimulating both the process of combination of monomer molecules into polymeric chains and the process of modification of synthesized (by any methods) one-dimensional polymer into a three-dimensional (spatial) cross-linked material. In the latter, adjacent molecules become chemically bound and cross-linked. The peculiarities of the cross-linking process are discussed in next section.

In some cases, polymerization can be initiated merely by irradiating a monomer for a short period of time for producing radicals and for "burning out" inhibiting admixtures. Polymeric chains continue growing even after the irradiation is discontinued (postpolymerization process).

Polymerization can be performed in liquid, gas, and solid phases and in solution and emulsion. In this case, the radiochemical reaction yield will reach $G \geq 10^6$. At some point in time, the polymeric chain stops growing for some reason or other and the

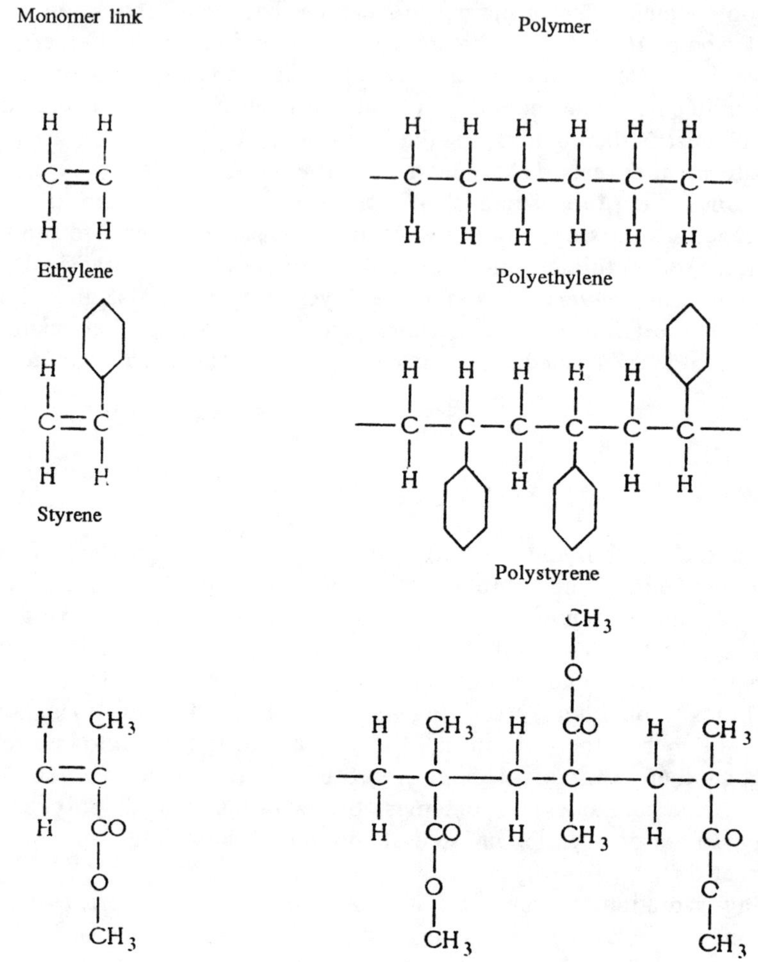


**Figure 3.1** Monomeric links and polymers formed from them.

chain is broken. At a high-intensity radiation dose, the important role in the process of molecular chain stopping may be played by high concentration of primary radicals produced under the action of primary radiation.

Reaction yield $G$ in polymerization in a general case is related to absorbed dose rate $P$ by exponential function $G \sim P^n$. For many processes $n = 0.5$. With a rise of dosage rate $n$ may tend to diminish due to increasing effects of chain stopping as a result of recombination of radicals. Radiation polymerization may be carried out at a temperature optimal for chain propagation, in contradistinction to catalyst-initiated reactions which, as a rule, require higher temperatures. At high radiation intensity, the concentration of produced radicals may be so great that the effect of any inhibitors contained in materials can be overcome. Some polymerization reactions requiring higher pressures without radiation proceed readily at normal pressures under radiation effect.

Of principal interest for industrial use is radiation polymerization of acrylamide, trioxane, fluoro-olefins, as well as copolymerization of ethylene with vinyl chloride, tetrafluoro-ethylene, acrylonitrile, carbon monoxide, sulfur dioxide, and some other processes.

### 3.2.1 Radiation Emulsion Polymerization

In this process, a monomer is dissolved in water, which substantially creates conditions for the reaction process. The produced polymer particles are suspended in water, which ensures a developed surface and effective cooling. Chain stopping due to recombination of microradicals is restricted by their spatial isolation. Rate of polymerization and molecular mass during a long period of the reaction process are independent of concentration of the monomer, which provides a highly efficient continuous process with a great degree of conversion of the monomer. Thus, for example, the rate of radiation polymerization of styrene in emulsion proceeds 100 times that of the process in a pure monomer; additionally, the molecular mass of the product is much higher. A similar gain is characteristic of emulsion reactions for polymerization of methylmethacrylate, vinyl acetate, acrylonitrile, and other compounds.

### 3.2.2 Radiation Graft Copolymerization

In this process, polymeric side chains are produced on free radicals or ions arising under the effect of radiation in main polymer chains. Depending on the process conditions, monomer grafting can be localized within the surface layer or be propagated throughout the full depth of the polymeric base. For localized grafting, it is common practice to change the surface properties, such as adhesion, colorability, resistance to light aging, rotting. Grafting is applicable not only to synthetic polymers but also to wood, paper, and other materials. There are several different methods of graft polymerization, namely:

1. In the process of irradiation the base polymer under processing is brought into contact with the liquid or gaseous monomer to be grafted.
2. Radiation is applied after preliminary contact with monomer (monomer is applied onto polymer surface or is mixed up with polymer).
3. Polymer base is first irradiated in inert atmosphere (in some cases at lower temperature for better conservation of free radicals) and then is brought in contact with monomer.
4. The polymer of the base is irradiated in an oxygen medium or in the air. Peroxide and hydroperoxide groups produced this time remain stable for a long time (up to several months). For the purpose of copolymerization reaction the contact with a monomer is provided at a higher temperature. Peroxide and hydroperoxide groups decompose, producing active radicals which initiate the process of production of a copolymer.

### 3.2.3 Radiation Polymerization in Heterogeneous Systems

This process is used as the basis for producing wood-polymer and concrete-polymer materials.

**Radiation Modification of Wood.** This process consists in impregnating wood articles in vacuum by a monomer (up to 50 to 70 percent in mass) with subsequent irradiation up to a dose of 10 to 15 kGy in a tight chamber in inert atmosphere or in monomer vapours at a pressure of the order of one MPa. On the basis of cheap wood species it is possible to obtain new materials to substitute oak, beech, hornbeam, etc. The properties of these materials depend both on initial wood to be processed and on kind of monomer and impregnation technique. This method enables fabrication of fire-proof, biologically resistant materials intended for use in aggressive media. One of the most common applications of this process is production of parquet.

**Radiation Modification of Concrete Articles.** This is similar to the process described above for wood. First, all the articles to be treated undergo drying at a temperature of 110° to 130°C and then are subjected for about 1 h to evacuation of up to 1500 Pa. These articles are then impregnated with a monomer at a pressure of approximately 1 MPa. Unlike the wood base, the amount of monomer in this process amounts to 4 to 7 percent in mass. The radiation dose also lies within 10 and 15 kGy. Tables 3.1 and 3.2 list some characteristics of wood-polymer and concrete-polymer materials [13].

**Table 3.1 Mechanical properties of different wood species before and after modification by polymethylmethacrylate**

| Strength characteristics | Oak, natural | Birch | | Aspen | |
|---|---|---|---|---|---|
| | | Natural | Modified | Natural | Modified |
| Cross-breaking strength, MPa | 100–300 | 100–110 | Up to 250 | 70–75 | Up to 150 |
| Shearing along fibers, MPa | | | | | |
| Tangential | 8.5 | 7.0 | 21 | 5.5 | 15 |
| Radial | 7.5 | 6.5 | 15 | 5 | 10 |
| Compression along fibers, Mpa | 50–52 | 40–45 | 100–125 | 35 | 100 |
| Detrition, mm | 0.66 | 0.70 | 0.23 | 1.15 | 0.55 |

The articles produced by means of radiochemical modification of wood and concrete are, as a rule, of large mass, and in their production process use is made of gamma radiation. The radiation dosage rate should be moderate to avoid heating the articles. In some cases, it is found preferable to first irradiate the articles at high doses for a short time so as to exclude evaporation and leakage

**Table 3.2 Some characteristics of concrete polymers**

| Properties | Concrete | Concrete polymer |
|---|---|---|
| Ultimate strength, MPa | | |
| Compression | 30–50 | 100–200 |
| Tensile | 2–3 | 10–14 |
| Cross-breaking | 5–6 | 20–30 |
| Compression modulus, $10^4$ MPa | 2.5–3.5 | 4.5–5.0 |
| Adhesion with reinforcement, MPa | 1–2 | 10–18 |
| Ultimate coefficient of moisture absorption, % of mass | 14.1 | 0 |
| Moisture permeability of specimen, g at $p = 0.3$ MPa | | |
| for 1 h | 173 | 0 |
| For 4 days | – | 0 |
| Cold resistance at cyclic freezing/defrosting | | |
| After 500 cycles | 20 | 6 |
| After 5000 cycles | Failure | 8.5 |

of monomer from the article. Concrete-polymer articles are generally large-sized, so that it is rather difficult to provide uniform irradiation. Cooling conditions in this case are in practice not inferior, because despite great masses, the specific content of monomer and, correspondingly, the heat release per unit volume in concrete-based materials are one order lower than those in impregnated wood materials.

## 3.3 RADIATION CROSS-LINKING OF POLYMERS

Radiation of polymers may radically transform their structure. One type of transformation consists in creating cross-links between linear molecules and in producing three-dimensional (spatial) polymers with higher molecular masses (cross-linking process). The other way is polymer chain termination and decrease of molecular masses (destruction process). (A third way may also be noted: change of unsaturation, i.e., in the type of double links.)

In the process of cross-linkage, negligible chemical variations of large molecules (for instance, 1 variation for $10^4$ and more links) cause pronounced changes of their properties. (Irradiation yields low-temperature products of radiolysis, that in turn causes evolution of gas from polymers under irradiation.)

Cross-linking and destruction generally occur simultaneously. However, the ratio between their rates tends to vary, depending on the polymer involved, radiation conditions, and presence of various impurities. The cross-linking process is preferred for irradiation of polyethylene (with destruction to cross-linking rate ratio of 1:3), polypropylene, polyvinyl chloride, polyacrylamide, polystyrene, some types of rubber, polyamide, and polyethers. The structure of a cross-linked polyethylene is diagrammatically shown in Fig. 3.2. The destruction process plays a leading role in irradiation of such polymers as polyisobutylene, polymethylmethacrylate, polytetrafluoroethylene, and cellulose.

Cross bonds in the cross-linking process may be created both between different molecules and also within one molecule. The number of isolated molecules tends to diminish. At a certain density of molecular bonds, a single space-network is formed in the material. One of the basic properties of this network is insolubility in hot organic solvents: the material does not become creepy at temperatures exceeding the melting temperature of nonirradiated polymer and offers great cracking resistance. Below are set forth some characteristics of radiation cross-linkage:

1. The degree of polymer cross-linking is proportional to absorbed

radiation dose and is almost independent of dose rate. When the process is performed in the air, the increase of radiation dose rate leads to a higher quality of produced materials since the exposure time is reduced, which makes for minimizing radiation oxidation of polymers under irradiation. This is beneficial because oxidation is an undesirable process competing with the cross-linking process and causing deterioration of mechanical and electrical characteristics of processed materials.

2. The degree of cross-linking is negligibly dependent on the kind of high energy radiation (electrons, photons, or heavy particles).
3. Radiochemical yield $G$ is substantially dependent on small admixtures of such substances as acetylene, thiourea, and oxygen.

As compared with cross-linkage initiated by chemical catalysts, radiation cross-linking as well as other radiochemical processes enable easy control of reaction rate and the finished product contains no catalyst remainings.

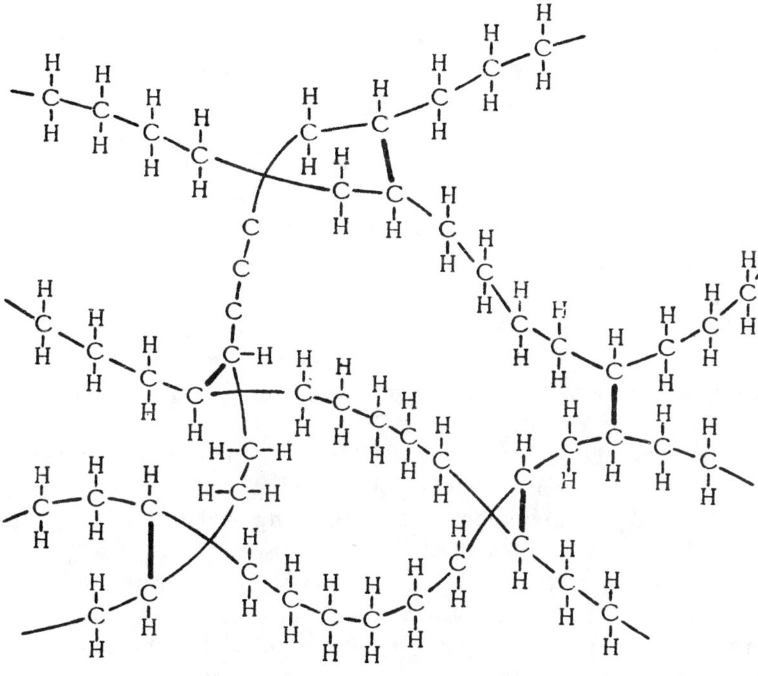

**Figure 3.2** Structure of cross-linked polymer. Intermolecular links are shown with thick lines.

### 3.3.1 Radiation Cross-linking of Polyethylene

This process is the most extensively used industrial radiochemical process. There are a number of applications of irradiated polyethylene: pipes and insulation of wires and cables operating in extremely hard conditions (high temperatures, aggressive chemical media, irradiation, etc.); sheet porous polyethylene used for thermal, noise, and electric insulation and as shock-absorbing pads; articles with "memory effect" utilized for enclosure of cables, for packing, and so on.

Radiation treatment is applied to a wide assortment of articles made of composites of polyethylene and other polymers: polyethylene of high and low density, copolymers of ethylene with vinyl acetate and ethyl acrylate, with tetrafluoroethylene, polyvinyl chloride, and others. For obtaining the desired performance characteristics use is made of additives such as antioxidants, sensitizers, fillers, and plasticizers [14]. The dosage rates in production of polymers cross-linked by irradiation effect usually lie within 50 and 300 kGy.

One of the problems arising in the course of service life of articles made of cross-linked polyethylene is protection against oxidation, which is particularly dangerous at high working temperatures. This problem is solved with the aid of thermal chemical structure stabilization. The antioxidants are selected for this purpose both with due regard for their chemical activity as an oxidation inhibitor and with account taken of their effect on the structure-formation process in polymer [15].

Applications that are most characteristic of irradiated polyethylene in cable industry are insulation of wires, power cables, communication cables, computers, and electric motors. Cables covered with insulation from radiation-modified polyethylene allow higher current loads and display proper serviceability at high temperatures (Table 3.3). The quality of such insulation will not deteriorate in a medium free of oxygen with long-term operations in the temperature range of 250° to 300°C.

To obtain a porous polyethylene (Table 3.4) [16], foaming agents are added to the initial material. Upon molding and irradiation, the articles are heated to disintegrate the foaming agents. As a rule, two-sided irradiation by an electron beam is used. A high degree of foaming corresponds to the quota of gel (unsolved fraction) in the irradiated material (i.e., 30 to 40 percent).

A porous polyethylene of natural white color has closed pores, can be readily dyed in desired colors, is perfectly workable, and allows impression and hot molding. Common applications of porous polyethylene are for thermal insulation of pipelines, sound insula-

**Table 3.3 Heat resistance of cross-linked polyethylene**

| | Operating temperature, °C | | |
|---|---|---|---|
| Polyethylene | Continuous | Up to 100 h/year | For short time |
| Chemically cross-linked | 90 | 130 | 250 |
| Cross-linked by radiation | 150 | 200 | 300 |

tion of air conditioners, insulating tapes, coating material of ceilings and walls, packing material, and sports protective facilities.

Other uses of irradiated polyethylene are based on the "memory effect." For example, when an article fabricated of irradiated polyethylene is heated, deformed, and then cooled down in deformed state the polyethylene proves to be capable of "memorizing" its initial size. If the said article is heated again to a temperature above 100°C, it will assume its original dimensions. This is accounted for by the fact that the polyethylene structure cross-linked under irradiation effect features equilibrium. As a result of heating with subsequent deformation, the crystallites melt and then, in the process of cooling, they fix the molecules in the deformed state, for the intermolecular force is superior to the restoring elastic force in cross-links. In the case of repeated heating of the polyethylene material, the crystallites remelt and the cross-linked structure immediately assumes equilibrium. Shrinkage depends on the degree of deformation and may amount to 50 to 300 percent. The shrinkage force reaches tens of kilograms per square centimeter. Thermo-shrinkable tubes enable cabling connections to be insulated promptly and reliably (see Fig. 3.3). Bags made from such material

**Table 3.4 Properties of some porous materials**

| Properties | Irradiated porous polyethylene | Porous polyethylene | Porous polystyrene |
|---|---|---|---|
| Density, $g/cm^3$ | 0.33 | 0.017 | 0.032 |
| Breaking strength, MPa | 0, 4, 3 | 0.1 | 0.38–0.48 |
| Elongation, % | 204 | 200 | 1 |
| Pressure hardness, $kg/cm^2$ | 0.36 | 0.02 | 0.6 |
| Working temperature, °C | 80 | 100–200 | 70 |
| Moisture absorption | Negligible | High | Low |

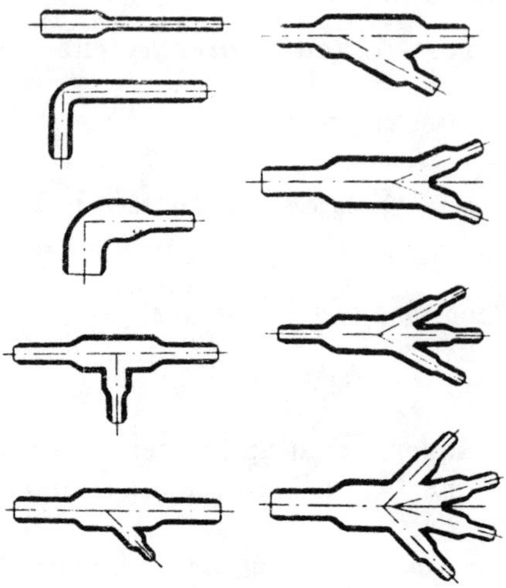

Figure 3.3 Versions of thermoshrinkable parts for insulation of cable connections.

are used for proper and inexpensive tight packing of goods. Owing to high strength of thermoshrinkable polyethylene and its close-fitting ability, it ensures tight packing of any article including, ones with sharp edges and projections. The properties of the polyethylene are exemplified by the characteristics of thermoshrinkable films: thermoplen and polyrad (Table 3.5) [17].

A new type of thermoshrinkable tubes cross-linked by irradiation and fabricated on the basis of fluorinated polyolefin has been described in [18]. The tubes are transparent, fire-resistant, and feature good resistance at high temperatures (up to 120°C).

### 3.3.2 Radiation Vulcanization of Rubber

In addition to cross-linking of polyethylene, radiation vulcanization of rubber also finds wide application. The cross-links produced by the vulcanization process are much more stable than the long sulfur bridges formed with the usual hot vulcanization. The rubber cross-linked by irradiation features higher heat resistance, resistance to aging and deformation at high temperatures, and much better wear resistance. Rubber used for fabrication of tires and other articles is a mixture based on a high-polymeric substance, an elastomer capable of elongating 10 to 15 times. Synthetic or natural rubber is used as an elastomer, with the addition of different fillers and antioxidants introduced into the rubber. If a need arises, some sensitizers are also applied for increasing the radiation vulcanization yield.

The radiochemical yield for different types of rubber lies within $G$ = 0.1 to 20. The process of cross-linking proceeds simultaneously with a destruction process. As with the cross-linking of polyethylene, oxidation under high-intensity radiation in the air is not essential. Therefore, it is expedient to use electron accelerators for radiation vulcanization. The radiation dosage amounts to 5 $\times 10^4$ to 10 $\times 10^4$ Gy, and the energy consumed directly in vulcanization may be many times less than by the thermochemical method. As the radiation vulcanization proceeds more rapidly, it seems to be practical to develop continuous production lines.

When fabricating molded articles (such as tires), a partial chemical cure is first required when the molded article is found capable of retaining its geometry unchanged, it is extracted from the mold and subjected to irradiation. In vulcanizing tires, it is necessary to irradiate the rubber layers of a few tens of millimeters that would demand the electron energy of 10 to 15 MeV and more. In this case, however, two problems arise: on the one hand, there is the too high cost of radiation of using electron accelerators, and, on the other hand, there is a probability of residual activity in materials subjected to treatment. The profitability of the process is improving yearly due to continuous reduction of cost of electron beam energy. As far as activation is concerned, the radioactive nuclides produced at electron energies of up to 30 MeV are characterized with half-life periods not exceeding 1 h, and therefore, the induced activity will decrease within several hours to a level approximating the natural radioactivity background.

Radiation vulcanization finds use in manufacturing rubber insulation of cables. Wide application is also found in the production of heat-resistant insulation tapes and rubber-glass textile on the base of polyheterosiloxane rubber. These materials ensure

**Table 3.5 Properties of radiation-modified thermoshrinkable and hardened films**

| Property | Thermoplen | Polyrad |
|---|---|---|
| Width, mm | Up to 3000 | Up to 3000 |
| Thickness, μm | 50–150 | 100–300 |
| Ultimate tensile strength, MPa | Up to 50 | Up to 20 |
| Elongation at rupture, % | 200 | 450 |
| Impact strength, g | 800–1000 | Up to 1400 |
| Shrinkage, % | Up to 70 | – |
| Shrinkage strain, MPa | Up to 3 | – |
| Resistance to cracking, h | 6000 | 6000 |
| Brittle point, °C | –70 | –70 |
| Penetrability, g/(mm²/day) | 1.5–1.7 | 1.5–1.7 |
| Working temperature range, °C | –70 – +100 | –70 – +100 |

a long period of service at 250°C and feature perfect autohesion, excellent insulating properties, and moisture and cold resistance. Radiation vulcanization is used in the cable industry for production of extruded insulation from siloxane rubber.

## 3.4 RADIATION DESTRUCTION OF POLYMERS

As noted in Section 3.3, irradiation of polymers can cause chain scission and formation of polymers with lower molecular masses. The destruction radically differs from depolymerization (often invited by heating of materials). Depolymerization is a chain reaction which results in converting the entire or parts of the chain into a monomer.

In the case of destruction, only some scissions of chains take place and chemical bonds undergo some transformation close to chain-opening points. No monomers will be produced, as a rule. The typical change of properties caused by the destruction are increase of solubility and decrease of viscosity.

The destruction process is used for production of substances that are unlikely to be obtained by the use of direct synthesis. Thus the destruction method enables, for example, production of spray treatment material from polytetrafluoroethylene and lubricant from butyl rubber. It is possible to produce sugar from cellulose, but it is uneconomical. The absorbed radiation doses in these processes amount to hundreds of kilograys. The destruction process prevails in irradiation of many polymers in extremely diluted aqueous solutions that in large measure are conditioned by involving products of water radiolysis in reactions with polymer radicals.

### 3.4.1 Sewage Treatment

One of the promising applications of the destruction process is for sewage treatment [19]. Pronounced increase in production and industrial and domestic utilization of surface-active synthetic substances, has exacerbated the sewage treatment problem, and contamination of water reservoirs has increased. These synthetic substances are basically used for producing detergents featuring much better properties than fatty-acid soap. The distinguishing feature of surface-active substances is a high stability to chemical interaction with solutions of alkalies and salts, slow biochemical oxidation, and high foaming capacity. The sewage of industrial enterprises also receives a number of substances used in various technological processes. Treatment of water by active carbon or ion-exchange resins may be effective, but from an economical standpoint, this is not expedient.

Radiation of surface-active substances promotes their destruction, i.e., decomposition into lighter and easily removable substances. Radiation treatment, in contrast to biological treatment, affects all organic compounds. In the presence of oxygen, organic compounds are oxidized. The treatment becomes much more efficient when sewage is subjected to effects of both radiation and ozonization. This fact is accounted for by transformation of $H_2O$ radicals into OH radicals which are highly oxidizing for most of organic compounds.

### 3.4.2 Radiation Reclaiming of Butyl Rubber

Radiation destruction enables raw materials with the required plastic and elastic characteristics to be obtained from waste tires. In sulfur vulcanization, when butyl rubbers (as distinct from a number of other kinds of rubber) are subjected to irradiation, the destruction prevails. By varying the radiation dose, it is possible to control the product: the higher the dose rate, the higher is the plasticity of the product. Table 3.6 contains the properties of reclaimed rubber produced by different methods [20].

## 3.5 COMBINATION PROCESSES

As can be seen from the description of the polymerization processes of cross-linking and destruction, it is seldom that only one of these processes takes place under irradiation. Irradiation usually causes various radiochemical reactions to occur, some of which promote while others impede achievement of desired properties of the final product. In some technological processes it has been possible to employ combined effects of different reactions.

**Table 3.6 Properties of reclaimed rubber produced by different methods**

| Method of reclaiming | Softness, mm | Tensile strength, MPa | Elongation, % |
|---|---|---|---|
| Radiation (100 kGy) | 2.1–3.2 | 6.0–10.0 | 350–550 |
| Water digestion | 2.5–3.3 | 4.5–6.5 | 340–500 |
| Thermomechanical | 2.7–3.2 | 4.0–5.5 | 300–425 |

### 3.5.1 Radiation Curing of Varnish and Paint Coatings

One of the most widespread technologies of this kind is radiation curing of varnish and paint coatings. In the case of the commonly applied method of thermochemical curing of varnish coatings, the articles are warmed to 60 to 70°C in drying ovens whose length reaches a few tens of meters. The articles are held in the ovens for several hours. In some cases, the varnished surfaces are polished, necessitated by sagging varnish coats and adhesion of dust.

In contradistinction to the traditional technology, radiation polymerization takes less than 1 s and requires no additional grinding and polishing of the articles. Three processes take place simultaneously: (1) polymerization of monomers (or low-molecular polymers), (2) cross-linking of produced polymeric molecules, and (3) development of chemical bonding between polymer and molecules of the substrate. Under the effect of irradiation, some harmful processes may also occur, such as oxidation or chain scission. However, these processes tend to decrease at a high dosage rate of ~10 kGy/s (which is considered a normal dose rate for today's accelerators).

When the purpose is to remove oxidized varnishes in the radiation zone, some equipment is used in an inert atmosphere, such as nitrogen, or some other method is used. The radiation-cured coatings feature high mechanical strength, chemical stability, and heat resistance.

Unlike traditional technology, explosion and fire safety of this process is much improved. The coatings polymerized at normal temperature conditions may be applied onto wooden articles, paper, fabric, etc. Most applicable to radiation polymerization are systems based on unsaturated oligoethers usually used in conjunction with monomers or in the form of mixtures of different oligomers. Irradiation of varnish and paint coatings made of polyurethane, fluoroplastic, polyxyloxane, unsaturated polyolefins also improves thermomechanical properties of coated surfaces. Depending on the desired compositions, the radiation dosage may vary in the range of 20 to 200 kGy.

For curing, graft polymerization and other processes in thin surface layers of material (generally within 500 μm) use is made of low-energy electron beams (150 to 300 keV). These processes also find application for fabricating thin, laminated materials, foiled dielectrics, electrical insulation coatings on different substrates, and packs of multilayer plates. Curing of binders is used in the manufacture of magnetic tapes, in polygraphic industry, and in some other processes. Curing of coatings deposited on the magnetic tape improves the quality of coating, reduces the cost of the hardening process and helps to achieve some other results not feasible by the common methods. For example, in case of curing, no heat treatments are used and, as a result, the strength of the coatings is improved. Curing of deep print ink on banknotes has also been investigated [21]. The obtained results bear witness to potential of this method to extend circulation time of paper money.

### 3.5.2 Radiation Modification of Textiles

This process is used for fabrication of crease-resistant, fireproof, hydrophilic, and hydrophobic fabrics, for improvement of resistance to contamination, to decrease of accumulation of static electricity, to increase of resistance to rotting, and so on. For radiation processing of fabrics, use is made of cross-linking, destruction, graft polymerization, and other processes.

For production of crease-resistant fabrics capable of remaining unwrinkled after washing, cross-links must be produced between hydroxyl groups in cotton or viscose rayon fibers. Radiation polymerization of fireproof preparations makes it possible to achieve incombustibility. Fluorinated acrylates and siloxanes are grafted as water repellents. Resistance to contamination is obtained by grafting hydrophilic polymers. Radiation may also be used for fixing dye with polymerized vinyl groups.

### 3.5.3 Curing Fiberglass Materials

The properties of materials obtained by the use of curing of fiberglass materials are determined both by the characteristics of the fibers and polymer and by the physical and mechanical phenomena at the interface. Use is basically made of polyether resins and vinyl polymers as binding polymers for fiberglass materials. Irradiation initiates simultaneous copolymerization and cross-linkage. Characteristics of the finished material include a relatively low density, high-impact strength, and chemical stability. This process is performed, as a rule, with electron accelerators, and polymerization is effected gradually, layer by layer for instance, when the fiberglass fabric is spooled up. The desired result will be achieved if the electrons penetrate to the approximate depth that the layers of fiberglass are impregnated with a binding agent. As a result of such a treatment, it is possible to obtain a substantial thickness of fiberglass material at a relatively low electron energy. The radiation dosage is 100 kGy. This method enables different layers of the article to acquire the predetermined properties. Table 3.7 lists properties of radiation-produced fiberglass materials.

## 3.6 PROCESSES IN GAS-PHASE MATTER

**Radiation purification of industrial gases** is a subject whose study is comparatively recent (approximately 15 years old) [23]. The most deleterious impurities produced by burning oil, carbon fuels, and other materials are sulfur dioxide ($SO_2$) and nitrogen oxides $NO_x$ (particularly NO). $SO_2$ contained in exhaust gases may be bound with alkaline solutions. Removal of $NO_x$ presents appreciable difficulties because of the low reactivity of these oxides. Effective industrial methods have not been yet devised. When gas is cleaned by irradiation, it is found necessary to add some ammonia ($NH_3$). The irradiation initiates reactions of $NO_x$ and $SO_2$ with $NH_3$. The produced particles of ammonia salts are entrapped by an electrical precipitator. At a gas temperature of ~70°C and a radiation dose of ~150 kGy, some 80 percent of $NO_x$ and about 98 percent of $SO_2$ can be removed. Industrial use of this process may be started in the near future.

Another radiochemical reaction in a gas medium that is of interest for industrial application is removal of hydrogen from

**Table 3.7 Properties of wound fiberglass materials produced by radiation method [22]**

| Binders* | Radiation dose, kGy | Content, % | | Ultimate stress, $MPa/cm^2$ | |
|---|---|---|---|---|---|
| | | Binder | Gel fraction | Compression | Interlayer shear |
| Oligomaleinatacrylate | 200 | 22.9 | 90.5 | 775 | 66 |
| Phosphorus-bearing oligomaleinatacrylate, higher elasticity | 200 | 22.5 | 98.5 | 693 | 59 |
| Phosphorus-containing oligomaleinatacrylate | 300 | 23.5 | 91.7 | 750 | 30 |
| Oligomaleinatacrylate with higher phosphorus content | 200 | 22.8 | 90.5 | 717 | 30 |

*Filler BMC-8–26 × 1 × 2-80

electrolytic chlorine [24] by the use of irradiation. The evergrowing demand for chlorine in various production processes necessitates its increased production and provision for its transportation over long distances. In order to transport the chlorine, it is liquefied. For liquefaction, it must be purified of some hydrogen admixtures to attain an explosion-proof level (i.e., the content of the residual hydrogen must be less than 4 percent of the chlorine volume). The electrolytic chlorine produced by means of mercury radiolysis contains ~1 percent hydrogen, which makes the chlorine explosion-hazardous in case of liquefaction to ~80 percent. The commonly known purification methods (thermochemical, photochemical, electrolytic discharge) do not find industrial application because of substantial consumption of energy and corresponding high cost.

Purification by irradiation results in production of atomic chlorine reacting with hydrogen and the following chain reaction is thereby initiated:

$$Cl_2 \rightarrow 2Cl$$

$$Cl + H_2 \rightarrow HCl + H$$

$$H + Cl_2 \rightarrow HCl + Cl$$

and so on. However, in addition to hydrogen, the chlorine contains water vapors and other impurities, so that a number of accompanying reactions occur simultaneous to hydrogen fixation, which cause chain opening, production of hydrogen (water radiolysis, in particular), and other effects that hamper the main process.

The feasibility of radiation fixation of atmospheric nitrogen is under study [25]. Under the effect of irradiation, nitrogen oxides are produced and entrapped. However, some competitive processes cause an extremely negligible output of useful product; therefore, this method has not yet been found profitable.

Relevant research is carried out to study radiation synthesis of organochlorosilane, hydrocyanic acid, phosgene, hydrazine, nitriles, and other compounds. It is found feasible to apply electron radiation for extraction of hydrogen from methane, for decomposition of carbon dioxide, and for production of vinyl chloride and styrene.

CHAPTER

# FOUR

# IRRADIATION OF BIOLOGICAL OBJECTS

## 4.1 PHYSICOCHEMICAL BASIS OF RADIOBIOLOGICAL PROCESSES

Research on the effects of ionizing radiation on living organisms was stimulated by discovery of roentgen radiation and was primarily aimed at determining its deleterious effects and developing radiation-shielding methods. While studying the processes adversely affecting irradiated objects, researchers succeeded in elucidating the recovery processes of irradiated biological objects on the molecular and cellular level as well as with respect to the organism as a whole. The purposeful application of ionization radiation for biological effect spawned numerous radiation-biological technologies in the past few decades.

Depending on the applied dose, different results of radiation processing may be obtained. These include growth stimulation, growth retardation, genetic changes, sterilization, cessation of vital functions, and chemical decay. Radiation sensitivity of biological objects is dependent first of all on the properties of the object under irradiation and on the environment—temperature, humidity, and gas composition [26, 27].

Radiation injury of organic cells, as a rule, comprises two

stages: (1) initial damage followed by (2) realization or recovery of latent defects. Initial defects in macromolecules are noted in the first stage. They are caused by direct or indirect effect of irradiation which consists in removal of electrons by active products of radiolysis of water and by low-molecular impurities. Water radiolysis gives rise to a number of chemically active products including $H^+$, $OH^-$, and hydrated electrons. A hydrated electron is essentially an electron surrounded by water molecules polarized in its electric field. The formation of this kind is highly persistent in water and offers high reactivity. An important part may also be played by products of radiolysis of low-molecular impurities contained in cells or specifically injected into the biological object.

The nucleus of the DNA molecule, to be more precise, is the main object affected by irradiation of a living organism. However, processes of injury in cytoplasm and, according to recent investigations, damage of biological membranes are also important processes that occur.

Defects in protein molecules occur at one site, as a rule, irrespective of whether the injury is caused by direct or by indirect irradiation. Energy injected into a molecule will migrate until a "weak" point is reached. The weak place is usually a disulfide bond due to great lability of electrons of sulfur atoms.

At the second stage of radiation injury, either realization or recovery of latent defects takes place in macromolecules. Development of the process at this stage may last for hours, days, and even months. One of the most essential factors responsible for modification of biological macromolecule is the presence of oxygen ("oxygen effect"). At the same time, oxygen is capable of deactivating the radicals arising as a result of radiolysis of water and low-molecular impurities, thereby minimizing the indirect effect of irradiation. Should the need arise to protect any biomaterials against radiation injury, it is common practice to add some low-molecular admixtures which can serve as electron donors or be adsorbed by macromolecules [27].

For protection against radiation injury, use is generally made of sulfur-containing compounds. With this in view, some sulfhydril compounds, such as cysteine and cystamine, are injected into living organisms before or after irradiation so as to substantially minimize the irradiation aftereffects. The number of radiation protectors grows steadily, and search for other effective substances against radiation injury is underway.

The approximate values of the absorbed radiation dose applied for various effects on living plants are:

| Process | Dose, Gy |
|---|---|
| Stimulation of growth (presowing irradiation of seeds) | 5–40 |
| Genetic changes* | $10–10^3$ |
| Retardation of growth | 50–150 |
| Cessation of vital functions | $10–1.2 \times 10^3$ |
| Chemical decay | $5 \times 10^5–10^6$ |

The dose rates required for obtaining the desired effect are reduced, as a rule, when irradiating organisms of higher organization.**

The following examples indicate the approximate dosage causing death of 50 percent of irradiated objects in different groups of organisms [26].

| Type of organisms | Dose rate, Gy |
|---|---|
| Mammals | 7 |
| Birds | 10 |
| Insects | $40–100^3$ |
| Angiosperms (seeds) | $50–1.2 \times 10^3$ |
| Gymnosperms (seeds) | 10 |
| Algae | 100 |
| Fungi | $50–5 \times 10^3$ |
| Protozoa | $100–3.5 \times 10^3$ |
| Bacteria | $20–5 \times 10^3$ |
| Viruses | $100–1.5 \times 10^3$ |

## 4.2 STERILIZATION OF MEDICAL MATERIALS

Sterilization of medical materials was the first radiation technological process adopted for industrial use and is now the most widely applied radiobiological technology. Successful application of radiation sterilization is favored by several factors, namely:

1. The dose rate suitable for reliably killing microorganisms (as a rule, 25 kGy) does not cause damage to most medical equipment:

*The values stand for the dosage rates used for radiation mutagenesis of plants. Genetic effects of ionizing radiation may also take place at other doses and are linearly dependent on the doses.

**This does not apply to chemical decay. The dosage of this process (~1 MGy) is similar for all biological objects.

surgical threads (catgut), dressing, syringes, surgical instruments, and other articles containing metal and plastic components, as well as rubber of some grades.

2. Sterilization of some materials by boiling is either forbidden or undesirable. For example, one-application articles fabricated of cheap thermoplastics can be subjected only to cold sterilization. Surgical catgut sterilized by irradiation features some 10 percent higher strength as compared to that sterilized in boiling water.
3. Articles can be sterilized by irradiation in tight enclosures made of cheap plastics.
4. The process is highly reliable. When sterilization is carried out in autoclaves or chemically with the use of ethylene oxide, it is essential that several parameters (temperature, time, pressure, humidity and concentration of ethylene oxide) be continuously monitored and maintained within the prescribed limits. In contrast, radiation sterilization required only the time and dose rate to be controlled.
5. Radiation sterilization can be performed continuously.

The mechanism of lesion of microorganisms depends on liability to radiation injury as well as on the surrounding medium. There are two basic ways of radiation injury: direct ionization of biomacromolecules or effect of chemically active products of radiolysis of cell sap. Unlike thermal sterilization, where microorganisms are killed immediately due to decay of protein, microorganisms survive radiation for a few hours. It has been found experimentally that nonsporogenous microorganisms tend to stop replicating at radiation doses of 5 to 10 kGy. The same effect is obtained for sporogenous bacteria at a dose of 15 to 20 kGy. The number of microorganisms on an article before sterilization may be $10^{10}$ and more; therefore, it is not inconceivable that after sterilization, some radiation-resistant organisms may still remain alive. Naturally, detection of microorganisms in a sterilized article will be less likely with an increase of radiation dose. For instance, for *Salmonella* bacteria, application of a dose of 6 to 9 kGy results in a population reduction by $10^7$ times. Increase of the dose by 1.5 times leads to decrease of population by $10^{11}$ times. The universally adopted criterion of sterility allows for presence of 1 nonsterilized article among 1 million sterile ones, or in case of some rigorous requirements, 1 nonsterile article for 100 million sterile ones. As a rule, this condition can be met with application of a dose of 25 kGy. However, this dose rate may be varied with due regard for the required degree of sterility, degree of contamination of articles, and some other factors. Radiation sterilization is effected, as a rule, by the use of gamma radiation of bremsstrahlung, which is required for the pronounced thickness of articles to be processed, particularly those confined in packing enclosures.

Radiation is also used for preparation of vaccines. In this case, the dose rate is less than for sterilization, so that the microorganisms are only weakened and then used as vaccine. Resistance of various medicines to radiation doses of 10 to 50 kGy has been investigated [28]. It has been shown that the overwhelming majority of solid medicinal preparations is radiation-proof, while medical preparation in the form of aqueous solutions are liable, as a rule, to pronounced decomposition. Most aqueous preparations can be stabilized by adding sodium pyrosulfite.

At present, 35 percent of all sterile medical facilities in the United States and Canada are processed by the use of the irradiation method. It is anticipated that by the end of the 1980s this amount will reach 80 percent [29].

## 4.3 PRESERVATION OF FOODSTUFF

Although radiation sterilization of food has great potential, this process has not yet found industrial application. On the one hand, long-term storage (without freezing) of meat, fish, and various meals ready to serve is economically profitable because there is no need to maintain low temperature and, on the other hand, radiation treatment provides better gustatory qualities of preserved food. Wide-scale adoption of this method is restricted mainly by two factors:

1. Application of high doses (25 to 70 kGy) for sterilization of food is required for killing spores of botulism pathogens (*Clostridium botulinum*) but causes oxidation of fats, decomposition of pigments, and other undesirable reactions, finally leading to unwanted change of color, smell, appearance, and structure of food.
2. Second, ferments contained in food tend to gradually decompose foodstuffs, resulting in thinning of food and loss of flavor. With thermal preservation or freezing, the ferments are made inactive and, therefore, are incapable of adversely affecting the foods. For inactivation of ferments by means of irradiation processing, untolerably high radiation doses (of the order of 100 kGy) must be used.

As a result of some 30 years of research, it has been found that the one of the most promising applications of radiation technology is sterilization of foodstuffs that can be ready for eating only after some culinary processing (for example, boiled or roasted meat or fish). Initially, the food undergoes thermal treatment (i.e., warmed to a temperature of 50 to 80°C) as a result of which the ferments contained are inactivated and the food is made ready for eating. Then the ready meal is frozen in a tight package without access to oxygen and sterilized by irradiation (with the temperature being maintained within the range of –40° to –10°C) [30]. Freezing state and absence of oxygen during irradiation treatment minimize physical and chemical changes of food. During subsequent storage of the treated food, the packing material prevents effects of oxygen, moisture, light, and microorganisms. For packing foodstuffs, laminated materials are often used: the inner layer is polyethylene, the next layer is aluminum foil, and the outer layer is also polyethylene or another film. The food prepared and treated in this way can be kept at room temperatures for months.

Successful tests have been made involving a broad variety of foods: smoked and fried sausages, ham, meat meals for breakfast, roasted potatoes, chicken, filleted fish, and so on. The same technique may be used for preserving large pieces of roasted meat, turkey, chicken, etc. Foods subjected to radiation sterilization are used for nourishment of cosmonauts and patients who need a sterile diet. Toxicological investigations of irradiated foodstuffs are being carried on involving many generations of animals. For example, in the United States, 5-year investigations of radiation-sterilized beef involved 1500 dogs, 27,000 rats, and 20,000 mice. Additionally, associated radiochemical research is also in progress for studying the physiological action of substances produced under irradiation.

The aforesaid technology is not applicable to foodstuffs that are eaten raw or that lose their quality when they are subjected to freezing and defrosting. Fresh vegetables and fruit juices may be successfully irradiated for preservation when contained in tight cans. First the cans of vegetables or juice are heated to 80 to 85°C for inactivation of ferments and then are irradiated at a maximum dose of 15 to 20 kGy.

Another use of radiation is to suppress the vital activity of microorganisms in fresh, perishable foodstuffs. This can extend their storage life, for example, by 1 or 2 weeks. Radiation with a dose of 1 to 3 kGy is applied to irradiate strawberries, raspberries, cherries, peaches, grapes, tomatoes, etc. Meat enclosed in polymeric bags irradiated with doses of 1 to 6 kGy can be successfully preserved at a temperature of 2 to 4°C for several weeks. In addition to radiation treatment of foodstuffs, of interest is sterilization of raw materials in the tanning industry to ensure preservation of materials during storage and shipment.

It is likely that radiation treatment of foodstuffs may be performed by the use of bremsstrahlung with an energy of up to 10 to 11 MeV, because the induced activity drops within few days to the level of the natural background of ionizing radiations [31]. Radiation of foodstuff for preservation is expected to become widespread in the near future.

## 4.4 DEINSECTIZATION OF GRAIN

It is well known that protection of crop from pests is one of the most important problems of agriculture. Apart from the insect control in tilled fields, the appropriate measures must also be taken to preserve the crop in storehouses. Pests are brought to storehouses together with grain and other crops and their population proves to be capable of destroying a noticeable part of cropped yield, for example, 5 to 10 percent of world yield of feed grains and some 50 percent of yield in tropical countries.

The main method of insect control at present is chemical treatment of grain (by use of methyl bromide, chlorpicrin, or other chemicals) both in storehouses and in transit (at definite transshipping points). The disadvantage of this method is its high cost and the danger of remaining toxic substances in products. Research in radiation preservation treatment of grain has disclosed that with a dose of 100 to 200 Gy, most insects become sterile and die within 2 to 3 weeks. Such treatment of grain is referred to as radiation deinsectization. Radiation doses of some few hundred Gy will not deteriorate the quality of wheat, maize, peas, buckwheat, rice, and other products. In many countries, public health services have granted official permissions for radiation treatment of grain with doses of 500 to 1000 Gy. In the Soviet Union, authorization has been given to perform radiation deinsectization of grain and dried fruit by use of ionizing radiation (with doses of 1000 Gy), dry food concentrates (with doses of up to 700 Gy), and industrial-scale irradiation of grain by electrons with an energy of up to 4 MeV [32]. Since 1980, a pilot installation has been operating in the seaport grain elevator in Odessa for radiation deinsectization (see Section 9.4)

Deinsectization provides partial protection of grain against repeated infestation for a period of time due to the fact that the fertility of newly brought pests is decreased as a result of copulation with sterile species. A similar method is used for suppression of pest populations in natural conditions by dissemination of sterile insects over a crop field. For example, in the state of Florida in the United States in the late 1950s, the program of destruction of the Callitroga flies was fulfilled over an area of 13,000 km$^2$. Fifty million insects were bred and sterilized every week at a special factory. About 500 people and 20 aircraft took part in the operation. The cost of the program exceeded 10 million dollars. Since 1979, operations aimed at destruction of this fly on the Amami and Okinawa islands (Japan) have been carried out. These operations are expected to be completed within a period of 10 years.

The advantages of radiation over chemical treatment of agricultural products are lower cost, noncontamination of products under treatment, and continuous process by use of production lines.

## 4.5 PRESOWING IRRADIATION OF SEEDS

Irradiation of plants and living organisms may accelerate the cell-division process (sometimes for a short period) and correspondingly enhance rate of growth or reproduction. There are several theories concerning the stimulating effect of small doses of ionizing radiation. According to one of the most popular hypothesis, radiation stimulates a number of physiological and biochemical processes, thereby making for more intense growth and development.

The stimulating effect of radiation doses of a few tens of grays has found wide application in plant production. Many agricultural plants were subjected to irradiation at this dose rates, including corn, vegetable crops, industrial crops, and fodder plants. One of the principal objects of presowing irradiation is stimulation of germination, growth, yielding ability, and quality [33].

## 4.6 INHIBITING GERMINATION OF CROPS IN STORAGE

The problem of unwanted germination occurs when agricultural crops are to be stored for several months. Apart from creation of low-temperature conditions and an oxygen-free medium, this problem can be solved by application of ionizing irradiation of the stored produce. For example, irradiation with a dose of 50 to 150 Gy prevents germination of potato for 5 to 10 months at normal storage temperatures [32]. Higher dosages may bring about undesirable effects, such as conversion of starch into sugar in cases of long-term storage. The radiation treatment is performed a few weeks after the harvest. When the crops are treated by the use of an electron beam, the dose may be increased to 300 to 500 Gy, for it is only the outer layer of tubers that is liable to radiation injury (location of tuber eyes). The storage-effective term after radiation treatment of the crops can be prolonged by several months. However, it should be noted that potatoes cannot be irradiated directly in containers, and that the growing points contained deeply in tubers are inaccessible for effects of electrons applied.

Onions can undergo radiation treatment with a dose of some 100 Gy immediately after the harvest. In some countries, potatoes and onions are allowed to be irradiated with a dose of 150 Gy.

## 4.7 RADIATION SELECTION

Ionizing radiations represent a powerful mutagenic factor that has made them very useful for breeding a large variety of new plants with valuable properties. Genetic alterations (mutations) are also characteristic of natural evolution under the action of various chemical factors, cosmic radiation, and other agents. The mutation rate tends to increase many times under the effect of irradiation. Disarrangement of structure of any section of a biomolecule is inadvertent in nature and generally may result in degrading the quality of a plant concerned. Only a small percent of mutation is considered to be potentially useful. Despite this fact, irradiation may give rise to new valuable characteristics of plants, which justifies the expenditures for selection of plants with useful properties. New methods have been devised for exerting purposeful effects on plants with a view to obtaining desired modifications. These methods cover particularly the following aspects: change of irradiation conditions (temperature, medium, humidity), alteration of contents of some elements in cells, storage conditions after radiation treatment, and selection of optimal period in development of plants.

The typical dose for radiation selection of plants is 100 to 300 Gy. Irradiation of seeds is found preferable. Undesirable mutations that arise in plants along with useful qualities can be removed by selection methods. Irradiation can also be used in conjunction with hybridization. This combination makes it possible to obviate difficulties of interbreeding of different plants, to succeed in crossing mutants obtained by means of irradiation effects, and so on. In some cases, a new quality shows up in the subsequent generation but not in the first one. For the purpose of selection, use is usually made of bremsstrahlung, gamma radiation, as well as fast neutrons, which sometimes prove to be more effective.

Radiation selection has allowed the breeding of high-yielding as well as precocious and non-drowning varieties of wheat. In some countries this method ensured the so-called "green revolution," i.e., breeding and distribution of much better, high-yielding varieties of wheat and other plants. For example, the yield of wheat in Mexico in the period of 1959 to 1970 rose by 4.5 times, in India from 1967 to 1969, by 1.5 times. According to the information presented by the IAAE and FAO, more than 400 varieties of crops mutated by radiation have been regionalized in the world, and this quantity is still growing. By the method of radiation selection in the USSR, new crop strains such as 'Novosibirskaya-67' wheat, 'Obsky' barley and other varieties have been bred. The irradiation method made it practicable to breed plants resistant to low temperatures and some diseases and noted for higher content of protein and other nourishing components.

## 4.8 RADIATION THERAPY

The ability of ionizing radiation to suppress vital functions of living cells can be successfully used for destroying malignant tumors. Irradiation may attenuate functions, suppress growth, and cause destruction of cells and some tissues of the pathological focus. The ideal result of radiation therapy is complete elimination of reproducible tumor cells. The main problem in this therapy is to minimize possible injury to sound tissue. This can be done by selecting the kind and energy of radiation and the method of application of ionizing flux to the place of pathology. For curing surface tumors, irradiation with electron beams is considered to be more rational. The ionization peak (see Fig. 2.4) should be in compliance with the depth of location of the tumor in question. With the use of special tubes, the electrons can be directed into some cavities of the patient's body (such as mouth and rectum). The electron energy used in medical practice reaches tens of megaelectron volts.

Irradiation with heavy particles allows for a higher degree of localization of ionizing effect of the beam. For instance, a proton beam moving practically rectilinearly releases energy at the end of its track in the region of the Bragg peak (see Fig. 2.3*b*). This requires use of more expensive accelerators (with energies of hundreds of megaelectron volts), and therefore application of protons involves some difficulties. Still higher concentration of the absorbed dose may be achieved by means of a beam of $\pi$-mesons (with zero rest mass of ~140 MeV). Their ionizing effect is negligible over most of the length of the trajectory. At the end of their path, they are captured by nuclei and impart their energy to these nuclei. As the result, the nuclei decay with emitting protons, neutrons, alpha particles, and heavier ions. High penetration of $\pi$-wave mesons enables deep irradiation (of the order of 10 cm). For obtaining a beam of $\pi$-mesons of adequate intensity, use should be made of accelerators of protons or electrons with an energy of the order of $10^3$ MeV, proton currents of 30 to 400 $\mu$A or electron currents of 1 to 10 $\mu$A.

When bremsstrahlung is used for radiation therapy, the patient is gradually turned over relative to the radiation source (or the source is rotated with respect to the patient) so as to minimize the dose rate imparted to the sound tissues. The irradiation zone boundaries are less defined than in the case of application of accelerated electrons. The radiation dose applied to the entire body of the patient is also greater.

To affect the tumor as required, it is necessary to perform several cycles of irradiation. The dose of each cycle is 1 to 10 Gy. The total dose of the complete treatment course comes to a few tens of grays. As a rule, radiation therapy is performed along with other treatment methods (medicines, vitamins, and others). The main object of ordinary therapy is to minimize harmful aftereffects caused by irradiation of sound tissues and to intensify injury of tumor cells. In some cases pathological cells feature higher radiosensitivity than healthy cells, and this factor facilitates the treatment process. Radiation therapy may result either in complete elimination of the tumor or in suppression of development of the disease so as to prolong the patient's life by several years.

At present the most universally used radiation therapy facilities are based on electron accelerators because of their acceptable costs and suitability for being mounted in ordinary clinics. Wider application is being found by heavy particles of beams of accelerators basically used for nuclear physics research.

## 4.9 BIOMATERIALS

Medicinal practice is in constant need of materials compatible with living tissue. Particularly, these materials are required in implantation of artificial vessels or other artificial elements. The effect of medicines may be materially intensified in many cases if they are well fixed with biologically compatible substances and are implanted into the patient organism for long periods, generally into the pathological site. As a rule, biomaterials are polymers that cause no undesirable effects from interaction with living tissue. One of the main properties of such polymers consists in liquid absorbability. The best implantation materials are hydrogels which are capable of absorbing and preserving a substantial water fraction (usually more than 10 percent). Moist hydrogels have a rubber-like consistency and resemble (in appearance) soft, living tissue. These materials cause minimum mechanical irritation and allow for diffusion of organic substances with small molecular mass. The important property of hydrogels is the low interfacial tension between the hydrogel surface and an aqueous solution. This factor enables minimum interaction with proteins and correspondingly minimum danger of thrombocytosis or initiation of various mechanisms of physiological incompatibility. In order to obtain the required form, the hydrogels are usually combined with a strong mechanical carrier (as a rule, a hydrophobic polymer). The required biomaterial is produced by radiation grafting of hydrogel to a carrier. By varying the proportions of hydrophilic and hydrophobic substances, it is found possible to change over a wide range the structure and composition of the materials subjected to graft polymerization.

The biomaterials are produced by polymerization, graft copolymerization of monomers and polymers, and cross-linking. The advantage of radiation polymerization is that the process is carried out at a room temperature, that there is no need for extraction of undesirable additives, and in some cases, sterilization occurs simultaneously. The biomaterials are fabricated in the form of fibers, tubes, films, membranes, microspheres, and molded articles. For producing the biomaterials, use may be made of such materials as aflon, Teflon, aflas with dimethylacrylic acid grafted to their surface. The grafting materially improves compatibility of blood with these polymers [34].

One of the first applications of biomaterials was fabrication of contact lenses. At present the biomaterials are in wide use for production of artificial blood vessels and coating of orthopedic joints and medical instruments. Biomaterials are also used as the basis for immobilization of biologically active substances—ferments, antibodies, antibiotics, antithrombogenic substances, etc. Mobility of immobilized medicines is limited. They shall be accessible for biomolecules of the organism or be capable of gradually dissolving and introducing into the organism.

## 4.10 STERILIZATION OF SEWAGE

Treatment of industrial and domestic sewage is a problem existing particularly in heavily populated areas. Sewage sludge contains useful organic and nonorganic substances and may be used as fertilizer in agriculture or as a nourishing addition to fodder. Apart from destruction of deleterious synthetic matter contained in sewage water (see Section 3.4), special treatment is required for killing pathogenic microorganisms. Concurrent with the commonly known method of disinfection, which is to hold the sewage for a few tens of minutes at ~70°C, research has been performed to determine the efficiency of radiation treatment of sewage with a dose of 30 kGy. In terms of their germicidal effect, both of the aforesaid methods are identical [35]. Experiments with the use of irradiated sludge of sewage have revealed that the sludge is actually similar to the fertilizers produced by means of heat treatment of sewage. The volume of sludge of primary waters is one or two orders less than the volume of the water delivered to decontamination plants, and therefore it requires comparatively low radiation power. As a matter of outlook for the future, radiation sterilization of sewage may prove to be much cheaper, considering that this method enables destruction of synthetic impurities.

CHAPTER

# FIVE

# THERMAL EFFECT OF INTENSE BEAMS ON MATTER

## 5.1 INTERACTION OF INTENSE ELECTRON FLUX WITH METAL

As noted previously, the major portion of the energy injected into an article being irradiated is eventually converted to heat. At high doses, radiation may result in pronounced heating of the article under treatment. For example, at a dose of 200 kGy, the temperature of water (or other material with a unit heat capacity) tends to increase by 50°C. Since substantial heating is generally undesirable, irradiation is performed gradually (specifically, in several passes) to allow the object or material to cool down. As a rule, the specific intensity of an electron beam with an ionization effect does not exceed 200 W/cm$^2$ and is determined both by technically practicable maximum moving speed of the objects under irradiation, when the heating of the objects lies within the tolerable limits, and by the conditions of cooling of the foil of the beam extraction window.

Under the thermal effect of intense electron beams on matter, the specific intensity is in the range of $10^3$ to $10^9$ W/cm$^2$, and the radiation treatment practically has no effect on the properties of the articles under treatment. The article to be processed is usually placed in a vacuum chamber. If the article is treated beyond the vacuum chamber, use is made of devices transparent to electron fluxes featuring high specific power (see Section 8.1).

When a metal surface is subjected to bombardment by accelerated electrons, the released energy amounts to $V \cong S\lambda_m$ (beam absorption zone, Fig. 5.1*a*) where $S$ is the beam cross section and $\lambda_m$ is the penetration of electrons (see Section 2.1). The beam power in terms of watts can be expressed as

$$P = IU$$

where $I$ is the beam current in amperes and $U$ is the voltage in volts, the latter being numerically equal to the kinetic energy of the particles expressed in electronvolts. In acceleration of electrons in a potential electric field, $U$ is the acceleration voltage. The specific power of the electron beam is

$$p = \frac{P}{S} = jU$$

where $j$ is the beam current density. The power released per unit volume is

$$p_v = \frac{P}{V} = \frac{IU}{S\,\lambda_m} = \frac{jV}{\lambda_m}$$

where $p_v$ stands for a power averaged for volume $V$. Since the beam is absorbed in this volume irregularly, different elements of the volume release unequal magnitudes of power. In the energy range for which the penetration is proportional to the electron energy $\lambda_m \sim U$, the intensity per unit volume does not depend on the energy of the beam and is determined solely by the current density: $P_v \sim j$.

In heat treatment by the use of a stationary beam, the energy released per unit volume of the article under treatment equals the product of $p_v$ by the beam effective time $\tau$. This time is determined by the speed of the article under treatment and the beam velocity with respect to each other or by the rate of evaporation of the metal (in some cases by the speed of escape of metal particles, by the rate at which the molten metal is poured off, etc.). It is common practice to use a notion, such as $P_v$ in thermal processes, unlike radiation processes where the measure of the effect of an electron beam on matter is determined by a dose and measured in grays. In the first case, the absorbed energy $(p_v\tau)$ is measured in joules per cubic centimeter, in the second case, in joules per kilogram.

The beam intensity is expended in heating the matter in volume $V$ as well as in providing the processes illustrated in Fig. 5.1*a*. The directions in which the beam intensity is taken are shown arbitrarily. Secondary electrons, macroparticles, and heat radiation propagate in the direction opposite to that of the primary beam within a spatial angle approximately equal to $2\pi$. Inward heat

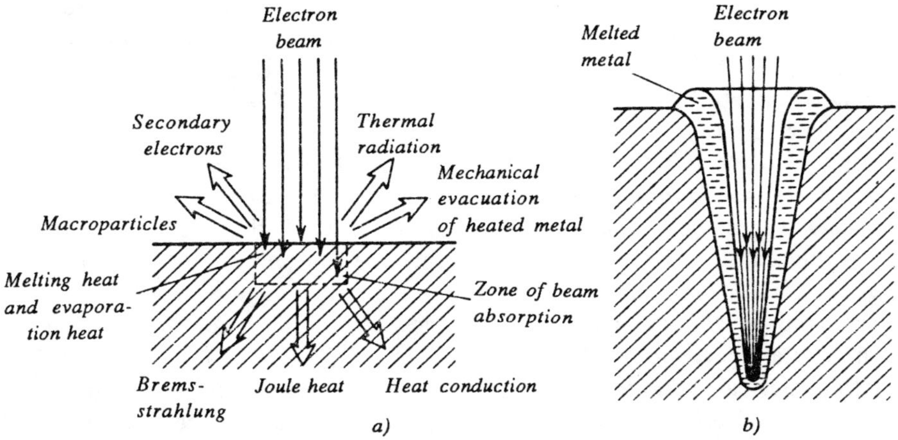


**Figure 5.1** Interaction of intense electron beam with metal. *a*) Extraction of energy from beam absorption zone. *b*) Penetration in metals by electron beam.

transmission at the expense of thermal conduction of the metal also occurs within a spatial angle of 2 π but in the direction of travel of the primary electrons. The loss in a vacuum due to thermal conduction and convection from the metal surface may be ignored. This loss may be appreciable if the matter under processing adjoins a dense medium, particularly in the case of forced cooling. The bremsstrahlung, in general, tends to propagate within an angle of 4 π. The energy brought by the electron beam may be consumed for melting and evaporation of metal. The illustration also shows heat removal in the form of kinetic energy of macroparticles (including drops and vapor), during mechanical evacuation (for instance, discharge of molten metal) and also expenditure of energy for heating and evaporation of metal. The metal is also energized with a current that is equal to the current of the primary beam (less the current of the secondary electrons and other carriers of charges escaping the metal), which results in joule heat.* The values of these losses are not essential, as a rule. Some negligible part of the energy of the accelerated electrons may be spent on changing the crystalline structure of metal.

Depending on the objectives pursued by electron bombardment of metal, the conditions are selected to facilitate a maximum

*Joule heat is directly proportional to the potential difference $\Delta U$ arising on the irradiated article when current flows across it. The beam electron energy will diminish respectively by a value of $e\Delta U$ when the electrons approach the article under radiation.

possible release of energy in the required process. To increasing the bremsstrahlung yield, every effort is taken to enhance the energy of the electrons and make use of targets fabricated of material featuring a maximum nuclear charge $Z$.

The energy transformed into bremsstrahlung in the range $W$ = 0.1 – 1 MeV is not above 1 to 3 percent. The energy consumed by the secondary electrons lies within 10 and 30 percent. In heat treatments of metals, the specific intensity $p$ of the electron beam is considered a basic parameter determining the nature of the process. At $p \leq 10^3$ W/cm$^2$, most of the energy from the heated zone is removed at the expence of heat conduction. With an increase of $p$, the metal starts to melt. Characteristic of $p = 10^3$ to $10^4$ W/cm$^2$ is a hemispherical fusion zone. Increase of the heat liberation rate up to $p = 10^5$ to $10^6$ W/cm$^2$ causes intense formation of vapor and evaporation of metal. From $p = 10^6$ to $10^7$ W/cm$^2$ an essential feature is ejection of metal under the action of volumetric effervescence and other effects. As a result, an extended fusion channel is formed (Fig. 5.1*b*). The channel depth may be 10 or more times its width. The process becomes unsteady state, for the beam periodically dissipates on the metal vapor, heating up the channel walls, and penetrates further inward releasing energy on the channel bottom. Such a deep penetration enables metal articles to be successfully welded or subjected to other treatments even though their thickness is many times the depth of penetration of accelerated electrons ($\lambda_m$). At the specific intensity of $p = 10^6$ to $10^9$ W/cm$^2$ it is possible to obtain through holes and also to perform dimensional treatment and cutting of metal.

Determination of regularities of the aforesaid penetration is one of the important problems of research concerning the electron-beam treatment of metals. Different approaches to consideration of this phenomenon has been described in [36]. Two kinds of oscillations are thought to take place in the channels, namely: (1) at a frequency of $10^4$ to $10^6$ Hz and (2) at a frequency of the order of $10^2$ Hz. The temperature of the channel varies substantially with height: 5000 K at the channel bottom and 2500 to 3000 K at the outlet. Some of the principal research goals are to find the best conditions for achieving a maximum depth of penetration, obtaining homogeneous (without cavities) welds, and elucidating the effects of the atmosphere, in which welding is performed, on the dimensions and quality of welds.

In some cases the accelerated electrons happen to be capable of passing through a metal whose thickness exceeds the depth of penetration of the electrons ($\lambda_m$) by a factor of hundreds of times. The work of Schvarts [31] treats the results of the experiments where an electron beam of 150 keV passes across a 25-mm-thick steel sheet. In this case the electron energy at the metal exit differs negligibly from the inital value. Upon cessation of the beam action, the hole is absent. This effect is attributed to conversion of a part of the material (under the action of the electron beam) into vapor whose pressure may be fairly low due to a high temperature. A small diameter of the beam is maintained due to gas focusing.

For treatment of surface layers, e.g., for hardening or changing the crystalline structure, use is made of pulsed heating of metal by electrons. The nature of this process is also determined by the energy release per unit volume, i.e., by the intensity liberated in a unit volume ($p_v$) and by the lifespan of the electron beam ($\tau$).

The temperature of the metal under adiabatic heating changes as $\Delta T \cong 4p_v\tau/c_v$, where $\Delta T$ is expressed in °C, $p_v$ in W/cm$^3$, $\tau$ in seconds, and $c_v$ is the heat capacity referred to unit volume in cal/(cm$^3$ °C). It is characteristic of the mentioned proceses that $p$ = $10^5$ to $10^7$ W/cm$^2$ and $\tau$ = $10^{-5}$ to 1 s. Single-pulse and repetitive-pulse beams are also used for making holes and for some other treatments. Fast heating results in ejection of metal caused both by melting and evaporation and by thermal strains. The nature of the process is affected not only by the parameters of the electron beams but also by the properties of the materials under treatment, such as density, melting temperature, heat conduction, and heat capacity.

## 5.2 TECHNOLOGICAL PROCESSES OF HEAT TREATMENT OF METALS BY INTENSE ELECTRON BEAMS

### 5.2.1 Electron Beam Melting

Electron beam melting is used for obtaining high-purity metals and ingots of refractory and high-activity metals. The main differences of this method from other methods of melting are high specific intensity of heating, easy control of electron beam power in space and in time, and accomplishment of the process in media with a minimum content of harmful impurities.

The high specific intensity of the beam is of particular value for such materials as niobium, tantalum, and other metals that have high melting temperatures. Low pressure in the melting chamber enables degassing and makes it possible to minimize absorption of gases during the melting process. Thus, in ferrous metallurgy, the process of removal of hydrogen, carbon monoxide, and nitrogen from melts at a pressure of 10 to $10^3$ Pa finds extensive application.

Figure 5.2*a* demonstrates a version of melting by use of accelerated electrons. Electron beam 1 melts down blank 2 and maintains metal 3 in liquid state at the end of ingot 4. The ingot is cooled down with the aid of crystallizer 5 whose intensity is removed by means of running water 6. The diameter of the beam at the point of contact with the metal amounts, as a rule, to 20 to 150 mm. In some installations, the electron gun shapes up an electron beam of a small cross section and the beam is scanned for maintaining a large surface of melt.

There are various types of electron guns in use and various arrangements with respect to the metal being melted (see, for example, [38]). One of the most important problems of the installations under consideration is attainment of adequate dielectric strength of the high-voltage accelerating gap in the electron radiation sources. Operation of the latter is impeded by vapors and splashes of metal, which is why every effort is exerted to reduce the accelerating voltage of the electrons. The parameters of the electron beam typical of the existing accelerating installations are as follows: energy of 15 to 40 keV, specific power of $10^3$ to $10^4$ W/cm$^2$, beam intensity (or total power of several beams) of 50 to 1000 kW.

Such high intensity at low electron energy hampers transport of the particles, and therefore the source of the electrons has to be located in the immediate vicinity of the melting zone. Thus, shielding the source against the abovementioned adverse effects is difficult. At the same time, an increase of electron energy allows an increase of the specific intensity and decrease of the beam divergence due to its natural space charge (see Section 1.3). It is possible that further development of accelerators of relativistic electrons with a power of hundreds of kilowatts may result in designing such melting units where the source of the electrons will be placed a few meters away from the melting zone and be conveyed

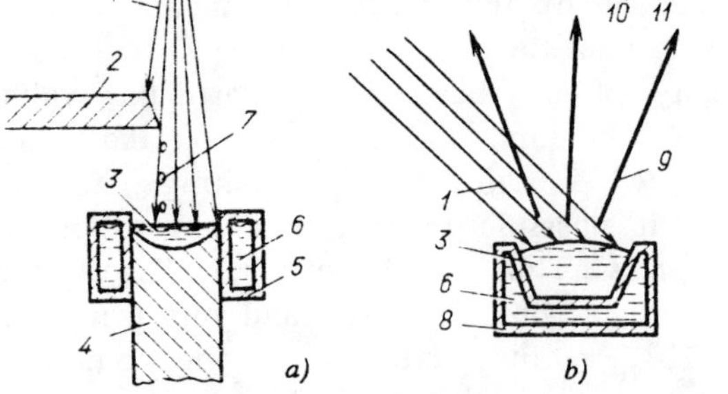


**Figure 5.2** Diagrams of processes of heat treatment of metals by intense electron fluxes. *a*) Electron-beam melting. *b*) Vapor deposition. *1*—electron beam; 2—blank; *3*—molten metal; *4*—ingot; *5*—crystallizer; *6*—running water; *7*—drops of molten metal; *8*—crucible; *9*—vaporizing metal; *10*—layer of metal coating; *11*—substrate.

to the point of application through a magnetic focusing channel. Such installations are likely to provide a means for varying the pressure and content of the atmosphere over a wider range.

### 5.2.2 Electron Beam Welding

Electron beam welding is a widely employed technological process where the electron beam serves as an energy carrier. A high specific intensity attained in the beam makes it possible to melt down limited sections of mutually adjoining parts. Upon solidification of the metal, a dead joint is obtained. In this process, energy consumption and temperature strain of the articles are minimized and new possibilities are opened up for welding refractory joints and for combining materials that are unlikely to be welded by other techniques. The electron-beam welding enjoys wide use in aviation and space technology for joining parts fabricated of titanium and titanium alloys. The ratio of the weld depth to its width may be much more than in other methods of flash welding, the thickness of the parts being appreciably more than 10 cm. Experiments with electron beam welding have also been carried out in cosmic space [39]. The performance characteristics typical of installations for electron-beam vacuum welding are: energy, 30 to 150 keV; power, up to 100 kW [40].

The value of the specific power $p$ at the welding site depends to a large measure on the properties of the electron optical system used for transport of the beam, particularly, on the focusing. The optimum focal point is generally located in the middle of the channel height. The dimensions of the fusing channel are dependent both on $p$ and on the rate of welding, as well as on energy of the electrons, pressure, and composition of the environment.

Figure 5.3 depicts the results of studies of effects of a stationary electron beam energy on weld depth in the range $W = 30$ to 125 keV at the pressure of the surrounding medium lying within $10^{-2}$ to $10^{4}$ Pa [41]. As the illustration shows, with a rise of electron energy there is an increase of fusing depth. This depth tends to diminish with increasing pressure, which stems, to a great extent, from scattering of the electrons and drop of the specific intensity $p$ in the course of transportation of the electron beam to the place of welding. In [42], for the sake of reduction of the 150 keV beam scattering in air, use is made of some lighter-than-air gas blown along the beam trajectory. The gas used may be helium or a mixture of hydrogen and oxygen which ignites directly on the welding spot. The increase of the metal fusing depth in open air with the use of a beam of up to 1.5 MeV has been investigated in [43]. Scattering of

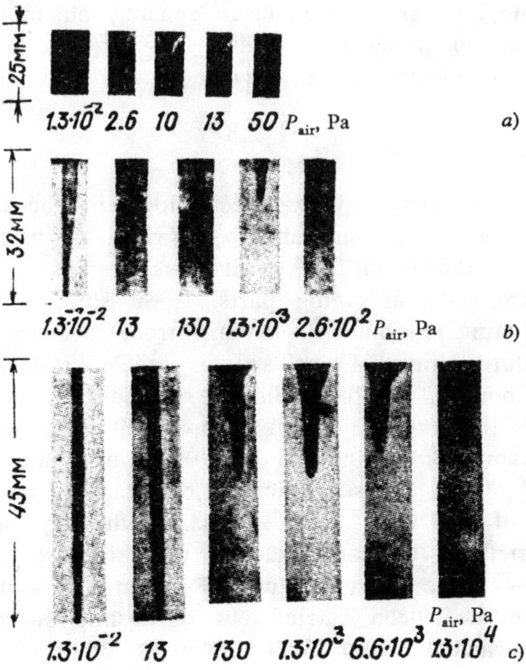


**Figure 5.3** Effect of electron energy $W$ and surrounding medium pressure $P_{air}$ on depth of fusion. Medium, air (pressure is indicated in each illustration), speed of beam motion relative to blank under treatment, 0.5 m/min. $P$ = beam power, $L_{air}$ = length of trajectory of electrons between electron source and blank (medium is under indicated air pressure). *a*) $W$ = 30 keV, $P$ = 6 kW, $L_{air}$ = 400 mm. b) $W$ = 75 keV, P = 8.4 kW, $L_{air}$ = 10 mm. c) $W$ = 125 keV, $P$ = 10 kw, $L_{air}$ = 10 mm.

a concentrated electron beam in air tends to decrease with rise of the temperature in the region of the beam transit and with respective decrease in pressure (see Section 5.4).

In industrial installations for out-of-vacuum welding, the energy of the beam does not exceed 200 keV, and the beam power lies within tens of kW. The article under treatment is at a maximum distance of a few centimeters from the beam extractor. No special problems are presented by out-of vacuum welding of articles with a minor ratio of the weld depth to its width (for example, welding of large-sized, thin-walled shells and containers [38]).

The method of welding by a beam of ~1 MeV extracted into the atmosphere is applied only in some experimental devices.

### 5.2.3 Dimensional Treatment

Dimensional treatment by an electron beam is used for obtaining shaped or cylindrical holes in hard-material parts such as block bearings and wire drawing dies. In a deep channel formed similarly to the channel formed in the welding process, the metal is melted down, and intense boiling under melt surface and splashing of metal occur. Energy is applied to the channel impulsively, as a rule. The pulse intervals of the accelerated electrons should suffice to enable vapors or metal particles to escape so as to avoid energy loss on the escaping metal by the electrons of the subsequent pulse. In deep channels (with a length-to-diameter ratio of 10 or higher), the cross section of the beam should be materially less than the dimension of the hole. The focal distance of the lens directing the electrons is varied, as a rule, as the hole under effect becomes deeper so as to keep the focus close to the channel bottom. The efficiency of the beam in deep holes approximates 100 percent. The specific energy consumption is determined by the energy expended for entrainment of a unit volume of the material. Thus, all other things being equal, this specific consumption is a function of the thermophysical properites of the material under treatment. As seen from Fig. 5.4, the specific energy consumption tends to increase with a rise in the melting temperature of the materials. The level of consumption value is intermediate between the values of the energies consumed for melting and for evaporation [44]. According to existing concept, most of the metal is ejected in a molten state and only a negligible percentage is in the vapor phase.

For dimensional treatment in the associated installations, the article to be worked is placed in a vacuum. The beam energy reaches 150 to 200 keV, the power generally lies within 15 kW.

### 5.2.4 Vapor Deposition

Vapor deposition in vacuum with materials heated by an electron beam is extensively used in industry for applying rust-proof, heat-

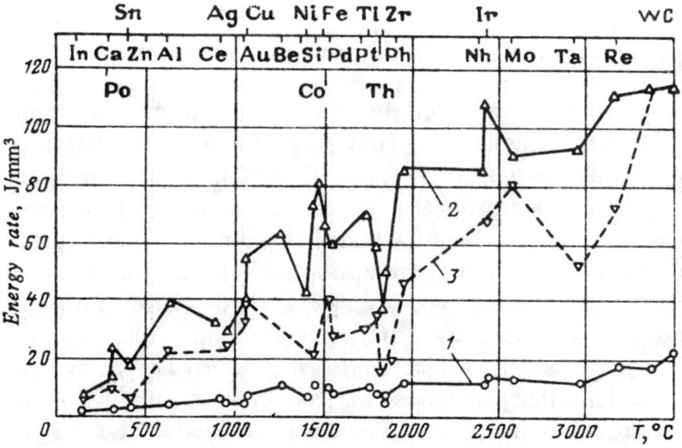


**Figure 5.4** Specific consumption of energy for melting (*1*), vaporization (2) and removal of material (*3*) under electron beam treatment for some materials, depending on their melting temperatures.

resistant (for example, blades of gas-turbine engine), light-reflecting, decorative and other coatings. The coatings are sprayed on surfaces of metals, plastics, paper, and other materials. The process of vapor deposition is diagramatically shown in Fig. 5.2*b*. The principal difference of evaporation by use of an electron beam from the other processing methods is the direct heating of the surface under vaporization. The crucible containing the metal is cooled, which completely prevents the crucible from evaporation and rules out reaction of the crucible material and the metal contained. Use is also made of crucibleless melting installations in which the material under processing is evaporated, for example, from a crater of molten material on the surface of a massive ingot. The thickness, composition of the deposited coating, and other factors can be varied by controlling the electron flux in space and time. High rates of vaporization, which can reach dozens of kilograms per hour in modern industrial installations, make it possible to provide high productivity, and enable, in particular, the spray-on coatings to be applied on fast-moving bands. Use is made of installations incorporating either one or several electron-beam evaporators. The beam intensity in each reaches hundreds of kilowatts; however, the electron energy, as a rule is not too much (10 to 20 keV).

A subject that is also under investigation is powder deposition by the use of an electron beam for coating metals to improve their corrosion resistance, thermal stability, etc. The coating can be sprayed on the metal surface out of vacuum by means of a beam extracted into the atmosphere. As compared to plasma spraying, this method provides higher productivity and a lower production cost [45].

### 5.2.5 Surface Hardening

Surface hardening of steel is effected by the use of a pulsed beam. The steel layer to be hardened is first heated to a temperature sufficient to obtain the austenite (i.e., ~1000°C). Then the heat energy is promptly transferred inward at the expense of heat conduction. A shorter time of the heated state as compared with other methods of hardening results in formation of particularly fine-grained austenite and then, upon cooling, very hard martensite. The cooling rate reaches $10^8$ degrees (°C) per second. The Brinell hardness of some special steels subjected to such a treatment is 100 to 200 units higher than that of the steels hardened by induction or other traditional hardening methods. To ensure speedy cooling of the steel layer being hardened, it is essential that the thickness of the material be considerably more than that of the layer. Since in electron-beam hardening only a thin layer of the material should undergo heating, warping of the metal under treatment tends to diminish. In some cases final treatment of the article may be completed before the hardening procedure, thereby providing a means for dispensing with expensive processes of treatment of hard materials. Seventy to eighty percent of the beam energy in the hardening process will be expended in the useful process, i.e., heating of the material under processing. If penetration of the electrons $\lambda_m$ is approximately equal to the thickness of the layer to be hardened, then it is found reasonable to heat the material for a short period of time to ensure that heat removal at the expense of heat conduction be negligible (adiabatic process). The main loss in the adiabatic heating is extraction of energy by the flux of the secondary particles. Should the thickness of the layer to be hardened exceed $\lambda_m$, the effective time of the beam should be extended in order to have the article under treatment heated as deeply as required due to the inherent heat conduction. As a rule, electrons are normally directed toward the surface to be hardened. However, inaccessible sections of the article may be treated by a beam directed at an angle, which will require somewhat longer duration of exposure. Relative ease of deflection of the beam by the use of the magnetic field allows for treatment of surfaces featuring intrinsic configurations. There are two ways of electron-beam heating of the surfaces to be hardened: (1) by a beam with a small cross section in conjunction with a quick magnetic scanning, or (2) by a wide beam whose cross section corresponds to the area of the surface to be hardened. In the former case, the specific intensity of the beam should be materially higher (approximately as a ratio of the area of the surface under treatment to the cross section of the beam). In order to produce a wide beam, use is made of accelerators with a cathode of a large effective area or defocusing of the electron flux.

In existing industrial installations intended for hardening by high-speed electrons, the energy of the particles does not exceed 150 to 200 keV, the power being 50 to 60 kW. The article under treatment is placed, as a rule, in a vacuum. For vacuumless hardening processes, it is necessary to increase the energy of the electrons and to use special beam extractors.

### 5.2.6 Electron Beam Refining Remelting

In electron beam refining remelting the surface layer of metal is heated for a short time by the electron beam to the melting point and then is promptly cooled down at the expense of inward heat transfer. The rate of crystallization in this case is substantially higher than in the process of solidification of ingots, resulting in finer microstructure and improved properties of the metal, viz., greater hardness, plasticity, corrosive resistance, etc.

### 5.2.7 Diffusion Alloying

The possibility also exists of subjecting metals to diffusion alloying. In this case, an alloy additive is applied to the surface of the metal under treatment. The additive diffuses over the metal melted under the action of the electrons. When the treatment is carried out with the aid of a highly intense beam out of vacuum use may be made of different active media interacting with the surface of the material under electron bombardment.

Heat treatment of the surface layer of metal in existing industrial installations is usually performed in vacuum. The energy of the electron beam is held in the same range as in the case of other kinds of heat treatments. Some difficulties are presented by charging the articles to be treated into the vacuum chamber. The articles are charged periodically or continuously through special sluiceways. As to some shapes of the articles, for instance, for endless bands (see Fig. 8.15), this problem can be easily solved. Relevant experiments study vacuumless treatment of surface layers of materials by electrons possessing energies exceeding 300 to 500 keV. The pulsed beams with the specific intensity required for this treatment are extracted through the outlet windows with a differential evacuation. It is conceivable that extraction of the beams through outlets opened for short lapses of time (see Section 8.1) or through a heat-resistant foil that is not destroyed by heating at every pulse will be more economical. The beam extracted into the atmosphere may be successfully used for treating some sections of surfaces of large-sized articles, including the treating operations in the process of repairs of assembled units.

## 5.3 TREATMENT OF DIELECTRICS BY ACCELERATED ELECTRONS

In the process of interaction of a concentrated flux of electrons with a dielectric, the energy is consumed for the same processes as in the case of heat treatment of metals (Fig. 5.1*a*). However, due to the difference of thermophysical and other properties of dielectrics, the distribution of energy consumption among the different processes is far from similar, and the heat treatment, as a rule, materially differs from that described above for treatment of metals. By virtue of poor heat conduction, the dielectrics feature great temperature gradients and respectively high thermal stresses. Bombardment of highly brittle dielectrics by an intense electron beam usually results in fissuring and chipping of the material. It is worth noting that maximum liberation of energy occurs inside the material (approximately by one-third $\lambda_m$). A quick rise of pressure in this zone and, after attaining a certain specific power $p$, vaporization of the dielectric may bring about microexplosion followed by splashing of the material. As in the case of the dimensional treatment of metals, in the process under consideration use may also be made of a pulsed beam. The interval between the successive pulses shall exceed the time of scattering of the material to ensure against consumption of the energy for additional heating of the destruction products. A high concentration of energy in the heating zone ensures generation of a stress field which propagates over the material under treatment. Due to this fact, cracking may also take place at some distance away from the heating zone. Shock waves may also appear due to formation of a solid-state nonequilibrium plasma deep in the material. Propagation of the shock wave is accompanied by change of density and temperature. In this case, chips can be observed on the rear of the material limited in thickness.

In treatment with the aid of a stationary beam, the dielectric starts melting and evaporating. The beam forms a cavity that grows until the thermal conduction through its walls becomes sufficient for takeoff of the power, as shown in Fig. 5.5*a* [46]. Most of the beam energy dissipates in the lower part of the cavity where the dielectric is in molten state. A small part of the energy is released by a vapor stream escaping the cavity. High temperature gradients arise in the dielectric. The form of the isotherms approximates a spherical shape. The high temperature gradients in some materials result in cracking and destruction (Fig. 5.5*b*). Energy consumption with such a method of destruction for granite, sandstone, and some other rocks amounts to ~250 J/cm$^3$, which is appreciably less than that characteristic of traditional methods [46]. Another advantage of the method in question is the absence of recoil forces. Therefore, in shaft sinking and tunneling, the electron generator (unlike mechanical processing equipment) should not be rigidly fixed in the working position. Schumacher and Taylor [46] discuss cutting granite and other rocks by means of an electron beam of 9 kw. The rate of movement of the beam is 10 cm/min. At higher powers and speeds of the beam, heating and chipping of thin layers at the rock surface may occur.

Another characteristic feature of treatment of dielectric materials by accelerated electrons springs from low electrical conduc-

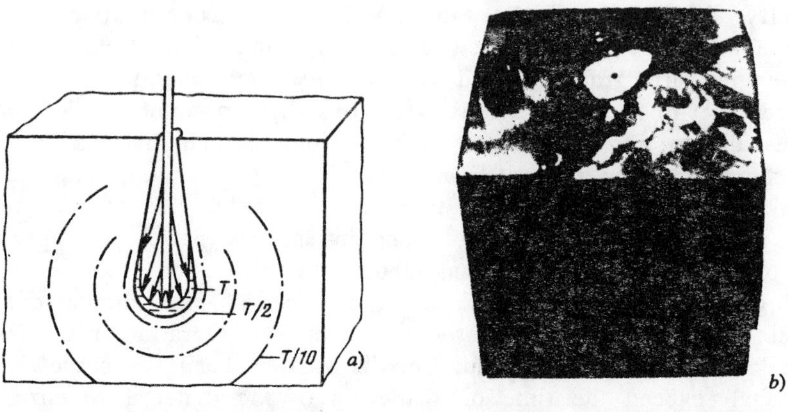


**Figure 5.5** Effect of stationary electron beam exerted on dielectric. a) Material is diorite (granitoid rock). Beam power is 9 kW, cavity depth is about 75 mm, cavity formation time is 30 s, $T$ is the melting temperature. *b*) quartzite (10 × 10 × 10 cm) destroyed by use of stationary electron beam of 150 keV, 9 kW. Operation time = 9 s.

tion. Charges accumulated in a dielectric cause breakdowns and also make for failure of the material. On the other hand, the negatively charged material defocuses and distorts the shape of the electron beam, thereby deteriorating the specific power $p$ at the point where the beam comes in contact with the material surface. Accumulation of the space charge decreases somewhat because of growth of electrical conduction of the dielectric and the environmental medium at the moment of irradiation (see Section 2.7).

## 5.4 PROPAGATION OF INTENSE ELECTRON FLUXES IN GAS

Interaction of electrons in a gaseous medium in cases when heating of matter may be disregarded is shown in Fig. 2.1, which illustrates scattering of an electron beam in the atmosphere. At high values of $p$, the heating of matter becomes essential, since it may lead to a considerable decrease of density and an increase of penetration of electrons $\lambda_m$. Additionally, most of the energy is removed from the region heated by the beam to high temperature as thermal radiation. The remaining, lesser amount of energy is extracted at the cost of heat conduction, convection, and bremsstrahlung. Calculation of the interaction between an intense electron flux and a dense gas can be found in a number of research works (e.g., [47, 48]). The increase of $\lambda_m$ reaches one order and above. A particular role in maintaining a high density of the beam is also played by ion balancing. An experimentally observed increase of $\lambda_m$ for an intense beam is presented in Fig. 5.6 [49].

Maintaining a high density of an electron beam in the atmosphere at a considerable distance from the extracting window is of crucial importance for a number of the technological processes, e.g., vacuumless welding, and crushing of rocks. As will be apparent from the stated data, the value of $\lambda_m$ may be increased not only by way of increasing electron energy $W$ but also by raising the specific power $p$.

## 5.5 COMPARISON OF ACCELERATED ELECTRON BEAM AND LASER RADIATION

The processes of heat treatment of materials with laser and electron beams have much in common. Both kinds of radiation feature high specific power, which allows for technological processes that are believed to be impossible by the use of other energy carriers. Lasers are also used in industry for welding, cutting, working of shaped metal articles, as well as for hardening and changing of structure of surface layers of materials. The specific intensity of radiation, as in the case of heat treatment by the use of electrons, lies within $10^3$ and $10^9$ W/cm$^2$.

The most significant advantages of laser radiation are: negligible dissipation and losses during transport of radiation in the atmosphere and simpler safety engineering. The advantages of electron beams are higher efficiency (from industrial network to object under treatment the efficiency is 70 to 90 percent as compared with the 5 to 10 percent characteristic of laser radiation efficiency), and better controlability. The considerable difference in efficiency is of no paramount importance in the case of low power of the installation; however, this may acquire decisive importance when the power is of the order of 100 kW and more. It should be noted that laser radiation may also be used to generate chemical reactions. The action of the laser provides selective resonance excitation of atoms and molecules of certain predetermined substances, which enables the reactiveness of the atoms and molecules to be radically im-

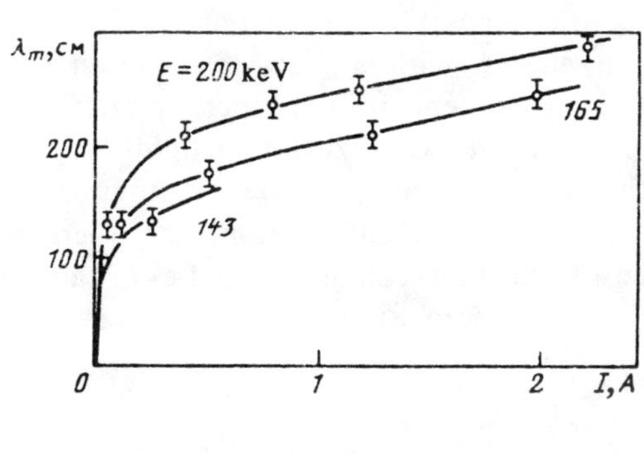


**Figure 5.6** Distance covered by beam in atmosphere versus beam power. Diameter of extracting window is 1 cm; pulse duration is a few seconds

proved. The fast-progressing laser atomic and molecular technology involves such processes as isotope separation, production of chemically pure substances, and chemical synthesis [50].

One application of laser radiation is for irradiation of biological objects. Laser equipment finds ever increasing use in medicine, primarily in surgery for dissecting tissues by thermal effect [51]. Besides, irradiation of the operated site causes coagulation of the tissue and prevents pathological formations. The therapeutic action of low-intensity radiation ($p = 10^{-3}$ W/cm$^2$) is also found useful for activating metabolic and regenerating processes. A problem under study is the possibility of selective effect on some components of biomolecules in order to investigate their structure.

## 5.6 CONCLUSION

In concluding the present review of applications of intense electron beams, it should be emphasized that by virtue of the tangible difference between radiation and thermal methods of treatment of materials, the two methods form independent and little-associated fields of industrial technology. While for the radiation method of treatment, use is made of electron beams developing energies of 150 keV and higher, in modern thermal treatment, electron energy does not exceed 10 to 175 keV. The specific intensity $p$ of the beams in the thermal treatment facilities exceeds the value of $p$ characteristic of the radiation treatment installations at least by one order. For a number of technological processes, this difference amounts to several orders. The absolute power of the electron beams in both types of equipment is approximately similar.

As a consequence of low energy of the electron beams in the heat-treatment installation, the problems of high-voltage insulation and radiation protection can be solved without great difficulties. The effective range of the energies, unitl very recently, was considered satisfactory for treatment of the entire nomenclature of the articles subject to processing by electron beams. The articles under treatment were placed in a vacuum, which in some cases is an indispensible condition for the technological process (vapor deposition, etc.). Desire to increase the beam energy with a view to extracting it from vacuum and performing heat treatment in the open atmosphere was aroused, first of all, with increase of size of articles to be treated and, consequently, with difficulty of positioning huge articles in vacuum chambers. One of the suggested solutions is to create a local evacuation with the use of vacuum chambers to be advanced along the welds (practicable only in case of simple shapes of articles under treatment). Another solution is to extract the beam from vacuum into the atmosphere. High-energy concentrated electron beams extracted from vacuum will be useful also in processes of crushing rocks (shaft sinking, tunneling, etc.). As far as the problem of radiation protection is concerned, it seems to be not very difficult with remote control of the production processes and should not be considered as principal one due to the tendency for automation of the technological cycles, particularly, electron-beam welding and melting.

CHAPTER

# SIX

# DIRECT–ACTION ACCELERATORS

## 6.1 ACCELERATION OF PARTICLES IN POTENTIAL ELECTRIC FIELD

Direct-action accelerators (or high-voltage accelerators) are devices in which charged particles are accelerated in a potential electric field. The increment of the energy of an electron is determined by the potential difference, viz.:

$$\varepsilon = e\Delta U$$

The circuit of a triode-based system for accelerating electrons in a potential electric field is shown in Fig. 6.1. This system comprises cathode 1, control grid 5, and anode 4. The triggering moment and the current magnitude are controlled by the voltage at control grid 5. Should no provision be made for a control grid (in diode-based systems) the time of acceleration of electron beam and the magnitude of the current are determined either by the accelerating voltage or by the cathode heater. The cathode in electron accelerators with potential electric field stays, as a rule, at high potential, while the anode is at the ground potential. With a view to enabling the particles to escape the accelerating gap, provision is made for a hole in the anode, and the subsequent transportation of the particles usually takes place in the equipotential space. In accelerators with accelerating potential electric field at a time-

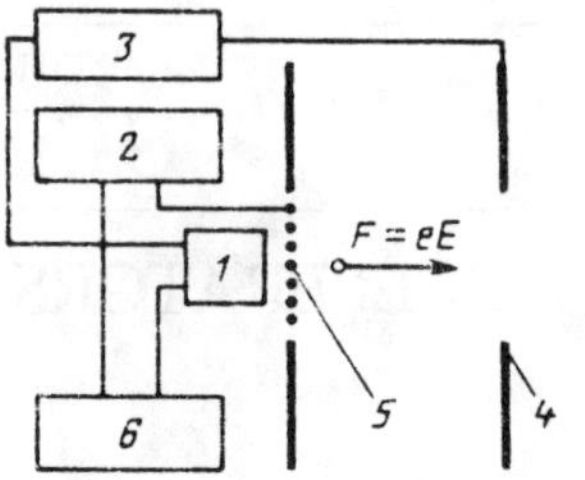


**Figure 6.1** Triode system for electron acceleration in potential electric field. *1*—Electron source (cathode); *2*—electron current control generator; *3*—acceleration voltage generator; *4*—anode; *5*—grid; *6*—cathode heater power supply.

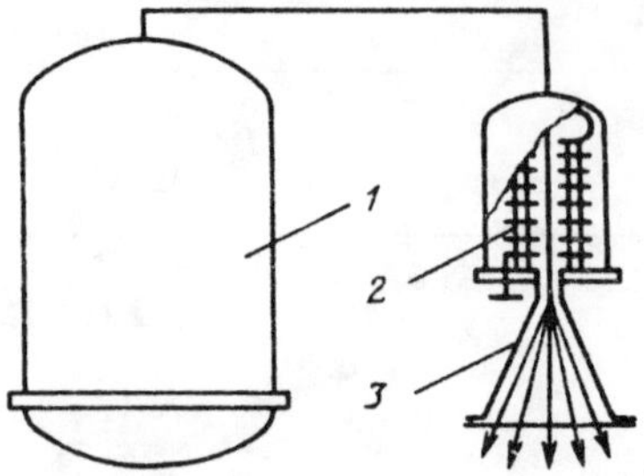


**Figure 6.2** Diagram of direct-action electron accelerator. *1*—HV generator; *2*—accelerating device; *3*—beam extractor.

constant voltage, more than 90 to 95 percent of the energy applied from the supply mains to generator 3 is likely to be imparted to the electron beam. However, in pulsed accelerators, this value actually lies within 20 and 70 percent.

Due to a high efficiency and relatively simple design features, these accelerators find application in the overwhelming majority of the relevant industrial installations with an average power exceeding 5 kW and an energy up to 3 or 4 MeV. In the case of higher energy, preference is given to the accelerators with vortex and high-frequency electric fields. This is associated with substantial complication of electric insulation of direct-action accelerators at voltages exceeding several million volts.

As illustrated in Fig. 6.2, a direct-action accelerator consists of three basic components: (1) a high-voltage generator, (2) an accelerating device, and (3) a beam extraction system (or internal target). Each of these elements may come in different modifications depending on the parameters and purpose of the particular accelerator. Some other compositions of the accelerators are also found. For instance, one generator feeds a number of accelerating devices, or the accelerating device is built in a high-voltage generator. Anode may also be at a high-potential, particularly when generating bremsstrahlung (or x-rays) on the internal target.

## 6.2 METHODS FOR GENERATING HIGH VOLTAGES

The main purpose of high-voltage power supply sources of continuous-action accelerators is multiple increase of the main voltage. The function of power supplies of pulsed accelerators is also to accumulate energy to considerably increase the power during the working (driving) pulse. Operation of high-voltage units used in accelerators is based on a number of main principles. Thus, in direct-current, low-frequency, high-voltage generators use is made of the following (see Fig. 6.3) [52]:

1. Mechanical transfer of electric charges. The charges from the electrode whose potential differs from the ground potential by several kilovolts (or tens of kilovolts) are transported by the use of a conveyer fabricated from an insulating material to a hollow, high-voltage electrode insulated from the ground.
2. Transformer. This contains a primary winding and a secondary winding with a common magnetic flux or is provided with a single winding (autotransformer).
3. Circuits containing capacitors and rectifiers for multiplication of voltage; in some circuits of this kind, the voltage is multiplied and rectifies at the same time.

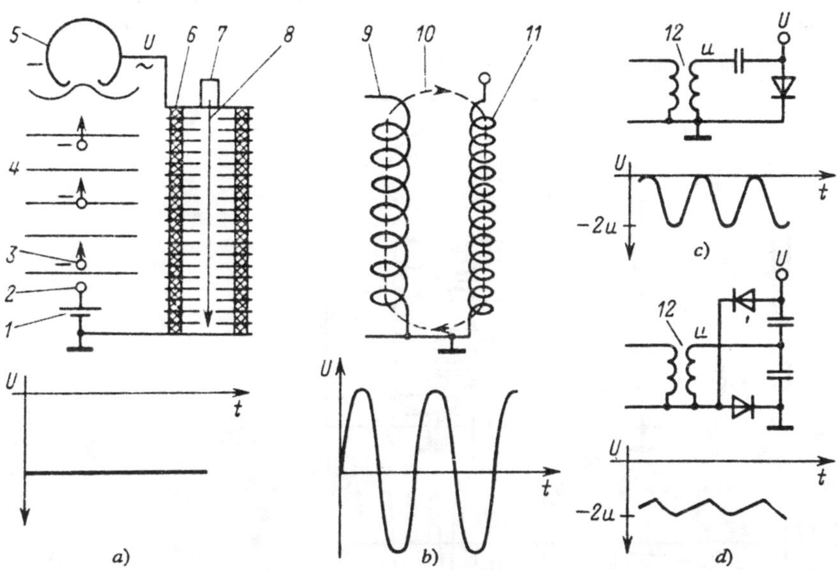


**Figure 6.3** Diagrams of direct and low-frequency, high-voltage sources and waveform of generated voltage at no load (in *b*, *c*, and *d*, voltage waveform under load will change). *a*) Mechanical transfer of charge (electrostatic generator). *b*) Transformer. *c*) Half-wave rectifier. *d*) Voltage doubler circuit. *1*—Rectifier; *2*—electrode under potential of some 10 kV; *3*—electric charge; *4*—electric field equipotential; *5*—high-voltage electrode; *6*—accelerating tube; *7*—electron source; *8*—flux of accelerated particles, *9*—primary winding of transformer; *10*—magnetic flux; *11*—secondary high-voltage winding; *12*—charging transformer; *U*—voltage generated by circuit; *u*—voltage of charging transformer.

The existing power supplies of the accelerators on the basis of mechanical transport of charges generate the voltage up to the value of the order of 30 MV. The voltage produced by the transformer, as a rule, is not in excess of 1 or 2 MV. Figure 6.3*c* and *d* illustrates a circuit diagram of a half-wave rectifier and a circuit diagram of a voltage doubler. The voltage generated by these circuits usually amounts to 100 to 200 kV. In order to obtain higher voltages (up to 3 to 5 MV), these circuits are united into stages. It is common practice to use combinations of different methods to increase voltage. The power supply sources on the basis of three-phase step-up transformers generate direct voltage.

The power supplies of industrial pulsed accelerators more widely employ the following circuits (Fig. 6.4):

1. Marx circuit. This contains capacitors connected in parallel (across resistors) during charging operation and in series (through spark gaps) during discharge.
2. Resonance systems. These represent a circuit formed by a multiturn high-voltage winding with its distributed capacitance. The systems are excited by a resonance-frequency generator.

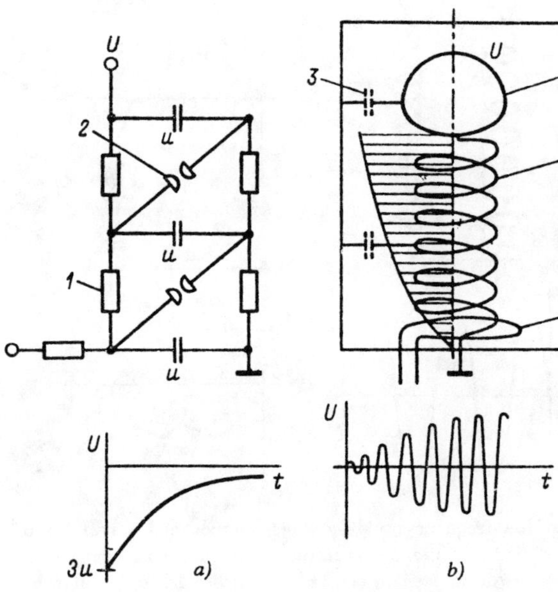


**Figure 6.4** Diagram of pulsed high-voltage sources and waveform of generated voltage. *a*) Marx's circuit. *b*) Multiturn resonance system. *1*—Charging resistor; *2*—discharger; *3*—distributed capacitance; *4*—high-voltage electrode; *5*—high-voltage winding; *6*—tank; *7*—excitation (field) turn; *U*—generated voltage; *u*—capacitor voltage. Distribution of voltage along high-voltage winding (5) can also be seen.

The voltage of one capacitor in the Marx circuit is usually equal to 10 to 100 kV, and some tens of stages can be included in the generator system. The Marx generator is one of the most widespread sources of high-pulse voltage in the range of pulse durations of $10^{-6}$ to $10^{-3}$ s. The time required to yield the rated voltage in the resonance systems is determined by the power of the supply generator and the degree of coupling between the resonant circuit and the exciting turns. As soon as the voltage reaches a preset amplitude, the particles undergo acceleration. This may be exemplified by Fig. 6.4*b*) which presents a resonator that is essentially a quarter-wave line. The ultimate voltage of these generators is within 2 to 3 MV.

For generating pulse voltage, use is also made of lines, inductive energy storage contrivances, and other devices. However, they have not yet found wide application in industrial accelerators.

## 6.3 ELECTROSTATIC GENERATORS

Mechanical transport of electric charges to the inner surface of a hollow electrode underlies the operation of electrostatic generators [53], which are considered one of the early sources of megavolt voltage. As Fig. 6.3*a* shows, charges (3) are conveyed from the side of the ground toward a high-voltage electrode (5). The source of charges is a rectifier (1) rated at a voltage of some tens of kilovolts. To transport the charges, use is made of mechanical conveyers. Figure 6.5 shows the circuit diagrams of different verisions of the electrostatic generators. The diagrams illustrated in Fig. 6.5*a*, *b*, and *c* demonstrate a conveyer in the form of an endless belt (5) made of an insulating material. For charging and discharging the conveyer, a corona discharge is produced from point 3 located adjacent to the conveyer surface. In the initial stage of operation, the load current (at the accelerating tube) is absent. This time the charges are accumulated on high-voltage electrode 8 and an increase of its potential $U = It/C$ occurs, where $U$ is the voltage (in volts), $I$ is the current (in amperes) carried by the conveyer, $t$ is the conveyer operating time (in seconds), and $C$ is the electrode capacitance (in farads). The steady-state current of the accelerating particles is equal to the current transported by the conveyer belt. Electron beam power $P = I \cdot U$ is equal, respectively, to the power consumed for surmounting the strength of the electric field during conveying of the charges by the transport belt. The conveyer current is determined by the density of the charge capable of holding on the conveyer surface, as well as by the width of $a$ the conveying belt and its speed, i.e., $I = \sigma a v$, where $I$ is expressed in

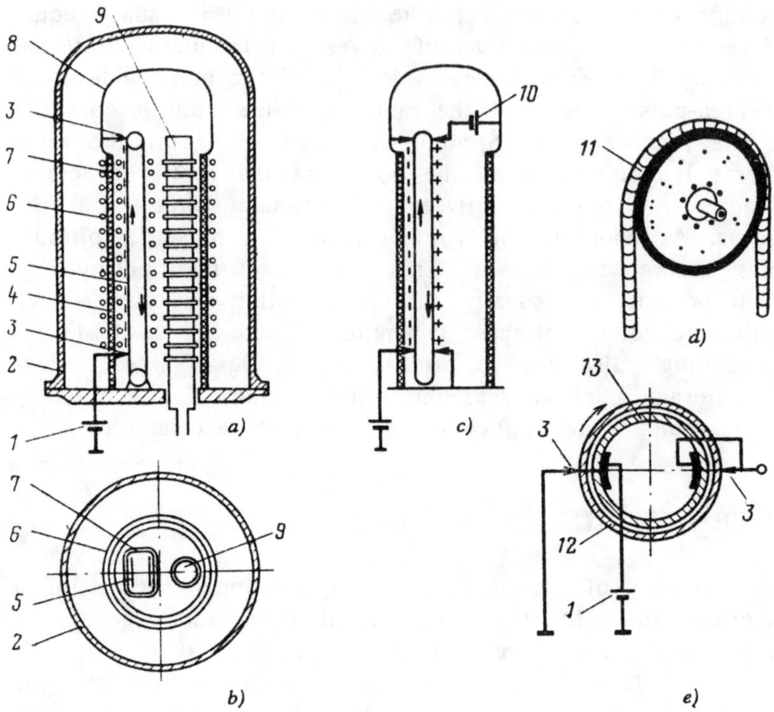


**Figure 6.5** Electrostatic voltage generators. *a*) and *b*) Construction diagram of generator with belt conveyor. *c*) Doubling of current by use of additional rectifier on high-voltage side of circuit. *d*) Chain-linked conveyor. *e*) Rotary generator. *1* and *10*—Rectifier for charging transpot belt; *2*—tank; *3*—corona points (brushes) for charging and discharging conveyor; *4*—insulating column; *5*—belt conveyor; *6*—gradient rings; *7*—gradient frames; *8*—high-voltage electrode; *9*—accelerating tube; *11*—chain-linked conveyor; *12*—rotor; *13*—stator.

amperes, $\sigma$ in C/m$^2$, $a$ in meters, and $v$ in m/s. The ultimate density of the charges under transport depends on the dielectric strength of the conveyer environment. For a flat transporting belt the ultimate density $\sigma_{max} = 2\,\varepsilon_0 \varepsilon\, E_{max}$, where $\varepsilon_0$ is the electric constant, $\Phi$/m, $\varepsilon$ is permissivity of environment surrounding the conveyer belt, and $E_{max}$ is the breakdown strength of the electric field in V/m. the maximum potential $U_m$ that can be obtained on the high-voltage electrode is also dependent upon $E_{max}$ of the surrounding medium. Thus, in the case of a spherical electrode with radius $R$ when the distance from the electrode to the part under ground potential is much longer, $R$: $U_m = E_{max} \cdot R$. The current of the electrostatic generators can be doubled by making use of an additional rectifier (10) inside high-voltage electrode 8 (see Fig. 6.5*c*). In this case, the conveyer (5) will transport not only negative charges to the higher potential, but also positive charges to the ground. Rectifier 10 as well as the other systems positioned on the high-voltage side are usually supplied with power by a voltage motor-generator located inside electrode 8 and driven to rotate by conveyer 5.

The present-day electrostatic generators with belt conveyers are placed together with accelerating tubes in hermetically sealed tanks [(2) in Fig. 6.5] filled with an electrically insulated gas (mixture of nitrogen with carbon dioxide, Freon, $SF_6$ gas, etc.) under a pressure of up to 1 to 2 MPa (Fig. 6.5*a*, *b*). For improving the dielectric strength of the structure, use is made of metal gradient rings (6) and frames (7), which are arranged on an insulating column (4) and also near to a conveyer (5) and accelerating tube (9), both mounted inside the column. Constant intensity of the electric field along the column is provided with the help of a divider connected to the gradient rings and frames (not shown in the Figure). The current transported by the conveyer in such machinery amounts to hundreds of microamperes. The width of the conveyer (*a*) usually lies within 0.3 and 0.4 m, the speed being 20 to 40 m/s. To increase the current of the accelerating particles, use may also be made of several belts operating simultaneously. However, this will require a more complicated construction of the generator and will deteriorate its reliability. In pelletrons (modification of electrostatic generators), the purpose of the conveyer is served by a chain (11) assembled of metal cylinders interconnected through insulating elements (Fig. 6.5*d* [54]). In some cases, electrostatic generators are positioned horizontally so as to materially reduce the height of the generator room. However, a cantilevered positioning of the column will necessitate a more intricate construction of the generator.

In addition to continuous operation, the electrostatic generators may also be used for accelerating great currents of particles in short pulses. In this case the current (and power) of an electron beam can be increased by a factor of several orders. The charge accumulated between successive pulses is carried away by the electron beam in a short lapse of time. The accelerating voltage at the pulse time is diminished by the value $\Delta U = I\tau/C$, where $I$ is the current of the accelerated particles and $\tau$ is the pulse length. In equipment operating in such working conditions, use is made of elongated high-voltage electrodes (see Section 6.8) with a view to increase capacitance $C$.

The substantial advantage of electrostatic generators with belt conveyers, which generate the load direct current, consists of a high stability of voltage (up to 0.01%). At the same time, the power of these generators does not exceed a few kilowatts, the efficiency being 10 to 20 percent. These performance characteristics are noticeably inferior to those of other versions of the voltage generators utilized in industrial continuous-action accelerators. Due to this fact, at present electrostatic generators with belt conveyers find application, for the most part, in research installations.

Another version of electrostatic generators is represented by a rotary generator. (Fig. 6.5*e* [55]). In this unit, the charges are transferred by the use of a fast-rotating insulating cylinder (rotor 12). Uniform division of the voltage is provided with the aid of a stator (13) fabricated of a semiconducting material. The current flowing across the stator amounts approximately to 10 percent of the load current value. The speed of rotation of the rotor reaches 3000 rpm. The generator is filled with hydrogen under a pressure of 1.5 to 2 MPa, which enables a high dielectric strength of the gaps to be obtained with a relatively negligible friction loss. The current in a rotor-type generator amounts to a few milliamperes. The rotor generators that were in batch production offered the following performance characteristics: voltage of up to 600 to 750 kV and power of 2.5 to 3 kW. Presently these generators are not widely used because moving parts incorporated in the generators significantly impair the competitiveness of these generators as compared with the voltage sources of other types, which are described below.

## 6.4 SINGLE-PHASE, HIGH-VOLTAGE TRANSFORMERS

Schematic diagrams of single-phase transformers are shown in Figs. 6.3*b* and 6.4*b*. These transformers, as a rule, are operated in two modes. The first mode of operation is continuous generation of sinusoidal high voltage and acceleration of particles at each half-wave of voltage of corresponding polarity. With a ferromagnetic core and at a high coefficient of coupling between the windings, the load power approximates the power applied to the primary winding of the transformer. The second mode of operation resides in the fact that the repetition frequency of the accelerating cycles is considerably less than the frequency of the feeding generator, and the pulse power of the electron beam considerably exceeds the average power of the generator. In this case, the transformer can be excited by the turns undercoupled with the secondary winding, and therefore, the ferromagnetic core may be omitted. With the object of obtaining high voltages and adequate efficiency in such a transformer, it is necessary to provide resonance between the frequency of the feeding generator and the natural frequency of the transformer. (The natural frequency of the transformer is determined with consideration for the inductance and capacitance of the load and the power supply feeder.)

Strong-coupling transformers are also used in various voltage sources incorporating rectifying elements and generating direct voltage. Description of such facilities can be found in the subsequent sections of this book. This section is devoted to industrial voltage sources containing single-phase transformers and generating sinusoidal (or approximately sinusoidal) high voltage. It should be noted that the strong-coupling transformer may also be employed for generating accelerated particles by single pulses, while light-coupling transformers are suitable for accelerating charged particles at each oscillation period of the high voltage. At present, such variants are used much more rarely, and generally only two modes of operation described above are of use in industrial accelerators furnished with single-phase transformers. However, in the process of developing industrial accelerators for generating electron beams with an average power of several kilowatts, it was also found reasonable to design and manufacture light-coupling transformers.

### 6.4.1 Resonance Transformers

The first single-phase transformers used in the industrial accelerators for generating sinusoidal voltage of the order of 1 MV were the resonance transformers manufactured by General Electric (U.S.). These transformers were commercially produced from the end of the 1930s until recently. Figure 6.6 shows the construction and circuit diagrams of one of the units, namely, an acclerator for flow detection, rated at 2 MeV with a mean current of 1.5 mA [56]. Primary winding 1 of the transformer is located near the bottom of tank 3. Secondary winding 2 consists of 243 flat sections. The coupling coefficient is about 0.2. Between the sections of the secondary winding there are gaps forming a passage for the gas. The equipment is cooled by forced circulation of the gas. As a result of dissipation, the intensity of the magnetic flux from the high-voltage end of winding 2 is appreciably less than in the area of primary winding 1. Therefore, for equalizing the gradient of the electric field potential over the secondary winding, the upper sections are arranged closer to each other. The transformer is supplied with a voltage of 180 Hz in frequency. The elements of the power supply circuit are shown in Fig. 6.6*b*. The natural frequency of the circuit constituted by the secondary winding and its effective capacitance to the tank walls also amounts to 180 Hz. This factor makes it possible to obtain the value of the transformer cos φ approximating a unity.

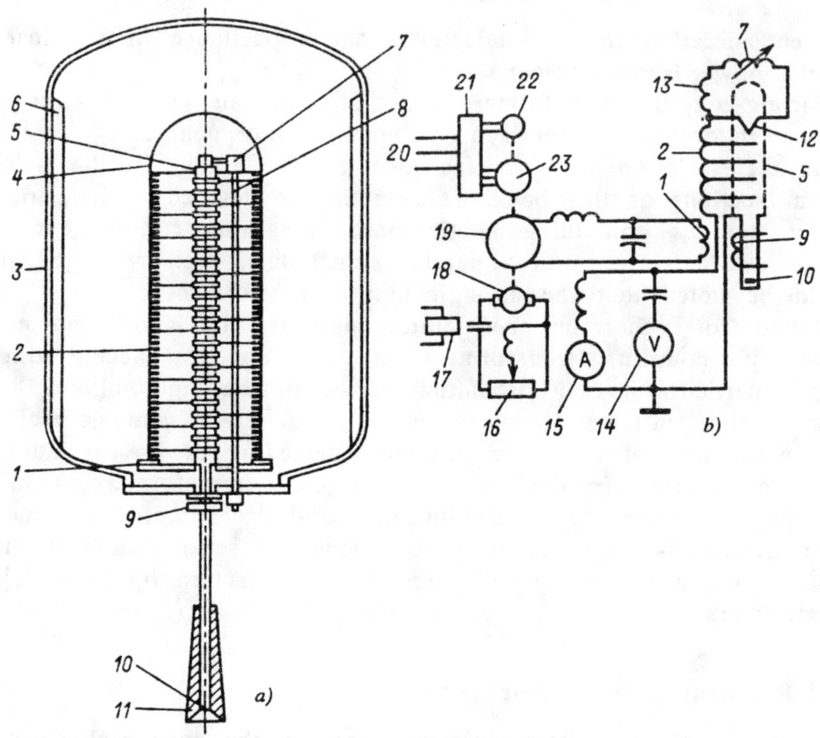


**Figure 6.6** Installation for industrial flaw detection based on resonance transformer. *a)* Construction diagram. *b)* Circuit diagram. *1*—Primary winding; *2*—secondary winding; *3*—tank; *4*—high-voltage electrode; *5*—accelerating tube; *6*—transformer steel; *7*—variable inductance; *8*—glass rod; *9*—magnetic lens; *10*—target; *11*—lead shielding; *12*—cathode heater; *13*—turns to feed cathode heater; *14*—megavoltmeter; *15*—milliammeter; *16*—rheostat with motor-drive control of high voltage; *17*—starting relay; *18*—generator exciter; *19*—single-phase generator, 440 V, 10 kW, 180 Hz; *20*—three-phase power supply network, 60 Hz; *21*—synchronous motor starter; *22*—motor exciter; *23*—synchronous electric motor, 15 kW, 1800 rpm.

The secondary winding is a quarter-wave resonant system on the top end of which voltage peaks of standing wave arise (see Fig. 6.4*b*). The voltages from some sections of the winding are fed to 24 electrodes of evacuated accelerating glass tube (5). To feed the heater of cathode 12, use is made of a part of turns (13) of the upper section of high-voltage winding 2. The heater current is controlled by means of variable inductance coil (7) which is set in motion by a glass rod (8).

Tank 3 is filled with Freon-12 under the pressure of 0.4 MPa. A ferromagnetic circuit is absent. To reduce the loss caused by eddy currents, the walls of tank 3 are coated with bands of transformer steel (6). The cooling system comprises a fan and a heat exchanger (not shown in the illustration). The accelerated electron beam is brought to internal target 10 for generating bremsstrahlung. The installation under consideration employs an accelerating diode system (cathode control grid is not available). The accelerating process takes place during every negative half-wave of high voltage within the range of zero to $U_{max}$. The overall dimensions of the tank are height, 2.5 m and diameter, 1.5m. The total length of the installation including the target is ~3 m. The transformer efficiency is 50 percent. Later modifications of the resonance transformers produced by General Electric incorporated systems for extraction of electron beams into the atmosphere through the outlet window of up to 30 cm in length. These transformers found application in production processes with radiation. In the early 1970s, manufacture of such equipment was discontinued due to development of new types of high-voltage generators featuring extraordinary power, efficiency, and much better monoenergetic characteristics of accelerated particles.

A similar principle has been used in resonance transformer, type RTE-1 rated at 1 MV and a power of up to 3 kW, on the average. This transformer was developed by NIIEFA (Leningrad) [57]. The natural frequency of the transformer is 500 Hz, the stability of the accelerating voltage amplitude lies within ±5 percent, the stability of the beam current is up to ±10 percent and the efficiency is 30 percent. The overall dimensions of the transformer are 1.1 m in diameter and 1.3 m in height.

In the early 1970s, a series of this type of accelerator was manufactured for flaw detection and radiographic inspection of materials.

### 6.4.2 Resonance Shock Transformers

Another variety of single-phase transformers is represented by a resonance shock transformer [58]. Unlike the resonance transformer discussed above, which is continuously excited by the resonance-frequency voltage source, the excitation of a shock transformer occurs periodically at the expense of discharge across the primary winding of a charged capacitor. An equivalent circuit diagram of a resonance transformer with shock excitation is shown in Fig. 6.7*a*. The natural frequencies of the primary and secondary circuits of the transformer are equal, i.e., $C_1L_1 = C_2L_2$, where $C_1$ is the capacitor contained in the primary circuit and charged by device 8, $C_2$ is the distributed capacitance of the transformer secondary circuit, and $L_1$ and $L_2$ are the inductances of the primary and secondary windings 1 and 2, respectively. The waveform of voltage $U_2$ arising on the

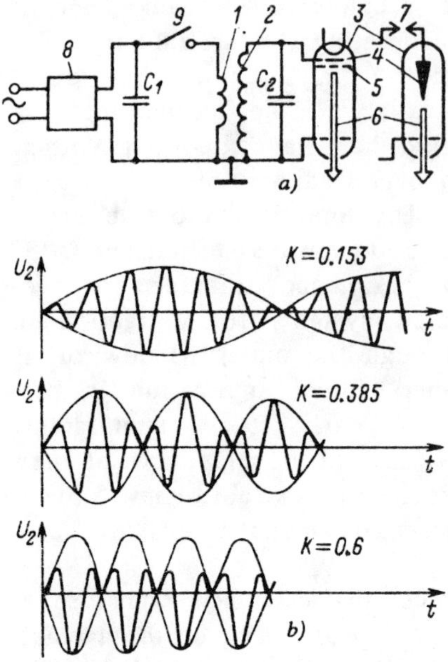


Figure 6.7 Resonance shock transformer. *a)* Equivalent circuit diagram. *b)* Waveform of generated voltage at different coupling coefficients $K$. *1*—Primary winding of transformer; 2—secondary winding; *3*—accelerating tube; *4*—cathode; *5*—control grid; *6*—accelerating particles; *7*—sparkgap; *8*—device for charging capacitor $C_1$; *9*—electron switch; $U_2$—voltage of secondary winding; $K$—interwinding coupling coefficient.

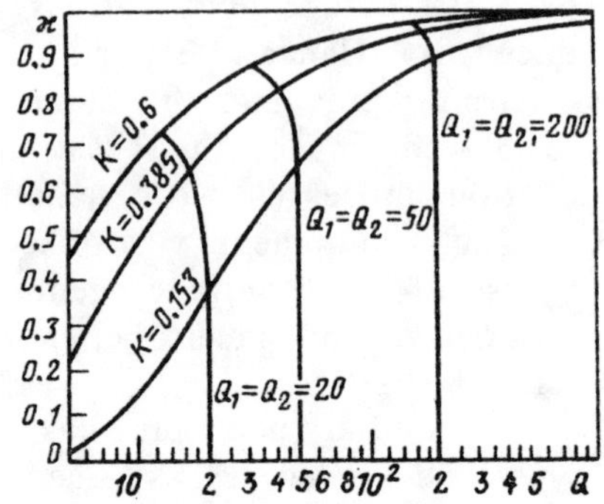


Figure 6.8 Energy transferred to secondary circuit versus coupling coefficient $K$ and $Q$ factor of system.

secondary winding when connecting charged capacitor $C_1$ to the primary winding through key 9 is illustrated in Fig. 6.7*b*. The greater the interwinding coupling coefficient $K$, the less is the number of oscillation periods required to attain the maximum value of $U_2$. As can be seen from Fig 6.7, at $K = 0.385$ this takes place at the third half-cycle of oscillations, whereas at $K = 0.6$ this is realized during the second half-cycle of oscillations. Generally, at the first maximum of $U_2$ the load is connected, and the acceleration of the particles is effected. Figure 6.7 also shows an accelerating tube (3), cathode (4), with control grid (5). The latter controls acceleration time and magnitude of the current of particles (6). In the absence of a load, the energy contained in $C_1$ is transferred to $C_2$ and back to $C_1$ until it dissipates in the form of ohmic losses. A number of beat periods is shown in Fig. 6.7*b*. Change of $U_2$, depending on the parameters of the transformer, takes the following form:

$$U_2 = u_1(0)\sqrt{\frac{L_2}{L_1}}\, e^{-t/\tau}\left[\sin\frac{\omega}{2}\left(\frac{1}{\sqrt{1-K}} + \frac{1}{\sqrt{1+K}}\right)t + \sin\frac{\omega}{2}\left(\frac{1}{\sqrt{1-K}} - \frac{1}{\sqrt{1+K}}\right)t\right]$$

where $u_1$ (0) is the voltage of capacitor $C_1$ at the initial moment and $\omega$ is the natural frequency of the circuits; $\tau$ is the attenuation time constant

$$\tau = \frac{4}{\omega}\frac{Q_1 \cdot Q_2}{Q_1 + Q_2}(1 - K^2)$$

where $Q_1$, and $Q_2$ stand for $Q$ factors of the circuits. From the above equation, it follows that $U_2$ reaches its peak value at $K$ = 0.6, 0.385, 0.28, 0.222, 0.18, 0.153, and so on. The part of the energy transfered from capacitor $C_1$ to secondary circuit capacitor $C_2$:

$$\kappa = \frac{C_2}{C_1} \cdot \frac{U_{2m}}{u_1^2(0)}$$

is presented in Fig. 6.8 where the value

$$Q = 2\frac{Q_1 \cdot Q_2}{Q_1 + Q_2} \cdot (1 - K^2)$$

is plotted along the $x$ axis. The value of the energy injected into the electron beam under acceleration depends on the modes of operation. Figure 6.9 illustrates several typical verisons of connection of load (accelerating beam) to the secondary winding of the transformer, namely,

(a) acceleration of high current in short pulses at the expense of the energy accumulated in capacitor $C_2$; in this case spark gap (Fig. 6.7*a*) usually is used.
(b) acceleration of comparatively small current in long pulses (of the order of one-fifth the duration of the oscillation period of $U_2$). The substantial portion of the energy taken away by the electron beam is supplied here from the primary circuit to the secondary circuit during the lifetime of the accelerating pulse.
(c) accleration of monoenergetic beam. Under the effect of the particular law of change of load (beam current) it is practical to attain the constancy of the voltage on the acclerating tube.

The radically new performance characteristics of shock transformers have been obtained by using a controllable key switch gap

(9) in the primary circuit (Fig. 6.7*a*) capable of both closing and opening the circuit [59]. The modes of operation of the installation with this key can be seen in Fig. 6.10. At $t_1$, when the residual energy of the transformer is accumulated again in $C_1$, key 9 becomes nonconductive, and energy is preserved in capacitor $C_1$ for the subsequent pulse. During the interval between the pulses, the voltage of capacitor $C_1$ is increased to $u_1$ (0) with the aid of device 8. The pulse repetition frequency may reach hundreds and thousands of hertzes, which allows for obtaining a high average power supplied to the load. The energy losses in such a mode of operation and at a high coupling coefficient ($K$ = 0.6 in Fig. 6. 10) are comparatively negligible, and the efficiency from the power supply mains feeding device 8 (Fig. 6.7) until the moment of loading may be as much as 60 to 70 percent.

The Tesla transformer was utilized in the early particle accelerators in the 1930s. As a rule, the coupling coefficient was not in excess of 0.1 to 0.2. A control grid in the accelerating device was not provided, and as a result, the particles were accelerated during all voltage half-cycles of the corresponding polarity. Oil or air served as insulators. The voltage was generated by emitting single pulses, and no recourse was made to recuperate the energy. Due to low efficiency, instability of particle energy, minor average power, large size, and other disadvantages, accelerators based on the Tesla transformer have given way to other types of equipment. A new series of resonance shock transformers dates back to the 1960s, when a number of modifications and improvements of both the construction and the electrical circuitry made it possible to obtain a comparatively simple and compact voltage source capable of generating voltage pulses at a high repetition rate and high efficiency.

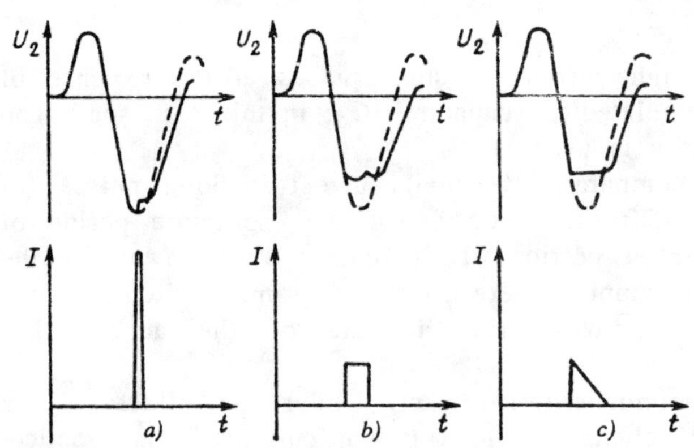


**Figure 6.9** Versions of acceleration of electron beam in resonance shock transformer. *a)* Acceleration in short pulse at the expense of energy accumulated in $C_2$. *b)* Acceleration in long pulse. *c)* Acceleration of monoenergetic beam.

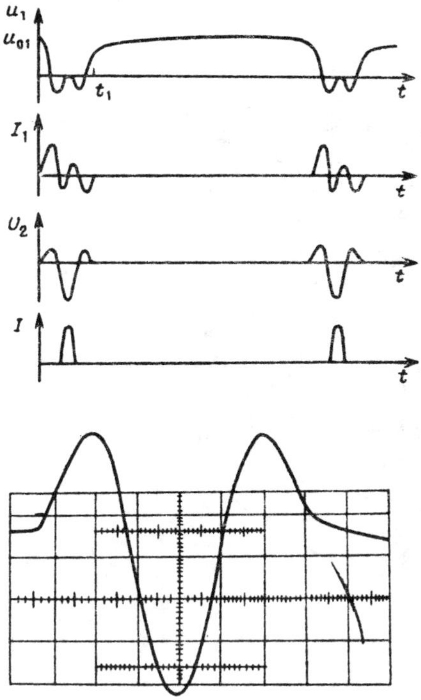


**Figure 6.10** Curves illustrating operation of shock transformer in energy recuperation mode ($K$ = 0.6). $u_1$—Voltage of primary stage capacitor $C_1$; $I_1$—current of primary stage; $U_2$—voltage of transformer secondary winding and load; $I$—accelerating particle current. $U_2$ is shown at no load and in energy recuperation conditions.

Figure 6.11 shows a construction diagram of an accelerator based on shock transformer Elita–1.5 (electron impulse transformer accelerator) rated at an energy of 1.5 MeV and an average power of 5 kW. Primary and secondary windings (1 and 2) are arranged coaxially and are approximately similar in height. A protecting screen (3) and high-voltage electrode (4) are transparent for magnetic flux. The coupling coefficient $K$ of the windings is equal to 0.6. In the transformers characterized by high natural frequency (tens and hundreds of kHz), secondary winding (2) is wound in one layer over insulating cylinder (usually made of Plexiglas). The generator, together with the accelerating tube (5), is placed in a tank (6) filled with a mixture of gases $SF_6$ and $N_2$ under a pressure of up to 1.5 MPa. The accelerating tube includes a cathode (7) with a control grid. The energy supply (8) for the cathode and control of the accelerating tube current is fed through the secondary winding wound with a double wire. The energy is applied from a source of low potential. To minimize the overvoltage effects arising on the secondary winding at the time of high-voltage breakdowns, provision is made for a shield (9) (for details, see Section 6.9). The energy of electrons in the installation displayed in Fig. 6.11 is controlled within the range of 0.4 to 1.5 MeV. The current can be

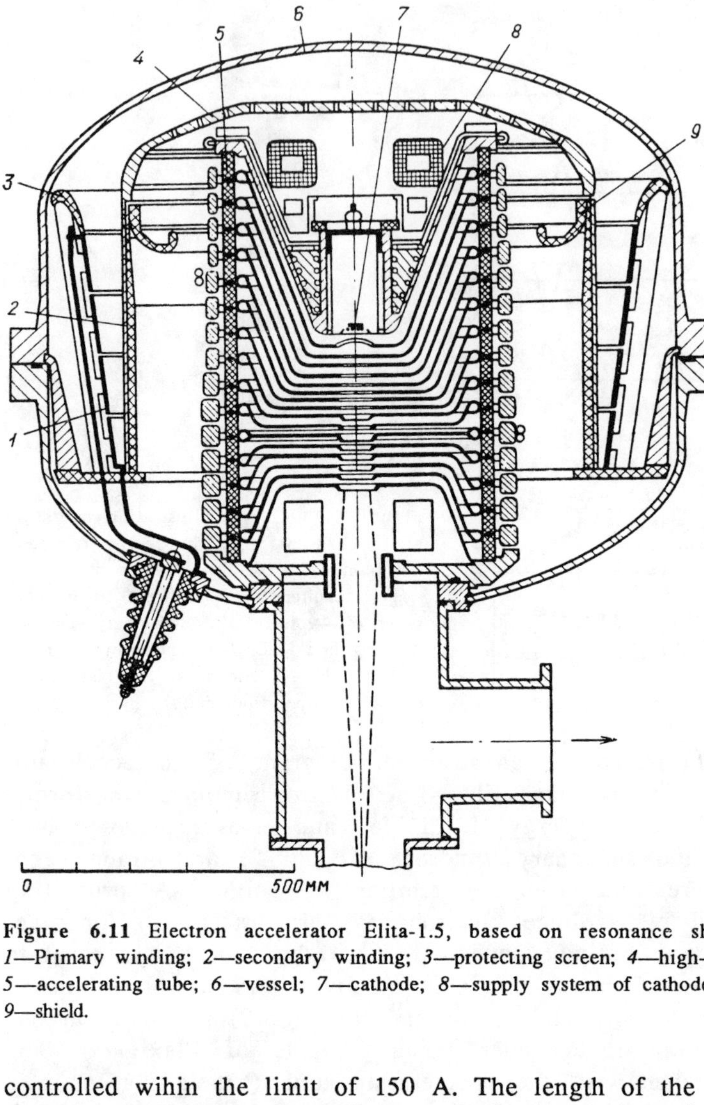


**Figure 6.11** Electron accelerator Elita-1.5, based on resonance shock transformer. *1*—Primary winding; *2*—secondary winding; *3*—protecting screen; *4*—high-voltage electrode; *5*—accelerating tube; *6*—vessel; *7*—cathode; *8*—supply system of cathode and control grid; *9*—shield.

controlled wihin the limit of 150 A. The length of the pulses can be varied within the range of $3 \times 10^{-8}$ to $2.5 \times 10^{-6}$ s. The pulse repetition frequency goes up to 100 Hz. With an energy of 1.2 MeV, the average power of the electron beam amounts to 5 kW. In the 1960s and early 1970s, the Nuclear Physics Institute (Novosibirsk) produced dozens of Elita units with different parameters. Some of this equipment has been used until the present time. Subsequently, similar high-voltage generators were developed in other laboratories (for example, see [60, 61]).

One of the important components of the installation under consideration is an electronic key (switch gap) in the primary circuit.

In the serial Elita units, use was made of hydrogen thyratrons TGI–I–2500/35 allowing operation with recuperation (Fig. 6.10) at a high pulse repetition frequency [62]. In the installation described in [60], a gas-blast spark gap was used as an electronic key. Achievements in the development of keys capable of not only closing but also opening the primary circuit and operating in the frequency-operation mode is a deciding factor that governs, to a large measure, the parameters of up-to-date accelerators based on shock transformers. The switching procedure in the primary circuit and the entire system of the primary power supply have been considerably simplified, with the natural frequencies of the transformer circuits reduced to a few kHz. (Specifically, this can be realized by way of increasing the number of the secondary winding turns and by converting from a single-layer to multilayer high-voltage winding). In this case, standard thyristors may be used as electronic keys, and capacitor $C_1$ can be charged directly from the industrial power supply mains.

Tesla transformers have also found extensive applications in compact x-ray pulse generators (e.g., MIRA) for industrial flaw detection (developed by Scientific-Industrial Objedinenije "Burevestnik," Leningrad [61] (Fig. 6.12). The Primary winding (1) of the transformer is wound by laying 6 to 10 turns of copper foil strips, 0.1 to 0.2-mm thick, with a width equal to the height of the secondary winding. The external turn is grounded. The turns of the secondary winding (2) are insulated from one another by means of insulating oil-impregnated paper. To improve the coupling effect, some ferrite rings may be fitted on lead (3). The coefficient of coupling between the primary and secondary windings of the transformer ($K$) ranges from 0.6 to 0.7. The voltage of the capacitor of the

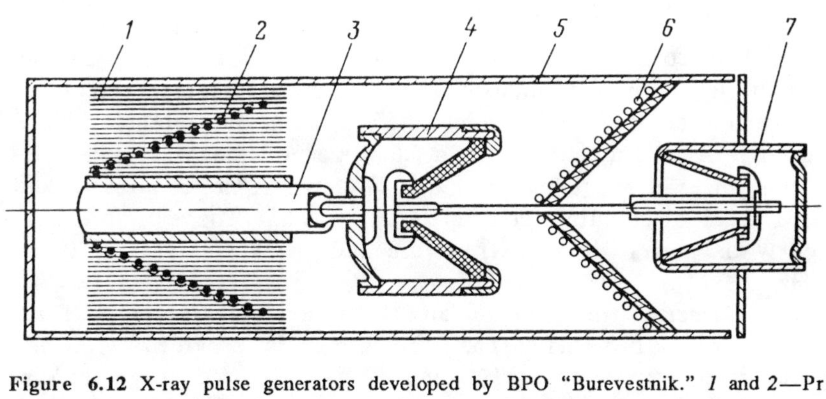


**Figure 6.12** X-ray pulse generators developed by BPO "Burevestnik." *1* and *2*—Primary and secondary windings of transformer; *3*—secondary winding lead; *4*—discharger-peaker; *5*—casing; *6*—bypass inductance; *7*—electron gun.

primary circuit reaches 30 to 50 kV, the transformation coefficient being 15 to 20.

The capacitance of the transformer secondary circuit is materially dependent on the size of the surface of casing 5 and the area of the elements being at a high potential. As a rule, generators of this type incorporate evacuated electron guns (7) with cold cathode. To reduce the duration of a pulse of the accelerating voltage applied to the electron gun, use is made of crowbar device (4) (see also 7 in Fig. 6.7*a*). Generally, these are serial spark gaps P-43, P-69, P-70 filled with a gas under a pressure to 2 to 4 MPa [63]. Self-breakdown of a spark gap takes place at the first half-wave of the voltage. Thus, the electrons are accelerated at the first half-wave of voltage $U_2$ as opposed to the Elita transformers in which charged particles are accelerated during the second half-wave of voltage $U_2$. This reduces by nearly one-half the voltage taken from the transformer, but considerably prolongs the service life of the spark gap. The switching time of the spark gap is 2 to 4 ns. The rate of the voltage rise on the accelerating tube amounts to $\sim 10^{14}$ V/s. The duration of the electron acceleration pulse and the length of x-ray radiation pulse equal $10^{-8}$ s. An inductance coil (6) bypasses the electron gun and thereby prevents acceleration of electrons until breakdown of the spark gap (4) takes place. The generators are rated approximately at $10^6$ pulses without replacing the component parts. Maximum voltage in the generators of this type reaches 500 to 600 kV, the pulse repetition frequency in different models ranging from 1 to 25 Hz. The mass of the generator (less the primary power supply and the control unit) ranges from several to tens of kilograms.

### 6.4.3 Electronic Transformers (ELTs)

Now we shall consider one more version of a single-phase step-up transformer used in the industrial accelerators. This is an electronic transformer ELT [64] which differs from the foregoing varieties in that it is furnished with a stabilizer of the voltage accelerating the charged particles, contains a ferromagnetic circuit, and possesses a high coupling coefficient $K = 0.85$. The transformer is supplied with power directly from the industrial mains of 220/380 V, 50 Hz.

The construction diagram of an electronic transformer ELT-1.5 [65] is shown in Fig. 6.13. The transformer is provided with primary and secondary windings (1 and 2), an external magnetic circuit (parts 5, 6, and 14), a core stacked of disks (3) separated with layers of insulation, a high-voltage electrode (4), and built-in accelerating tube (8). Primary and secondary windings contain, respectively, $w_1 = 94$ and $w_2 = 100.000$ turns. The magnetic circuit and the core are fabricated from transformer steel 343, 0.35-mm thick. Core disks (3) and magnetic circuit component parts (6, 13, 14, and 16) are radially laminated, and stuck together with epoxy resin. Then their surfaces facing the high-voltage gaps are properly finished. The transformer is confined in a tank (15) filled with a compressed gas (freon or a mixture of $SF_6$ gas and nitrogen). The maximum pressure of the gas is 1.5 MPa. The potential of each disk approximates that of the secondary winding arranged on top. The insulation between adjacent disks is provided by the use of gas gap $\delta = 0.7$ cm. The full operating voltage is applied to the high-voltage gap $a \cong 13$ cm. The accelerating tube is in electric contact with every disk of the core. The dimensions and number of turns of the secondary winding are selected so that the natural frequency of the transformer is close to 50 Hz. This minimizes losses in the transformer, and the power factor (cos $\varphi$) approximates unity.

The injector of the accelerating tube has a grid and grid voltage control system. The injector and the control system are supplied from a part of the secondary winding turns arranged at the high-voltage end of the winding. As indicated in Fig. 6.14, the accelerating tube voltage $U_2$ is maintained unchanged during acceleration of the electron beam. This can be achieved by modulation of the accelerating tube current I (electron beam current) during the pulse lifetime with the aid of a control system with feedback. The duration of the electron current pulse can be varied from the control board within the range of 0 to 5 ms by changing the instant (phase) of the beginning of the current flow. The pulse frequency corresponds to that of the supply mains and equals 50 Hz. In the case of maximum width of the pulse, the inverse voltage $U_{2m}$ exceeds the accelerating voltage by 10 to 15 percent. The essence of the applied method of stabilization of the accelerating tube voltage resides in the fact that during the life of accelerating pulses, the load (accelerating tube) resistance is varied by the law enabling the load voltage to be held constant. The law of electron beam current variation $I\ (t)$ during the acceleration pulse (i.e., $t_1 < t < t_2$) which provides the unchangeability of voltage $U_2$ on the accelerating tube, takes the following form (Fig. 6.14) [65]:

$$I = \frac{U_{2m}}{\omega L_s}\left[\left(\frac{w_1^2}{w_2^2} + \frac{L_s}{L}\right)\cos \omega t - \frac{U_2}{U_{2m}}\omega\left(\frac{w_1^2}{w_2^2} + \frac{L_s}{L}\right)(t - t_1) - \frac{w_1^2}{w_2^2}\cos \omega t\right]$$

where $L$ is transformer inductance, $L_s$ is primary winding leakage inductance, $t_1$ and $t_2$ are instants of switching on and switching off current under acceleration (see Fig. 6.14).

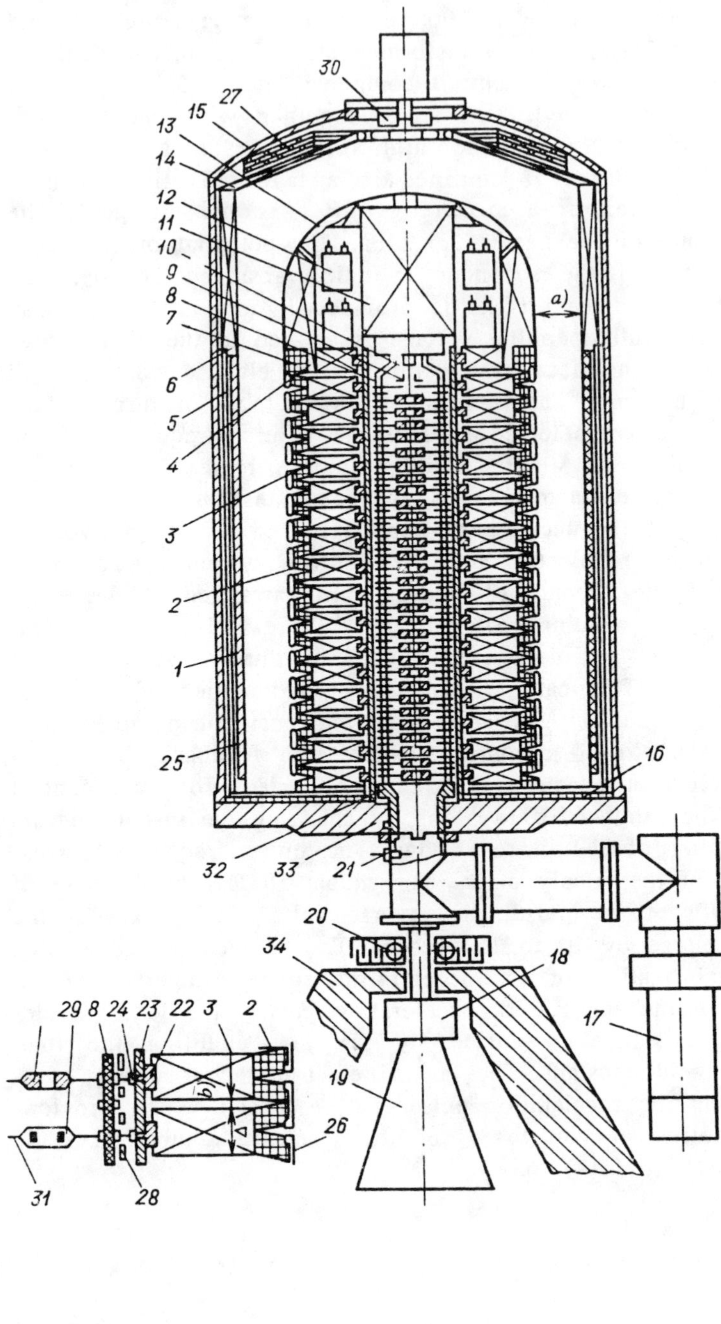
30
27
15
13
14
12
11
10
9
8
7
6
5
4
3
2
1
25
16
a)
32
33
21
20
34
18
17
19
29 8 24 23 22 3 2
б)
Δ
26
31
28

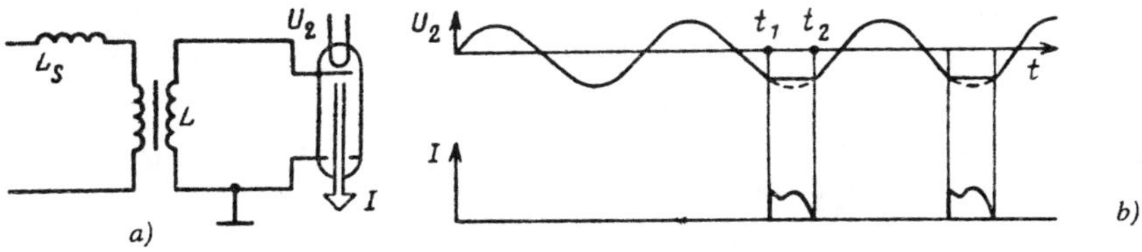


Figure 6.14 *a)* Simplified circuit diagram of electron transformer. *b)* change of voltage $U_2$ and current $I$ of accelerating tube in time. $L$—Transformer inductance; $L_s$—leakage inductance of primary winding; $t_1$ and $t_2$—switching-on/off moments of accelerating beam: $U_2$ and $I$ curves show initially no-load condition. Current in accelerating tube and, correspondingly, voltage stabilization are absent.

It should be emphasized that this equipment makes it practicable to realize various laws of energy distribution of accelerated electrons, particularly for providing a more uniform exposure dose of irradiation in depth of the object under processing.

Figure 6.15 shows a block diagram of the automatic control system of the ELT-1.5 transformer. Provision is made here for control of the accelerating voltage in the range of 400 kV to 1.5 MV. The voltage feeding the cathode and injector control circuit is automatically maintained at a constant level. The power at a preset accelerating voltage is controlled by changing the duration of the accelerating pulses ($t_2 - t_1$). The maximum power obtained during the tests of the accelerator was 390 kW. The power of the accelerator operated for many years was, as a rule 12 to 15 kw, the efficiency being ~90 percent. The ELT-1.5 is applicable to different beam extraction systems with a sweep covering the length of the outlet window (from 0.6 to 2.5 m) and with a differential pumping for extracting a concentrated beam of some 3 mm in diameter (see Section 8.1). The dimensions of the accelerator are: height of the tank, 2.1 m and diameter, 1.2 m. The unit mass is 8 t. About 10

---

**Figure 6.13** Electron transformer ELT-1.5. *1*—Primary winding of transformer; *2*—secondary winding section; *3*—disk of magnetic core; *4*—high-voltage electrode of magnetic circuit; *5,6* and *14*—parts of external magnetic circuit; *7*—coil of high-voltage electrode *4*; *8*—accelerating tube; *9*—control electrode; *10*—injector; *11*—power supply system of injector; *12*—capacitor unit; *13*—capacitance-type sensor; *15*—vessel; *16*—base of magnetic circuit; *17*—vacuum pump; *18*—scanning magnet; *19*—horn with extraction window; *20*—Rogovski belt; *21*—probe; *22*—copper rings; *23*—supporting cylinders; *24*—elastic contacts of accelerating tube; *25*—shield of primary winding; *26*—shields of secondary winding sections; *27*—cooler; *28*—ohmic divider; *29*—magnetic lenses; *30*—fan; *31*—electrodes of accelerating tube; *32*—spherical bearing of accelerating tube; *33*—sealing of accelerating tube; *34*—radiation shielding.

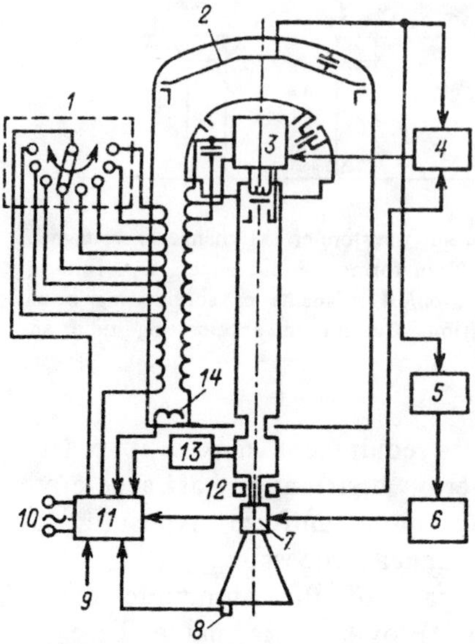


**Figure 6.15** Automatic contol system of ELT-1.5 transformer. *1*—Selector switch; *2*—capacitance-type voltage pickup; *3*—supply and control system of injector; *4*—power control and stabilizing circuit; *5*—circuit of sweep control and stabilization; *6*—sweep generator; *7*—scanning magnet; *8*—foil temperature pickup; *9*—from tank pressure pickup; *10*—mains; *11*—power supply network switching-off system; *12*—Rogovski belt; *13*—vacuum pickup; *14*—breakdown pickup.

ELT-1.5 accelerators were produced during 1965 to 1970 in the Nuclear Physics Institute in Novosibirsk. Some of these accelerators are still being used.

### 6.4.4 Cascade Transformers

In addition to the transformers in which the required voltage is obtained in one stage (cascade), use is made of voltage sources comprising several series-connected single-phase transformers. One of such sources widely used industrially in the USSR is the cascade transformer RUP-400, which generates sinusoidal voltage with an amplitude of 400 kV. The circuit and construction diagrams of the RUP-400 transformer intended for radiation are shown in Fig. 6.16 *a* and *b*. The transformer contains three stages. Each stage incorporates a toroidal core (1) with primary winding (2), high-voltage winding (3), and transmission winding (4). An evacuated accelerating tube (7) is built in the transformer and connected electrically to the taps of windings (3). The cathode is supplied with the aid of additional winding (5) placed in the final high-voltage stage. The transformer is placed in a Bakelite pipe (8). The latter is confined in a metal tank (9) filled with transformer oil. Shields (10 and 11) are shaped so as to be suitable for removal of corona and breakdowns from the parts energized with high voltage. The RUP-400 transformer is powered from the industrial power supply mains of 200/380 V, 50 Hz. The power requirement amounts to 4 kW. The accelerating tube receives a sinusoidal voltage of 400 kV. The transformer operates as a diode. The accelerated electrons are characterized with a broad range of energies. The average power of the electron beam reaches 2 kW. The installation illustrated in Fig. 6.16 is provided with an external cooled anode (13) for producing bremsstrahlung. The external diameter of the tank is 0.55 m; its length, including the external anode, is 1.6 m. Such dimensions make it a transportable apparatus of 600 kg in weight suitable for flaw detection. It is under serial production at the Aktubrentgen

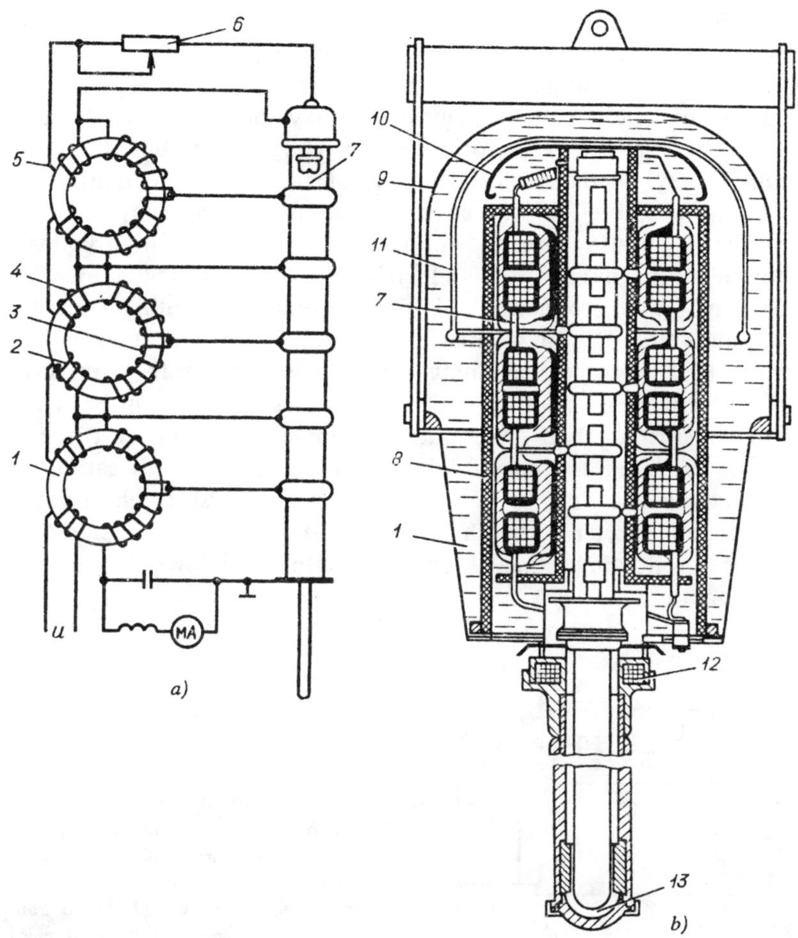


**Figure 6.16** Circuit *a)* and construction *b)* diagrams of x-ray apparatus RUP-400. *1*—Toroidal core; *2*—primary circuit; *3*—high-voltage winding; *4*—transmission winding; *5*—cathode heater supply winding; *6*—cathode heater rheostat; *7*—accelerating tube; *8*—Bakelite cylinder; *9*—metal tank; *10* and *11*—shields; *12*—focal lens; *13*—remote anode.

plant (Aktubinsk). There is also a modified version of the apparatus with extraction of electron beam into the atmosphere for radiation processing of materials [66]. The NIIFKH (Kiev) has already produced more than 20 modified versions of the RUP-400 sets.

### 6.4.5 Cascade Shock Transformer

The circuit diagram of a cascade shock transformer is shown in Fig. 6.17 [67]. The accelerating voltage in this generator is produced by several transformers connected in series. The energy is supplied to the high-voltage stages through the secondary windings of the transformers (the windings are wound with double wires). The transformation coefficients in each stage may be substantially decreased; therefore, in the case of preset primary voltage, the secondary windings may contain a less number of turns. On the one hand, this makes it possible to increase the natural frequency of the resonant circuits and in turn to diminish the length of high-voltage pulses, thereby improving the dielectric strength of the unit. On the other hand, the production of the secondary winding can be simplified, since the windings may be wound in one or two layers (see Section 6.9). Another advantage of shock generators with several stages is that the duration of the current pulses can be controlled for ensuring their adequate monoenergetic property, including cases of acceleration of high currents from cold cathodes. With this objective, a voltage is generated in the first (high-voltage) stage in a much shorter time than the duration of the accelerating voltage in the following stages, which practically allows for operation as a triode.

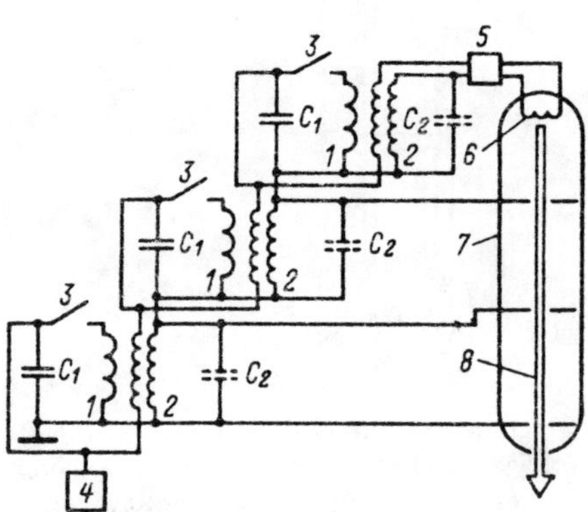


**Figure 6.17** Circuit diagram of cascade shock transformer. $C_1$—Capacitors of primary circuits; $C_2$—distributed capacitances of secondary circuits; *1*—primary circuits; *2*—secondary circuits wound by double wire; *3*—switches (control systems of electron switches are not indicated in illustration); *4*—charging system of capacitors; *5*-cathode supply system; *6*—cathode; *7*—accelerating tube; *8*—electron beam.

### 6.4.6 Conclusion

As the description of single-phase transformers showed, the main problems in their development and designing are dielectric strength of the construction and stability of the accelerating voltage. For solving the first problem, the units are placed in a high-quality insulating medium, making use of specially designed secondary windings (see Section 6.9). The constancy of energy of electrons is ensured by limiting the time of acceleration of particles or by modulating the electron beam current. Pronounced potentialities in decreasing the energy dissipation of accelerated particles are offered by application of rectifiers in the transformers under consideration.

## 6.5 DIRECT-VOLTAGE GENERATORS BASED ON SINGLE-PHASE TRANSFORMERS

### 6.5.1 Transformer with Insulated Core

The insulating core transformer or ICT is one of the earlier of these types of transformers. It was developed in the 1960s and manufactured on a commercial scale since then by the U.S. firm High Voltage Engineering Corp. [68]. The construction of the transformer is shown in Fig. 6.18*c*. For comparison, Fig. 6.18 shows the electron transformers Elita (Fig. 6.18*a*), ELT (Fig. 6.18*b*), and an electron transformer-rectifier ELV described below (Fig. 6.18*d*). The primary winding (1) and sections of the secondary winding (2) of the insulating core transformer are arranged coaxially. A magnetic flux passes along the external magnetic circuit composed of elements 3 and 4, and also over the central core stacked of disks (5) insulated from one another by the use of solid insulation (6). Located on the high-voltage side of the core is a magnetic circuit component (8). Each section of the secondary winding (2) is fitted on and connected electrically to disk 5 and is coupled to voltage doubler circuit (7) (see also Fig. 6.3*c*) incorporating rectifiers and capacitors. All the voltage doubling circuits are connected in series. Potentials of disks 5 change in compliance with changes of the secondary winding potential. This circumstance radically alleviates the problem of insulation of the winding turns from the core of the magnetic circuit. Accelerating tube 9 is mounted in common tank 10 filled with compressed gas.

An advantage shown of the foregoing high-voltage source is a 50 percent reduction of the voltage on the secondary winding sections,

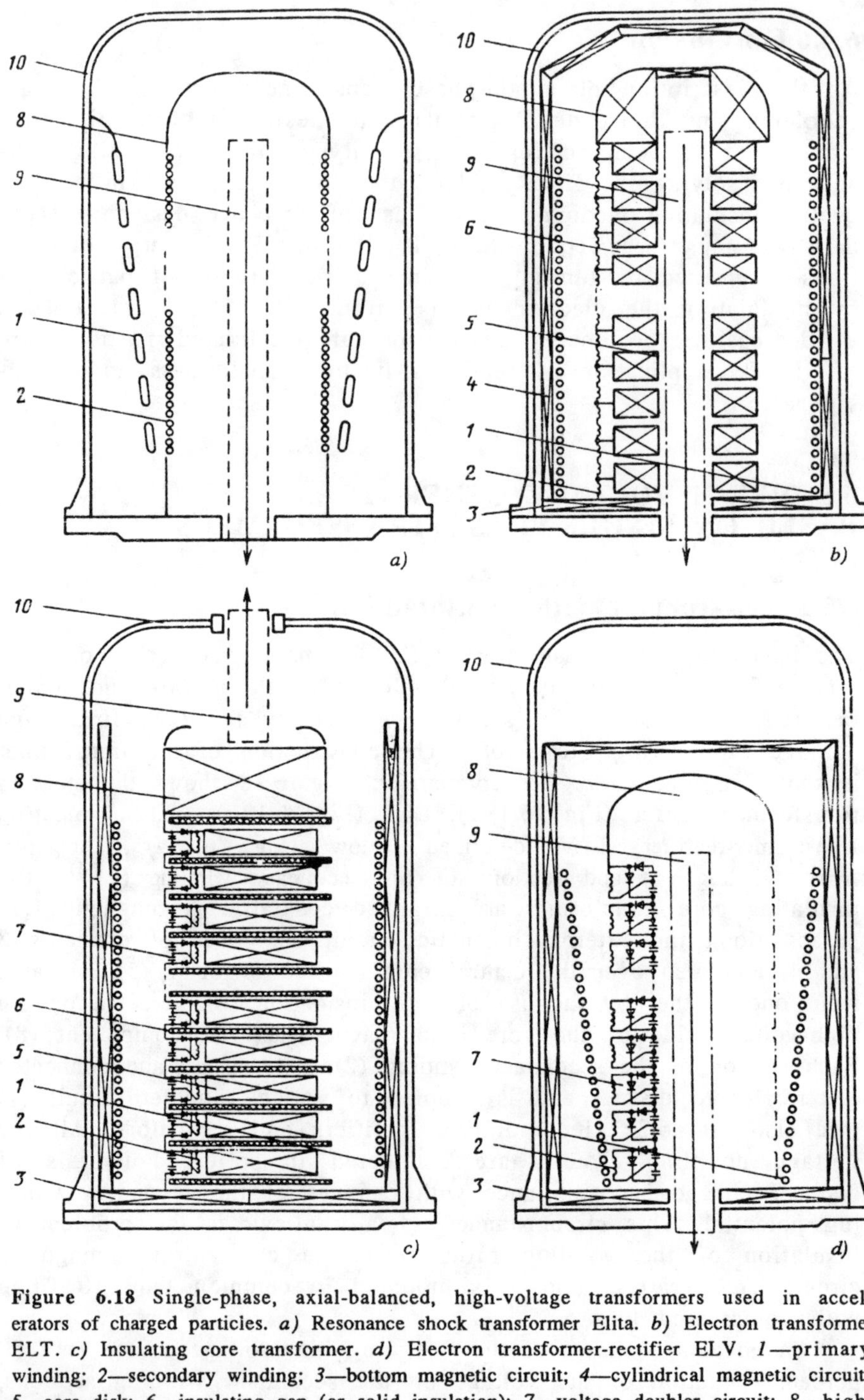


**Figure 6.18** Single-phase, axial-balanced, high-voltage transformers used in accelerators of charged particles. *a)* Resonance shock transformer Elita. *b)* Electron transformer ELT. *c)* Insulating core transformer. *d)* Electron transformer-rectifier ELV. *1*—primary winding; *2*—secondary winding; *3*—bottom magnetic circuit; *4*—cylindrical magnetic circuit; *5*—core disk; *6*—insulating gap (or solid insulation); *7*—voltage doubler circuit; *8*—high-voltage electrode; *9*—accelerating tube; *10*—vessel.

however, the circuit incorporates numerous rectifiers and capacitors. The approximate parameters of the high-voltage generators of this type are: voltage, 300 to 500 kV and power, about 10 kW. Following the development of a single-phase insulating core transformer, some more powerful three-phase transformers using the same principles have been designed and have found extensive application. Three-phase transformers are described in Section 6.7.

### 6.5.2 Electron Transformer-Rectifier

Similar in principle of action to the ICT is the electron transformer-rectifier ELV [69], developed and commercially manufactured since the mid-1970s in the Nuclear Physics Institute (Novosibirsk). The circuit diagram of the rectifier is shown in Fig. 6.18*d*. Sections 2 of the rectifier secondary winding are also connected to voltage doubling circuits 7 connected in series. As distinct from the foregoing generator, the ELV electron rectifier dispenses with the central ferromagnetic core. The elements of the voltage doubler circuits are accommodated inside the secondary winding; accelerating tube 9 is built in the generator. To improve the coupling coefficient, the primary winding (1) has been made in the shape of a cone. Figure 6.19 shows the diagram of a modification of the electronic rectifier ELV [70], where, in order to increase the intensity of the electron beam, provision is made for two high-voltage rectifiers connected in parallel with the accelerating tube. The top and bottom ends of the central core containing the sections of the secondary windings and voltage doubling circuits are at ground potential. The middle part of the core in Fig. 6.17 is at high potential. Both rectifiers have a common magnetic flux closing across the external magnetic circuit (through parts 4, 6, and 11). The height of the primary winding (5) is equal to that of the central core, while the length of the accelerating tube (9) is approximately half as much as the height of the core. The tube injector is energized from the coil located in the high-voltage portion of the core and is in inductive coupling with the primary winding. The apparatus is accommodated in a tank (3) filled with $SF_6$ gas under a pressure of 0.5 MPa.

The energy of accelerated electrons is 0.5 to 0.7 MeV. The power in the beam is 50 KW. The energy stability lies within ±2.5 percent. The efficiency of the accelerator is about 80 percent. The mass of the equipment amounts to 8 t. The frequency of the voltage applied to feed the primary winding of the transformer is 400 Hz. Several different versions of the ELV rectifiers rated at voltages in the range of 0.5 to 2 MV and a power from 20 to 50 kW have been developed and are in serial production.

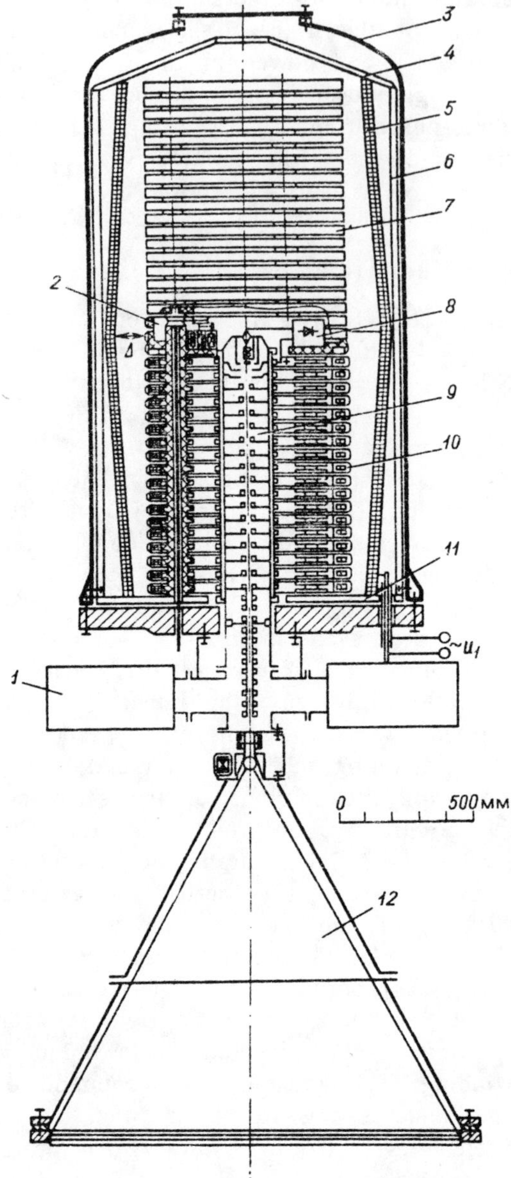


**Figure 6.19** Construction diagram of electron transformer-rectifier ELV-3. *1*—Vacuum pump; *2*—high-voltage electrode; *3*—vessel; *4*—tapered magnetic circuit; *5*—primary winding; *6*—cylindrical magnetic circuit; *7* and *10*—rectifying sections; *8*—control unit; *9*—accelerating tube; *11*—bottom magnetic circuit; *12*—extractor.

### 6.5.3 Transformer with Bridge Rectifier Circuit

Another version of a direct high-voltage source with a single-phase step-up transformer, is a transformer with a bridge rectifier circuit, type Elektron-3 (Fig. 6.20) [71, 72]. The generator contains two series-connected primary windings (1) and a secondary winding consisting of 200 sections (2). Unlike the foregoing transformers with voltage doublers, winding sections (2) in the transformer with a bridge rectifier are connected to the bridge rectifier circuits. The value of the rectified voltage equals the total sum of the amplitudes of the voltages on sections 2. The secondary winding in this circuit experiences higher intensity than in generators ICT and ELV; however, a substantially lower number of capacitors is used here. As far as design features are concerned, this generator differs from the preceding ones in introduction of closed O-shaped magnetic circuit (3) (which is at ground potential) and in using of solid insulation and insulating oil. The magnetic circuit is fabricated of a band of steel E350, 0.08 mm thick, 3.2 × 3.2 cm in cross section, and 1.5 m in height. Provision is made for a joint required for assembly of the unit. The total number of the turns of the primary and secondary windings are 140 and 440.000. The height of each winding is 1 m. Voltage is rectified by the use of selenium rectifiers (4), type TVS-7-19 M. The frequency of the supply voltage is 400 Hz. Sections 2 are insulated from the primary winding and the magnetic circuit by means of insulation (5) made of successive layers of polyethylene film and capacitor paper.

The high-voltage generator is installed in the upper part of a tank (6) filled with oil. An accelerating tube (7) is located underneath the generator in the same tank. The total height of the tank is 3 m, the tank diameter is 0.86 m. To provide a uniform division of the voltage over the height of the tank use is made of equipotential rings (8). The parts energized with high voltage are insulated from the tank wall by means of a cylinder (9) fabricated by winding up many layers of film insulation. For feeding the cathode of the injector (11) a special winding is used which is located on the high-voltage side of the magnetic circuit. The beam current in the accelerating tube is controlled by way of changing the heater voltage of the injector with the help of an autotransformer (10) controlled by an insulating rod. Figure 6.20*b* shows, among other things, a voltage divider (12), magnetic lens (13) for focusing the electron beam, magnet (14) for scanning the beam, horn (15), and outlet window (16) (1000 × 200 mm). The total height of the accelerator with the bell mouth is 7 m. The arrangement of the accelerator in localized shielding and the general view of the equipment are illustrated in Fig. 6.21.

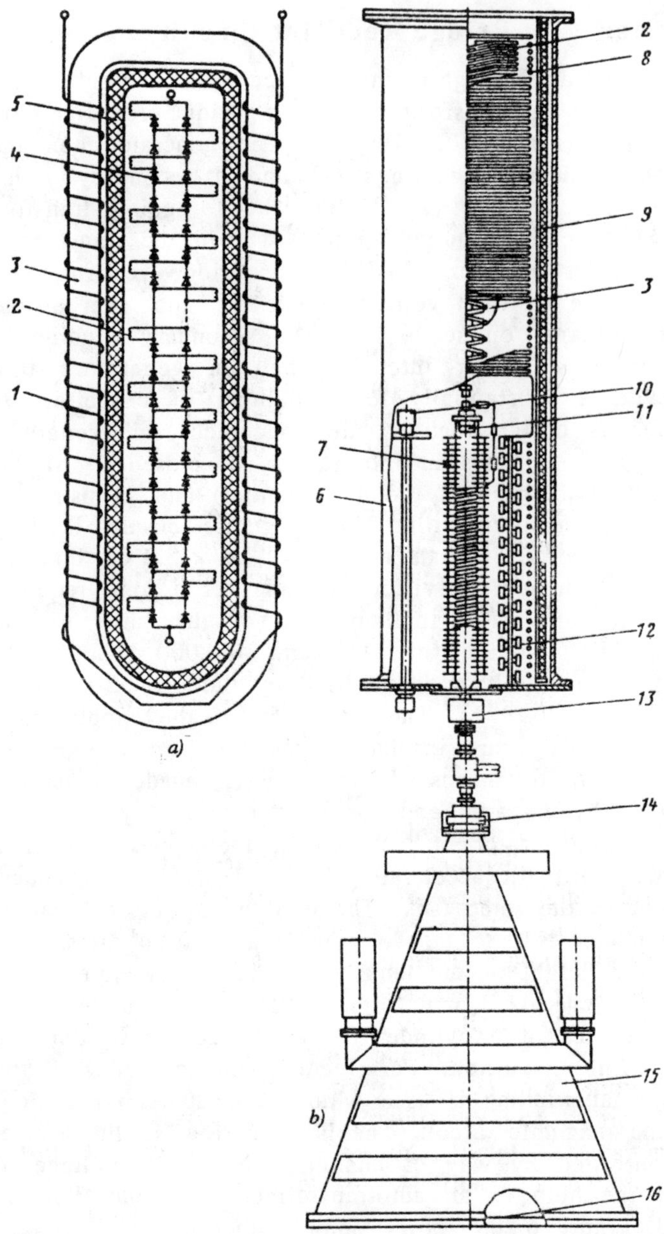


**Figure 6.20** Transformer with bridge rectifier circuit, type Elektron-3. *a)* Circuit of high-voltage generator (enlarged view). *b)* Construction diagram of accelerator. *1*—Primary winding; *2*—section of secondary winding; *3*—magnetic circuit; *4*—selenium rectifier; *5*—laminated insulation; *6*—vessel; *7*—accelerating tube; *8*—equipotential rings; *9*—insulating cylinder; *10*—autotransformer for injector heating; *11*—injector; *12*—voltage divider; *13*—magnetic lens; *14*—scanning magnet; *15*—horn; *16*—extracting window.

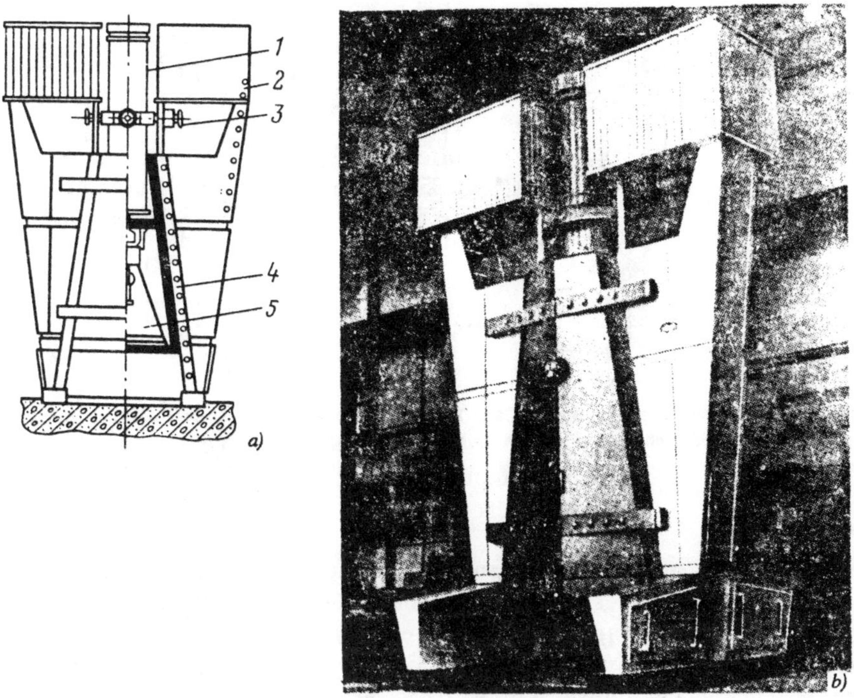


**Figure 6.21** Layout of electronic transformer accelerator Elektron-3 in localized radiation shielding *(a)* and general view of installation *(b)*. *1*—Metal vessel with high-voltage source and accelerating tube; *2*—servicing platform; *3*—adjusting appliance; *4*—radiation shielding; *5*—horn with extraction window.

The parameters of the electron beam generated by the Electron-3 accelerator are: energy, 0.5 to 0.7 MeV; energy stability, ±5 percent; efficiency, 80 percent. A number of these units were manufactured at NIIEFA (Leningrad) in the 1970s.

## 6.6 DIRECT-VOLTAGE, CAPACITANCE-COUPLED CASCADE GENERATORS

The general principle underlying different versions of cascade generators consists in a series connection of many voltage-generating sources, with the power supply of each of them derived from a source that is at ground potential.

In some generators, each subsequent stage is supplied from the preceding one. The energy is conveyed to the remaining stages via the stages closest to the power supply source (and correspondingly to the ground potential). In other generators, each stage is directly

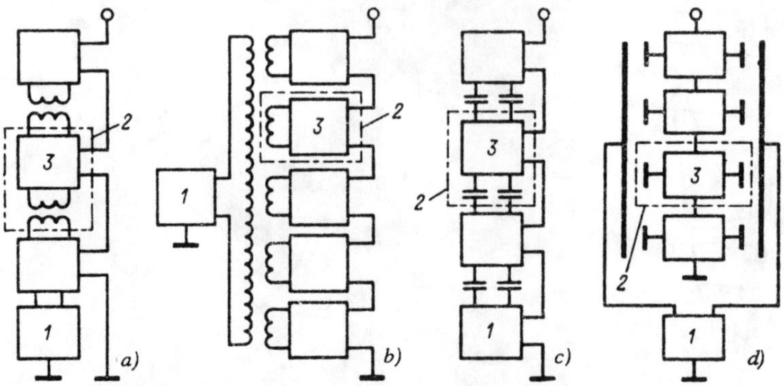


**Figure 6.22** Cascade generators. *a)* and *b)* Inductance-coupled generators. *c)* and *d)* Capacitance-coupled generators. *a)* and *c)* Generators with series supply of stage; *b)* and *d)* Generators with parallel supply of stages. *1*—Power supply; *2*—stage; *3*—voltage rectifying and multiplying circuits.

coupled to the power supply source. In the first case, a series supply of the stages takes place, while in the second case, it is effected in parallel. For transmission of energy through the insulation spaces, use may be made of both inductive and capacitive coupling. The schematic diagrams of four basic versions of the cascade generators are presented in Fig. 6.22. It is seen that in versions *a* and *b*, the stages are inductively coupled with the power supply source; in *c* and *d*, a capacitive coupling is shown. Figure 6.22 *a* and *c* shows a series power supply of the stages, while *b* and *d* illustrate a parallel method of power. Stages (2) may include different circuits (3) for rectification and multiplication of voltage. The construction features of the generators employing inductive coupling have been considered in the preceding section. Figure 6.22*a* corresponds to generator RUP, while *c* conforms to insulating core transformers as well as to installations ELV and Elektron-3.

## 6.6.1 Cockcroft–Walton Generators

The stages in Cockcroft-Walton generator (Fig. 6.23*a*) are supplied in series. The voltage is rectified by the use of a half-wave circuit. The value of the voltage at each capacitor is equal to the voltage *u* generated by the power supply source, i.e., as a rule, by a step-up transformer located at the input of the generator. This voltage usually ranges from 20 to 200 kV. The rectifying elements are rated at double inverse voltage. In order to reduce the dimensions of the capacitors, a voltage of a higher frequency (several kHz) is used. In Cockcroft-Walton generators, kenotrons served for a long time as rectifiers. The problem at the time was application of the energy for heating their cathodes. With the advent of solid rectifiers, construction of these generators was significantly simplified. Problems still existing in the Cockcroft-Walton generators are related to the fact that with the increase of the number of the stages, the operation of the generator tends to deteriorate. The voltage of the loaded generator drops, as compared to no-load voltage, and the voltage instability $\delta U$ tends to rise [72, 73]:

$$U \cong n \left[ 2u - \frac{I}{fC} \left( \frac{2}{3} n^2 + \frac{n}{2} + \frac{1}{3} \right) \right]$$

$$\delta U = \frac{nI}{2fC} (n + 1)$$

where $U$ and $u$ are voltages in terms of volts (supply voltage, $u \cos \omega t$; frequency $f$); $I$ is the load current in amperes; $C$ is the capacitance of each capacitor in farads; $n$ is the number of the stages.

An important role is played by spurious capacitance of the generator parts to the ground, or, more precisely, the ratio of the longitudinal capacitance of the generator to its transverse capacitance. Cockcroft-Walton generators were usually installed in the open air. The number of the stages reached several tens of pieces. The characteristic voltage and current were, respectively, 1.5 to 2 MV and ~5 mA. At present, generators of this type have been replaced by cascade generators balanced with respect to the grounded point.

## 6.6.2 Balanced Cascade Generators

In balanced cascade generators with a series supply of stages (Fig. 6.23*b*) [74], energy is transmitted through two extreme rows of

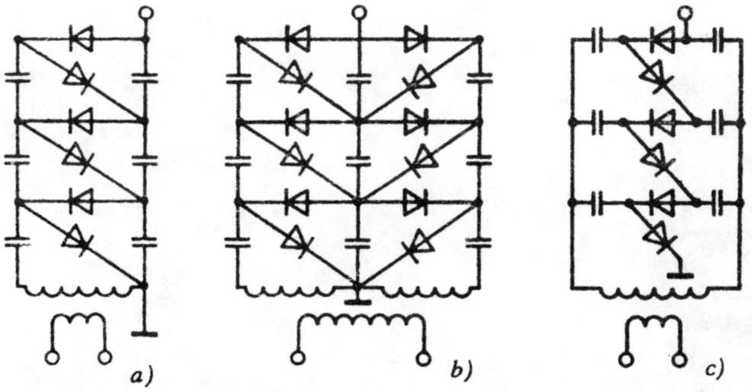


**Figure 6.23** Circuit diagrams of capacitance-coupled cascade generators. *a)* Cockroft-Walton generator. *b)* Balanced cascade generator. *c)* Generator with parallel supply of stage (dynamitron).

series-connected capacitors. The intermediate row of the capacitors is intended for filtration of the obtained direct voltage. The circuit of this kind makes for reducing the fluctuation of the rectified voltage both under load and at no load. The load voltage $U$ and ripple voltage $\delta U$ are [75]

$$U = n\left[2u - \frac{I}{fC}\left(\frac{n^2}{6} + \frac{n}{4} + \frac{1}{3}\right)\right]$$

$$\delta U = \frac{nI}{2fC}$$

The notation in this formula is similar to that stated above. Balanced generators are manufactured by several firms, including Emile Haefely Co. (Switzerland), Hitachi Ltd. (Japan), and Applied Radiation Corp. (U.S.). The power of these sets is 100 kW and greater, the voltage being several million volts.

Figure 6.24 shows a balanced cascade generator KGE-2.5 developed by NIIEFA (Leningrad) [75]. Each of two charging columns and the filtration column contains 12 stages of 250 kV each. Capacitance of each stage is $8 \times 10^3$ pF. The capacitor units are placed in hermetically sealed housings made of epoxy resin and are flooded with capacitor oil. The rectifiers assembled of selenium elements are also confined in casings filled with oil. The entire

**Figure 6.24** Cascade generator KGE-2.5 (without tank).

generator is accommodated in a tank filled with a mixture composed of $CO_2$ (25 percent) and $N_2$ (75 percent) under a pressure of up to 1.4 MPa. The KGE-2.5 generator is supplied with a voltage of 8 kHz from a dynamoelectric converter. The output voltage is controlled within the limits of 0.25 to 2.5 MV. The power amounts to 25 kW, the voltage stability being ±0.25 percent. Provision is made for changing polarity of the generated voltage. The dimensions of the generator are 3 m diameter and 5.8 m height.

### 6.6.3 Cascade Generators with Parallel Supply

The circuitry of the cascade generator with parallel supply of the stages (Fig. 6.23*c*) is employed in the accelerator dynamitron. This circuit was suggested as far back as 1919 [76], but it was not considered a promising version because the voltage of its capacitors tended to grow until the full operating voltage was obtained. With the advent of a high-quality gas insulation and saving primary high-frequency power supply sources, this type of generator was used as the basis for developing a comparatively simple and reliable accelerator in the late 1950s [77]. Used as high-voltage capacitors in this accelerator is construction capacitance $C$ between equipotential semirings (1) and high-frequency electrodes (2) which are essentially semicylinders approximately as high as the central core (Fig. 6.25). Arranged inside the semirings (1) are rectifying elements (11) and an accelerating tube (3). The potential of electrodes (2) is approximately equal to zero. The potentials of the semirings (1) go high with approach to the high-voltage electrode up to the full operating voltage. The voltage from the high-frequency generator (12) is applied to the high-frequency transformer (8) whose output is connected to high-frequency electrodes (2). The transformer is made in the shape of a toroid or a solenoid. The general view of the dynamitron of 4 MeV in the process of assembly is indicated in Fig. 6.26*a*.

Due to a large insulation gap (up to tens of centimeters), the coupling capacitances of the dynamitron are negligible, and the frequency of the supply voltage is ~100 kHz. The efficiency of the dynamitron depends, to a large measure, on the efficiency of the supplying generator and does not exceed 60 to 70 percent. The energy accumulated in the generator capacitors is minor, which makes it fairly stable and easy to service. In principle, the dynamitron differs from a generator with series-connected stages in that, at equal coupling capacitances, all the stages are in similar conditions and their quantity is restricted chiefly by the rise of the absolute potential difference, whereas in the case of series connection, the resistance of the charging circuit tends to increase with ascending ordinal number of the stage.

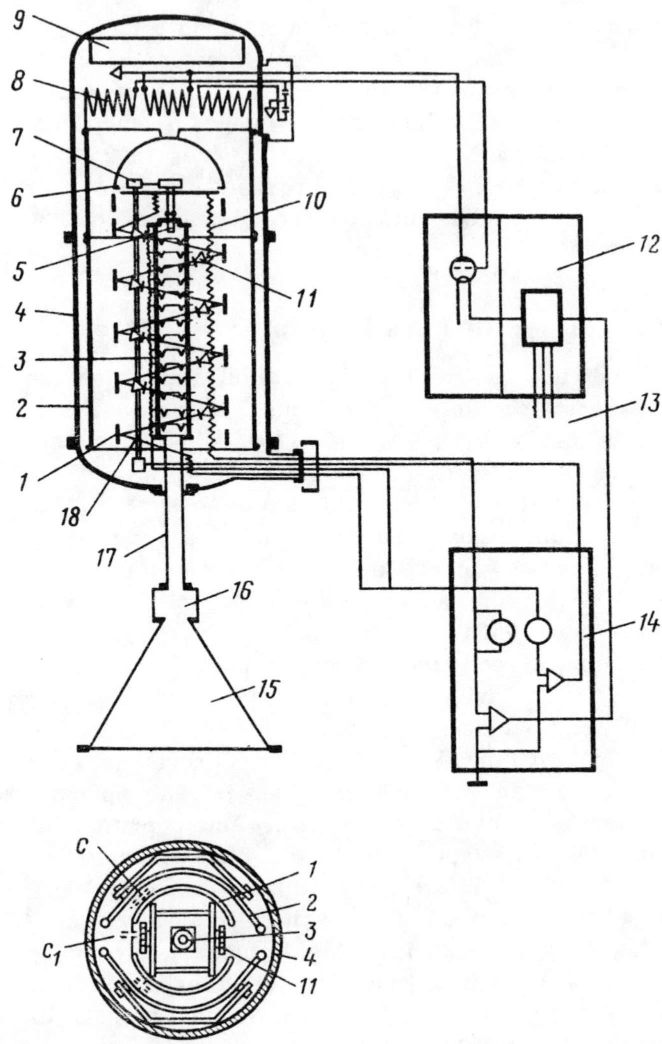


**Figure 6.25** Diagram of construction, power supply, and control system of dynamitron. *1*—Equipotential semirings; 2—high-frequency electrodes; *3*—accelerating tube; *4*—vessel; *5*—electron gun; *6*—high-voltage electrode; *7*—power supply system of electron gun heater; *8*—high-frequency transformer; *9*—cooling system; *10*—voltage divider; *11*—rectifying elements; *12*—high-frequency generator; *13*—power supply of high-frequency generator, 400V; *14*—control panel; *15*—horn with outlet window; *16*—beam scanning system; *17*—path of electron transport; *18*—electron gun heater control rod.

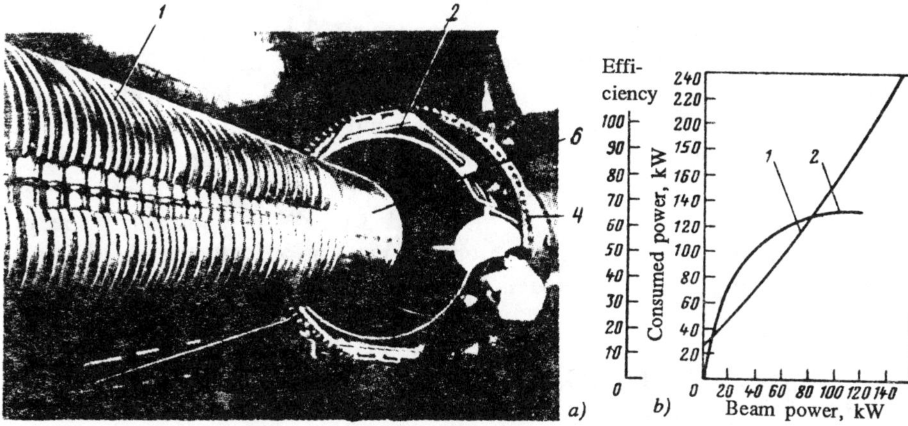


**Figure 6.26** *a)* Assembly of dynamitron produced by Radiation Dynamics (energy 4 MeV). Designations are the same as in Figure 6.25. *b)* Relationship between beam power, power consumption and efficiency. *1*—Power; 2—efficiency.

The no-load voltage of the dynamitron [78]

$$U = \frac{un}{1 + uc_1/C}$$

where $u$ is the amplitude of the voltage between high-frequency electrodes 2; $n$ is the number of the stages; $C$ is the capacitance between semirings (1) and high-freqency electrodes (2); and $C_1$ is the capacitance between semirings (see Fig. 6.25). At load current $I$, the voltage of the accelerating tube is

$$U = \frac{un}{1 + 4C_1/C}\left(I - \frac{I}{fCu}\right)$$

where $f$ is the frequency of the supply voltage. The value of the voltage ripples is independent of the number of the stages:

$$\delta U = \pm \frac{1}{2fC}$$

The relationship between the beam power, the consumed power, and the efficiency of the installation with a maximum energy of 3 MeV and a beam power of 150 kW is illustrated in Fig. 6.26*b*. For excitation of no-load voltage, a power of 24 kW is required. The efficiency at the power within the range of 70 to 150 kW approximates 60 percent.

The dynamitron is among the most extensively used industrial electron accelerators. The U.S. firm Radiation Dynamics, which developed these accelerators, has manufactured more than 150 sets since their business has been established. The range of the parameters of the dynamitrons are: electron energy 0.4 to 5 MeV and power, 30 to 150 kW. The generators are available both with the accelerating tube built in the generator and with the tube positioned externally. According to the long-term plan, Radiation Dynamics expects to increase both energy and power of electron beams. Preparation of an accelerator of 6 MeV and 300 kW is underway. One of the functions of this accelerator is generation of bremsstrahlung for irradiating massive objects.*

## 6.7 DIRECT-VOLTAGE GENERATORS BASED ON THREE-PHASE TRANSFORMERS

At a power exceeding 50 to 100 kW, some advantages are shown by generators with three-phase transformers. They provide a balanced consumption of power from industrial supply mains, feature high efficiency, and cause negligible ripples of rectified voltage. These installations are practically free of any restrictions in power. Now we shall consider three types of the generators with three-phase transformers which find application in industrial electron accelerators.

The circuit diagram of a three-phase insulating core transformer manufactured by the High Voltage Engineering Corp. (U.S.) is shown in Fig. 6.27. Primary windings (1) are arranged near the lower part of the magnetic circuit, which is at ground potential. The cores of magnetic circuit 3 consist of many separate elements (see also Fig. 6.28) insulated from one another through solid insulation (4). The sections of secondary winding (2) are placed on the elements of the magnetic circuit and connected electrically with them through one of the leads. Each section of the secondary winding is coupled to a voltage doubler circuit (5). All circuits (5) are connected in series. A direct high-voltage lead (6) is illustrated in the upper part of the generator. Correspondingly, the top part of the magnetic circuit is energized with high voltage. Also here are secondary windings (7) used for power supply of the high-voltage devices—the heater unit and control circuit of the accelerating device injector. The entire generator is accommodated in a tank (15) filled with compressed gas. The accelerating device is mounted beyond the generator either in the same tank or in a separate container. In the latter case, a lead (6) is connected to the accelerator by the use of a high-voltage feeder, that is with a cable coated with solid insulation when the voltage lies within 0.5 and 0.7 MV, or with a gas-filled coaxial line for higher voltages. The uppermost

* See new material added in proof on page 155.

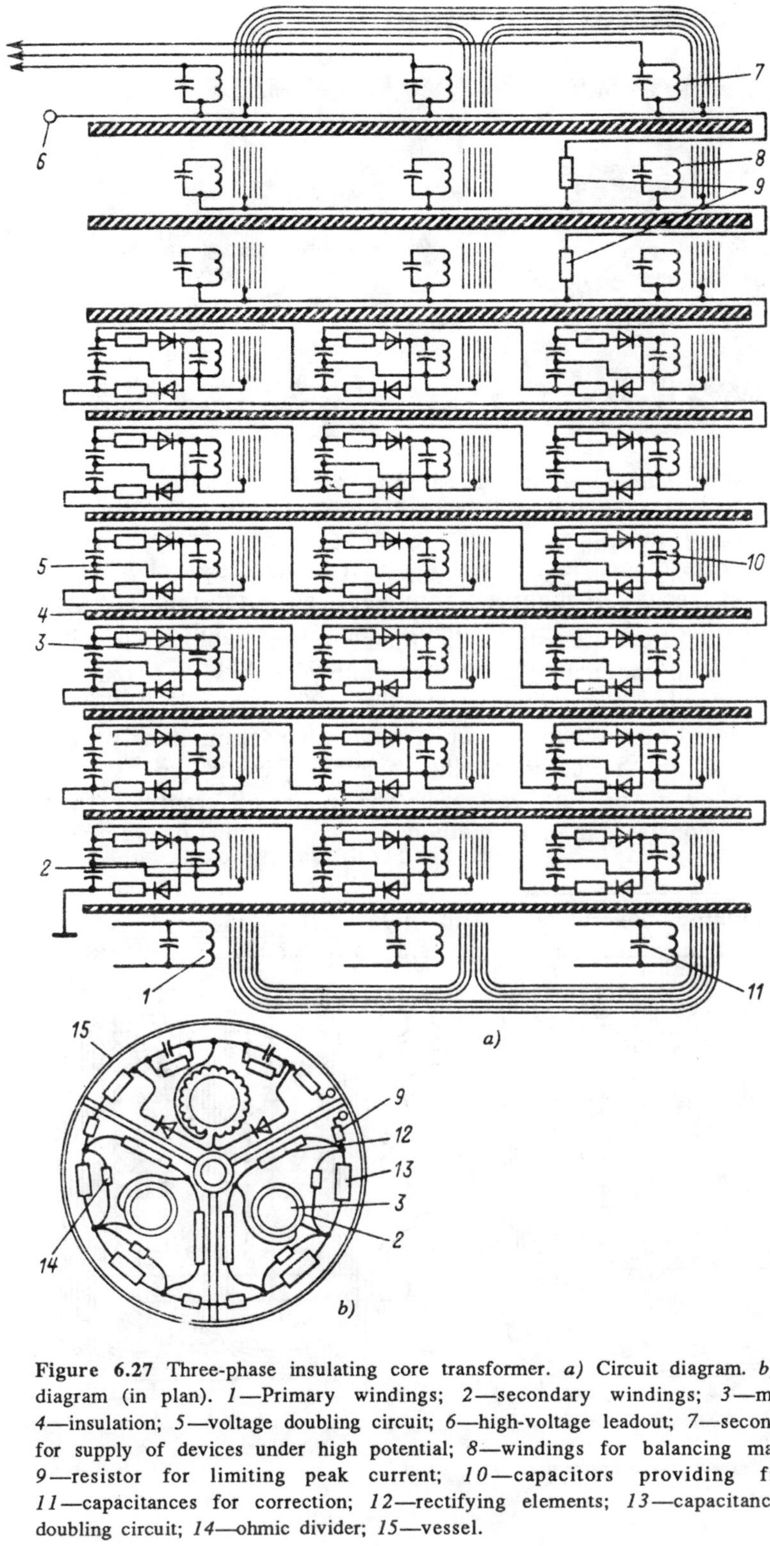


Figure 6.27 Three-phase insulating core transformer. *a)* Circuit diagram. *b)* Construction diagram (in plan). *1*—Primary windings; *2*—secondary windings; *3*—magnetic core; *4*—insulation; *5*—voltage doubling circuit; *6*—high-voltage leadout; *7*—secondary windings for supply of devices under high potential; *8*—windings for balancing magnetic fluxes; *9*—resistor for limiting peak current; *10*—capacitors providing field current; *11*—capacitances for correction; *12*—rectifying elements; *13*—capacitances of voltage doubling circuit; *14*—ohmic divider; *15*—vessel.

Figure 6.28 Three-phase transformer ICT at voltage of 1.5 MV.

voltage generated by the apparatus of this type is 3 MV, with a power of 100 kW. The power is supplied directly from industrial power supply mains, with an efficiency of about 90 percent.

Figure 6.29 illustrates two diagrams of three-phase transformers with a solid magnetic circuit. In both versions, the magnetic circuit is at ground potential and the secondary windings are insulated from the magnetic circuit (and also from the primary windings placed on the cores of the magnetic circuit) so as to be nonconductive even for full voltages. The main difference between these circuits consists in the method of connection of rectifying

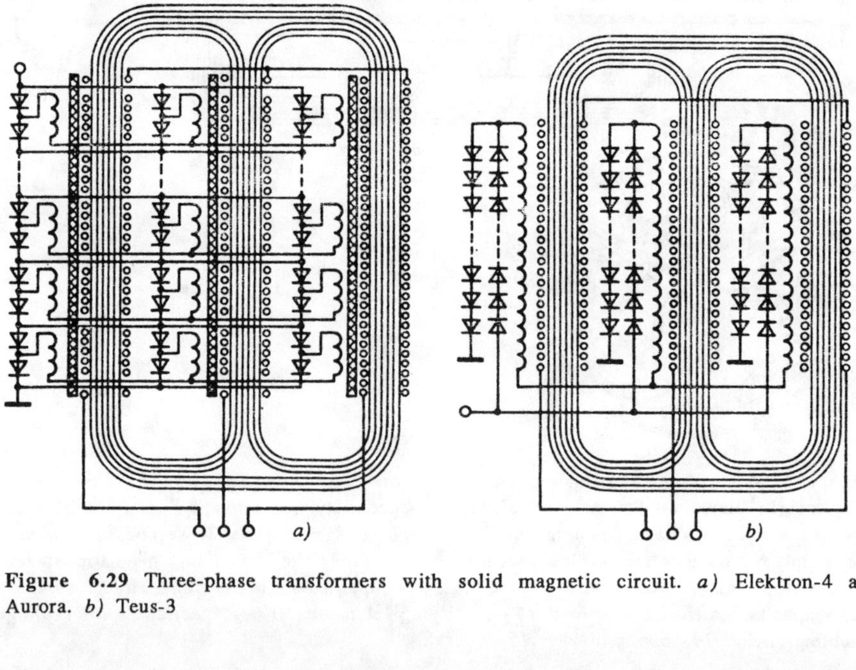

Figure 6.29 Three-phase transformers with solid magnetic circuit. *a)* Elektron-4 and Aurora. *b)* Teus-3

elements. In the version shown in Fig. 6.29*a*, the rectifiers are connected to each section of the secondary winding. The voltage undergoes rectification in each stage of the circuitry. The stages are connected in series. In the circuit indicated in Fig. 6.29*b*, the rectifiers are assembled into integral units rated at full operating voltage. Six units are connected to the high-voltage windings of the transformer employing a single-bridge rectifying circuit. As a rule, the three-phase transformers used in the industrial accelerators operate at frequencies of 50 to 60 Hz and are provided with three-dimensional balanced magnetic circuit composed of plates of transformer steel 0.35 mm thick.

The version shown in Fig. 6.29*a* is used in accelerators Elektron-4 and Aurora (developed by NIIEFA, Leningrad) [79]. Figure 6.30 is a construction diagram of the generator of installations Elektron-4 and Aurora-2. Three single-layer primary windings (7) are placed on three cores of magnetic circuit (5). The secondary winding comprises star-connected sections (8). The latter are insu-

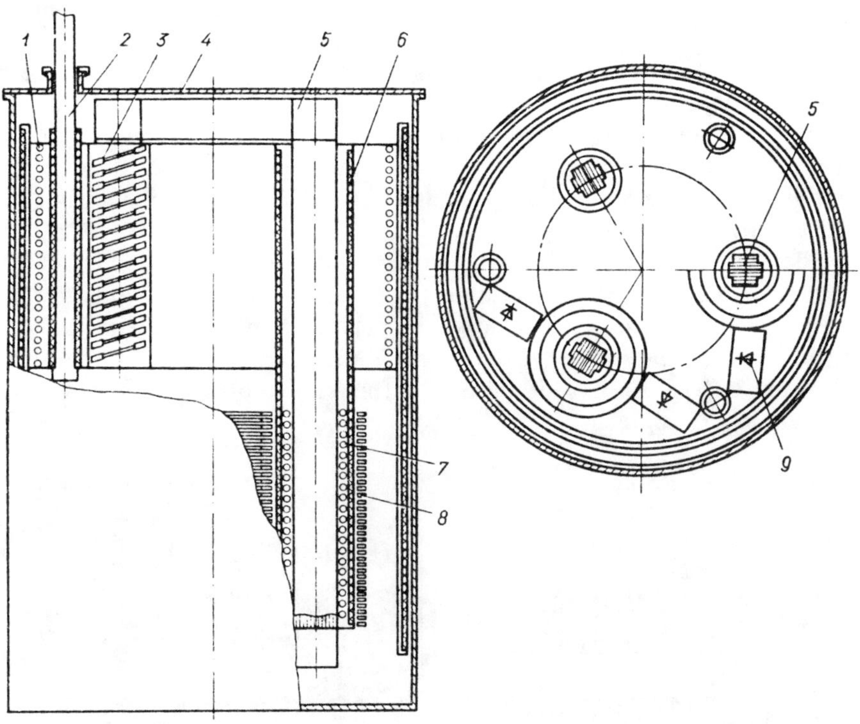


**Figure 6.30** Construction diagrams of high voltage generators of Elektron-4 and Aurora-2. *1*—Equipotential rings; *2*—high-voltage cable leadout; *3*—divider; *4*—tank; *5*—magnetic circuit rod; *6*—laminated insulation; *7*—primary winding; *8*—secondary winding section; *9*—rectifier.

lated for high voltage from the magnetic circuit by means of oil-impregnated laminated insulation (6) fabricated from polyethylene film and capacitor paper. The transformer is designed for supplying power to three accelerators through individual power cables (2). The generator is positioned in tank 4 filled either with capacitor oil or with transformer oil. The secondary windings are wound up as those in the Elektron-3 accelerator (see Section 6.5). The nominal voltage of the generator is 500 kV, the load current is 0 to 50 mA, the efficiency being 90 percent. The overall dimensions of the set are diameter, 1.23 m and height, 2.3 m. Gas insulation is used in new modifications of Aurora accelerators, which reach a voltage of 750 kV and a power (with three operating radiators) amounting to 100 kW. The electric power is supplied from industrial power supply mains of 220/380 V, 50 Hz.

A modified version of the circuit shown in Fig. 6.29*a* is implemented in generator ESU rated at a voltage of 500 kV and a power of 5 MW and intended for supplying power to experimental installations (developed in the Nuclear Physics Institute, Novosibirsk) [80]. To decrease the ripples of the high voltage, the secondary winding sections are alternately delta-connected and star-connected (Fig. 6.31). Three sections of the secondary winding and rectifiers constitute a stage. The generator contains a total of 22 series-connected stages. Mechanically, the stages are positioned one above the other and are secured to the shelves of a carrying support made of insulating materials. The main insulation is effected by means of $SF_6$ gas under a pressure of up to 0.6 MPa. The installation allows just a short-time operation (a few seconds); however, it is capable of continuous operation at lower voltage and power. The generator is assembled in a tank 2 m in diameter and 3.3 m in height. The power supply voltage is drawn from three-phase mains of 6.3 to 10 kV, 50 Hz. The fluctuation of the output voltage lies within ±5 percent.

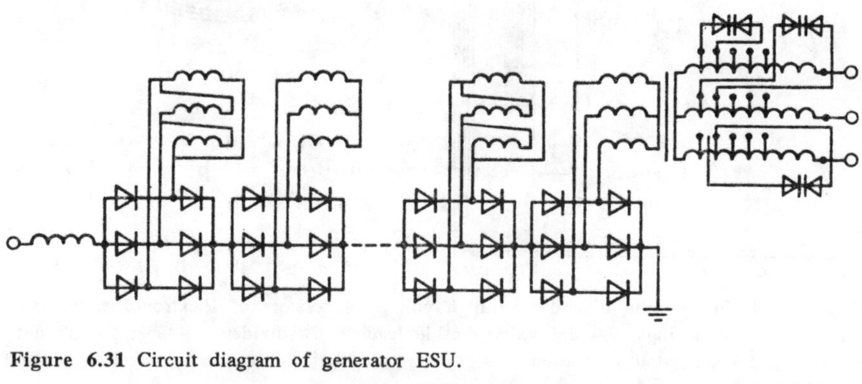

**Figure 6.31** Circuit diagram of generator ESU.

The circuit indicated in Fig. 6.29*b* is used in generators Teus (three-phase electron accelerator). The construction and schematic diagrams of the generator Teus-3 rated at a voltage of 300 kV and a power of 75 kW are illustrated in Fig. 6.32 (developed at the High-Temperature Institute, Academy of Sciences, Moscow) [81]. The sections of the secondary winding consist of flat or tapered layers (see Section 6.9) impregnated with an epoxy-based compound. Each phase winding (7) contains 10 sections of 3000 turns in each wound of a wire of 0.44 mm in diameter. The height of the phase windings is about 0.35 m. The insulation of the sections from primary winding (5), magnetic circuit (1), tank (9), and other components is effected with the aid of a mixture of $SF_6$ gas and nitrogen under a pressure of up to 1.5 MPa. For supplying the power to the devices that maintain high potential, provision is made for additional windings (13). The circuits for conversion and control of the voltage taken from windings are positioned inside an electrode (10) under high voltage. The circuits are controlled by light signals through a channel (14). For protection of the secondary winding in the event of high-voltage breakdowns, use is made of inductance (18) and lines (15). Each line comprises a conductive turn (16) (which makes part of the line lead) and ferromagnetic rings put on it. Rectifying units (2) (two units per phase) are rated at full voltage and represent sets of series-connected diodes (3) with a capacitive divider (4). The voltage is transferred to the load over one high-voltage cable KPV-7/300 (12) or through several cables. The installation is intended for continuous operation. It is supplied with a voltage of up to 230 V, 400 Hz from dynamoelectric convertor PSCh-100 (19). The efficiency is approximately 90 percent. The dimensions of generator Teus-3 are: diameter, about 1 m; height, about 1.8 m. Several generators were manufactured in 1974–1975 and are still in use.

Three-phase generators also find application in power supply sources of accelerators Electrocurtain and Electrocure (Energy Science Co., U.S.), EOL (USSR), and some other industrial-purpose accelerators.

## 6.8 PULSED VOLTAGE GENERATORS

Until recently, surge (pulsed) voltage generators were used in industrial installations mainly for obtaining single bursts of bremsstrahlung for radioscopy of materials, determination of their radioresistance, and investigation of quick processes. The pulse duration was $10^{-7}$ to $10^{-6}$s, the intervals between the working cycles amounted to several minutes or longer. Traditional use of pulsed generators

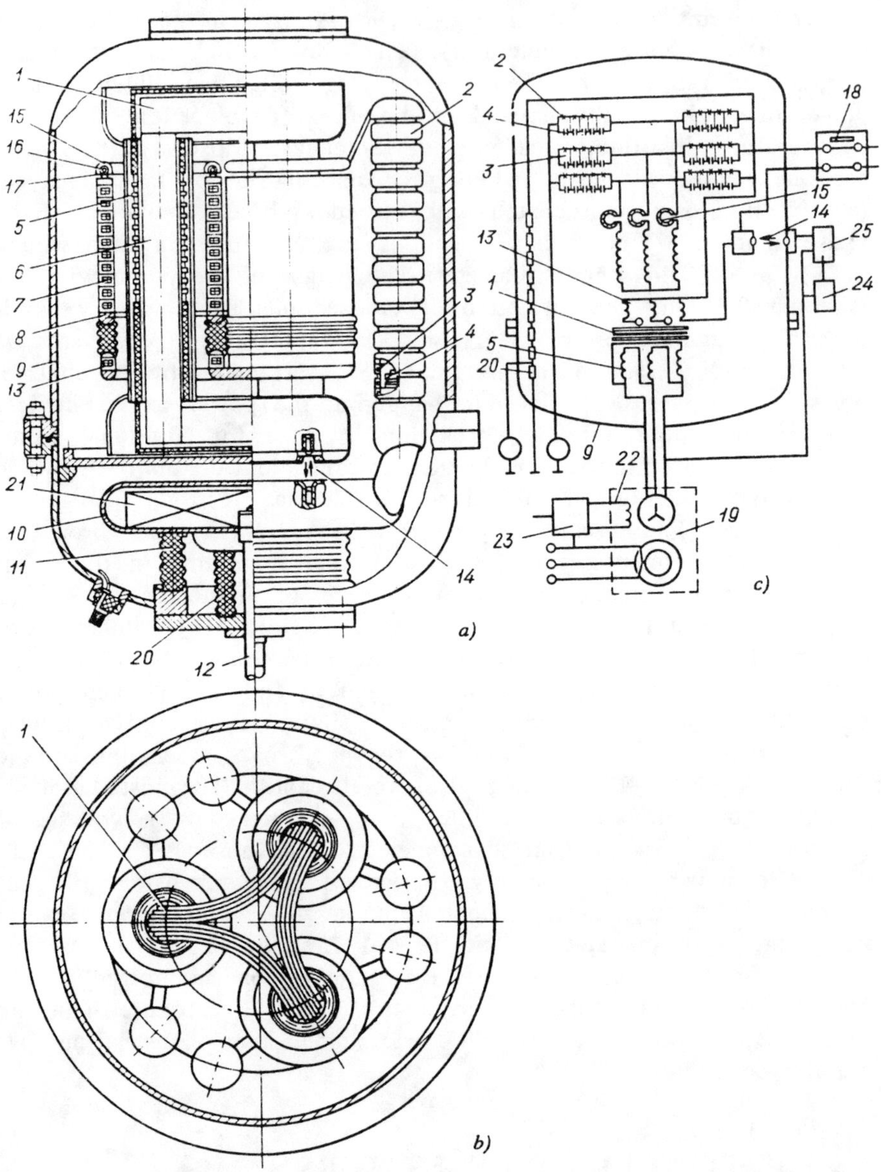


**Figure 6.32** Construction and circuit diagrams of generator Teus-3. *1*—Magnetic circuit; *2*—rectifying units; *3*—diode KD 209; *4*—capacitative voltage divider; *5*—primary winding; *6*—magnetic circuit core; *7*—secondary winding; *8*—supporting plate; *9*—vessel; *10*—high-voltage electrode; *11*—supporting insulator; *12*—high-voltage cable; *13*—additional windings; *14*—luminous control channel; *15*—line; *16*—conductive turn of line; *17*—ferromagnetic rings; *18*—protective inductance; *19*—frequency converter PSCh-100; *20*—voltage divider; *21*—controllable stabilized power supply source for feeding cathode heater, with luminous control; *22*—field winding; *23*—voltage control unit; *24*—auxiliary rectifier of cathode heater supply unit; *25*—common control device of cathode heater sypply unit.

in the sphere of electrical engineering is testing of elements of various high-voltage equipment. The pulse duration in this case reaches $10^{-3}$ s. In the last few years, there is a tendency to design pulse accelerators with a high-frequency repetition rate and an average power of tens of kilowatts. It is expected that some industrial accelerators of this type will allow for cheaper radiation as compared with continuous-action accelerators. In other cases, it is expected that great pulse power of radiation dose will facilitate some production processes. Pulse transformers used for acceleration of electrons are discussed in Section 6.4. The following are descriptions of some more pulse generators.

The circuit diagram of one of the early sources of high pulse voltage—the Marx generator [82]—is presented in Fig. 6.4*a*. The capacitors of the generator (capacitance of each $C$) are charged across resistors (1) approximately at a level of equal difference of potential $U$. To obtain a high voltage $V = Un$ (where $n$ is the number of the stages), the capacitors are connected in series through spark gaps (2). The capacitance of the generator in this operating condition is equal to $C/n$. To facilitate the starting of the generator, one or several spark gaps of the stages closest to the ground are fitted with controllable triggering circuits. The remaining spark gaps are characterized by development of multiple overvoltage that causes quick breakdown of the gaps.

The Marx generator rated at 0.5 to 1 MV are in batch production by a number of industrial firms (for example, the U.S. firm Maxwell, see Appendix, Table A.13. In order to change pulse wave form (e.g., reduction of length or increase of amplitude) provision is made in some accelerators employing the Marx generator for a pulse shaping line arranged between the generator and the load. In the most ordinary case, the coaxial line 1 is used. It is separated from the load (accelerating tube, 3) by an additional space (Fig. 6.33*a*). The accelerating tube employs a cold cathode (4). Upon charging the shaping line by a Marx generator (7), spark gap (2) is actuated and the voltage is applied to an accelerating tube (3).

The illustration indicates that the installation is accommodated in a tight tank (5) filled with a high-quality dielectric medium. Generators and pulse shaping lines in many research installations are placed in different media with an insulator placed between them.

Similar principles are characteristic of pulsed voltage generators based on electrostatic generators and shock transformers (Fig. 6.33*b* and *c*). High voltage produced by an electrostatic generator or transformer (see Sections 6.3 and 6.4) is applied to charge a coaxial line (1). The latter is then connected to an accelerating tube (3).

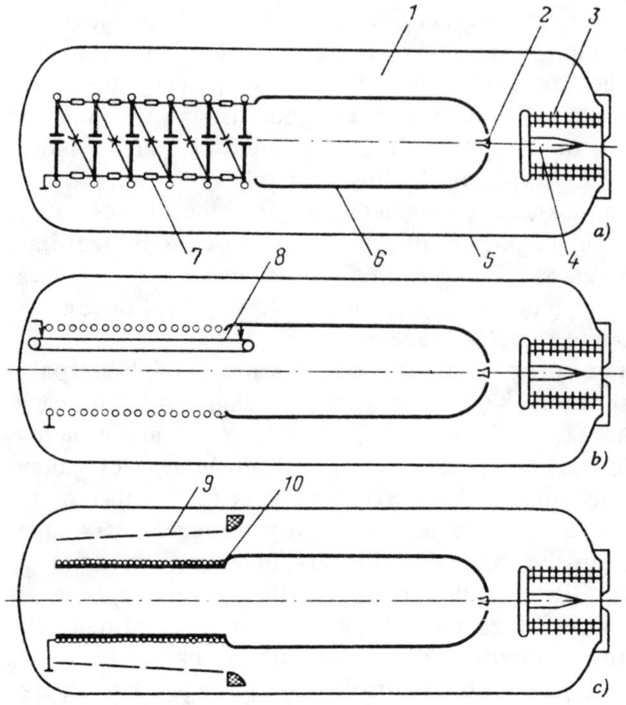


**Figure 6.33** Pulse voltage generators with coaxial pulse shaping line (with different charging circuits) *a)* charging by Marx generator. *b)* Charging by electrostatic generator. *c)* Charging by shock transformer. *1*—Coaxial line; 2—discharger; *3*—accelerating tube; *4*—self-emission cathode; *5*—tank; *6*—high-voltage electrode; *7*—Marx generator; *8*—charging belt of electrostatic generator; *9* and *10*—primary and secondary windings of transformer.

This principle was applied in the first serial sources of high-intensity bremsstrahlung, namely, febetron-705 (Field Emission Corp., U.S.) and plant FX-45 (Ion Physics Corp., U.S.). The energy and current in these installations were respectively 2 MeV, 5 kA and 4 MeV, 20 kA. The pulse length was ~30ns. The basic dimensional characteristics of the febetron-705 are tank length, 4.2 m and diameter, about 1.8 m.

A resonance shock transformer has been used in the Rius-5 accelerator (Nuclear Physics Institute, Novosibirsk) [83]. The Rius-5 accelerator ensures acceleration of electrons up to 4 or 5 MeV at a beam current of up to 30 kA with a pulse length of ~40 ns. The primary winding (9) contains four turns, the secondary winding comprises 10 to 580 turns (Fig. 6.33*c*). The natural frequencies of the primary and secondary circuits amount to approximately 32 kHz, the interwinding coupling coefficient $K$ = 0.4. The capacitor of the primary stage ($C$ = 5 μF, $u$ = 50kV) is connected to the primary winding by means of a spark gap. The amplitude voltage of the line may reach 7 MV. As soon as the voltage appears to be approximating the peak value (the third sinusoid half-wave, Fig. 6.7), a high-voltage electrode (6) is connected to an accelerating tube (3) by means of a spark gap (2). A tank (5) is filled with the gas ($SF_6$ + $N_2$) at a pressure of ~1.5 MPa. The approximate dimensions of the installation are diameter, 2 m, length, 5.5 m. A general view of the installation is shown in Fig. 6.34.

Several Rius-5 type accelerators were manufactured in the late 1960s and early 1970s. These installations have been used until the present. The Rius-5 generator was used, among other things, for performing experiments on acceleration of two electron beams at an interval of ~3 ms between successive pulses. In so doing, use was made of two separate primary circuits. The parameters of the beam were somewhat less than the nominal values.

When using the Marx circuit and a shock transformer, it is essential that the energy be stored in the capacitors and the pulse shaping line be charged during the lifetime of a short pulse. In order to raise the energy accumulated in the line, it is common practice to involve a dielectric with a large value of ε (for example, water with ε = 81, or glycerine with ε = 40). The dielectric also features a substantial dielectric strength during a limited lapse of time (of the order of $10^{-6}$s). The charging generator and the pulse shaping line in this case may be placed in different insulating media, and the electron beam power may be considerably higher than when compressed gas serves as a dielectric.

The advantage of short-pulse, high-voltage sources combined with Marx generators and shock transformers over installations

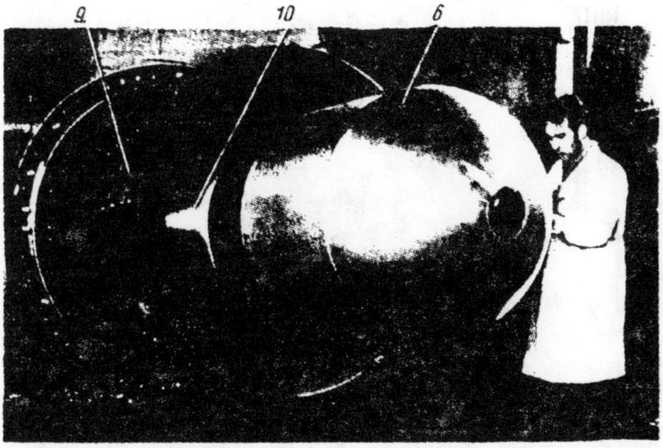


**Figure 6.34** Rius-5 generator (with vessel and accelerating tube removed). designations are as for those in Figure 6.33.

based on electrostatic generators is the absence of rather unreliable mechanical conveyers. As compared with a shock transformer, the Marx generator contains much more spark gaps, yet makes it possible to accumulate greater energy and inject it into the electron beam in a shorter lapse of time.

In recent years, pulse voltage generators have been developed on the basis of the Marx circuit principle to obtain pulse repetition frequencies amounting to tens and hundreds of hertz. These generators, as a rule, feature lesser pulse power as compared to the generators of single pulses. The spark gaps must first be modified to ensure the required frequencies. The next problem to be solved is prolongation of service life of accelerating devices in case of utilization of cold cathodes.

For the sake of acceleration of electrons by emitting short pulses, use is also made of a high-frequency resonator. Such voltage source has been used in the ILU-6 accelerator shown in Fig. 6.35 (developed at the Nuclear Physics Institute, Novosibirsk) [84]. The

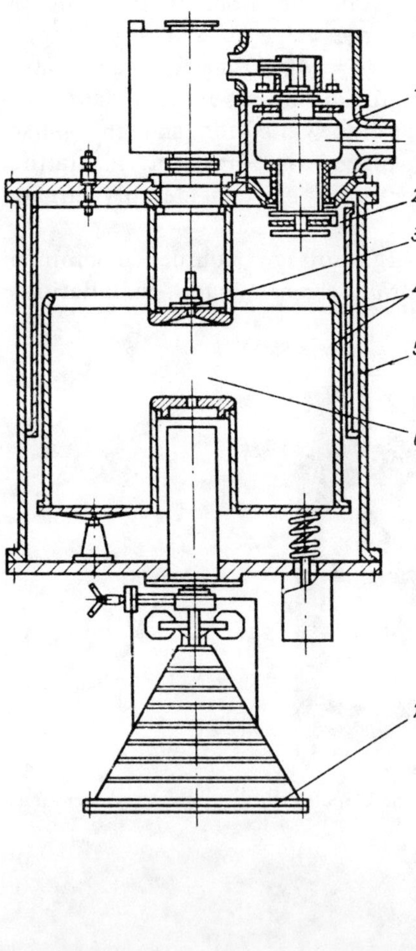


**Figure 6.35** ILU-6 accelerator. *1*—High-frequency generator; *2*—coupling loop; *3*—electron gun; *4*—resonator; *5*—tank; *6*—accelerating gap; *7*—outlet window.

high-frequency generator (1) and resonator (4) are combined into an integral unit. The resonator consists of two halves partially entering each other and is situated in the interior of the vacuum tank (5). High-frequency current flows from one-half of the resonator to the other through the capacitance formed by the gap between them. The lower part of the resonator is insulated from the tank and receives a voltage of 5 to 10 kV for suppressing the resonance discharge and drain of ions from the accelerating gap (6). The high-frequency generator is built around triode GI-50A on the principle of a common-grid circuit. The anode circuit is coupled to the resonator through a loop (2). The operating frequency of the generator is about 100 MHz. The maximum pulse length is 700 $\mu$s, the pulse repetition rate of 2 to 75 Hz. The maximum energy of accelerated particles is 2 MeV, the average power lies within 20 kW, and the amplitude of the electron beam current is 1 A. The stability of the beam energy within $\pm 10$ percent is ensured by selecting the integral between high-frequency voltage phases during which the beam is injected. The accelerator efficiency is ~20 percent. The accelerated beam is scanned with the aid of a magnetic field and is extracted into the atmosphere through the outlet windows measuring 75 × 980 mm or 75 × 1500 mm. Use of other types of beam extractors is also possible.

The ILU accelerators, in which charged particles undergo acceleration effect practically in a potential electric field, are a type of direct-action accelerators. (This holds true since the high frequency field varies negligibly during transit of a particle through the accelerating gap.) The principal disparity between ILU and other types of accelerators (discussed above) is the use of fine vacuum as insulation means in the high-voltage generator. Due to this, the particles can be accelerated in the same vacuum space, which simplifies the installation.

Figure 6.36 illustrates the electron accelerator Raps-1 where voltage is also produced in high-frequency resonator (developed in the VNIIOFI, Moscow) [85, 86]. High-voltage winding (1) and metal casing (2) form a quarter-wave coaxial resonator. Winding 1 is a single-layer coil excided from primary winding 9. Windings 1 and 9 are shown in Fig. 6.36 arbitrarily. The more correct location of these windings is indicated in 5 and 7 in Fig. 6.4. The accelerator employs an evacuated accelerating tube comprising a metal casing (2) and glass insulator (8). The cathode (4) is at ground potential, the target (6) is at the high-voltage end of winding 1. The installation contains two forced-cooling circuits. Windings 1 and 9 and the adjacent parts are cooled by transformer oil. The direction of the oil flow is shown with solid arrows. Denoted with 13 is a centri-

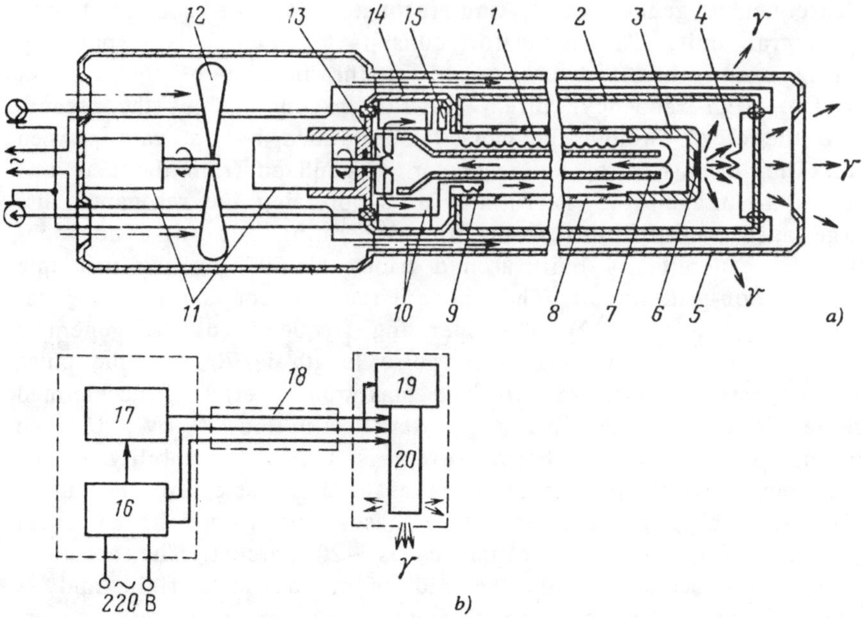


**Figure 6.36** RAPS-1 x-ray generator. *a)* Construction and *b)* block diagrams. *1*—High-voltage winding; *1*—metal casing of accelerating tube; *3*—target electrode; *1*—cathode; *5*—radiator housing; *6*—target; *7*—oil line; *8*—glass insulator; *9*—primary winding; *10*—oil cooling radiator fins; *11*—electric motors; *12*—fan; *13*—centrifugal pump; *14*—radiator for cooling oil; *15*—measuring resistor; *16*—control panel; *17*—high-frequency generator; *18*—cable; *19*—accelerator cooling system; *20*—electron accelerator; γ—bremsstrahlung.

fugal pump driven by an electric motor. The accelerating tube and the cathode fastenings are cooled with air (indicated by dashed arrows).

The block diagram of the Raps-1 installation is illustrated in Fig. 6.36*b*. The amplitude of the output voltage of the high-frequency generator (17) is 3.8 kV; the voltage frequency is 800 kHz. The generator employing valves GMI-27B is coupled to the resonator through a cable (18) 50 m in length. The duration of the working cycle is 50 μs; the cycle repetition rate is 300 Hz. The length of each pulse at the half-amplitude level amounts to 0.2 to 0.3 μs.

Electrons striking the target (6) produce a bremsstrahlung whose intensity reaches 15 R/min at a distance of 0.5 m from the target. The diameter of the focal spot is about 3 mm. The average power of an electron beam is some 0.6 kW. The overall dimensions of a portable x-ray radiator are 750 × 180 × 180 mm, the mass being 10 kg of which 7 kg falls to the share of the cooling system.

## 6.9 COMPONENTS OF HIGH–VOLTAGE GENERATORS

As is apparent from the description of high-voltage sources, to obtain similar parameters of electron beam for industrial use (i.e., energy of the order of 1 MeV and power of some 10 kW), different facilities can be used to generate accelerating high voltages, even though they differ from one another both in circuitry and in design features of principal components. Selection of reasonable circuitry of the generator, type of insulation, construction version, etc., is very dependent on the components under commercial production or those specially designed for the particular generator (rectifiers, capacitors, etc.). Improving the performance characteristics of any of these components, along with insulation or ferromagnetic and miscellaneous materials concerned, will allow for creation of a new modification or enable any one of the known high-voltage facilities to gain an advantage over the others. The competitiveness of any particular version of generator can be judged only after several years of operation of the facility.

Before considering aspects of generator component design, I shall present some current principles applied in developing direct high-voltage sources in direct-action accelerators.

1. Whenever the power lies within 50 and 100 kW (and, rarely, up to 200 kW), use is made of single-phase devices. In cases of higher power usage, it is common practice to use three-phase systems.
2. At voltages of 1 MV and higher, axial-balanced transformers in single-phase accelerating installations are used. The primary winding in these cases is arranged on the outside of the secondary.
3. In systems with voltage doubling, the frequency of supply voltage does not exceed 400 Hz.
4. At voltages of the order of 1 MV in the direct-voltage generators, as a rule, compressed gas serves as the main insulation. Installations with oil and solid insulations are usually rated at voltages not exceeding 0.5 MV. Supply voltage from generator to accelerator may be transmitted over high-voltage power cables with solid insulating coatings.
5. Multiturn secondary windings (of the order of 10,000 or more turns) are composed of many sections, as a rule.
6. Tight coupling between primary and secondary windings in ordinary transformers is characterized by coefficient $K \cong 0.9$. The same coefficient for resonance transformers $K \leq 0.6$.

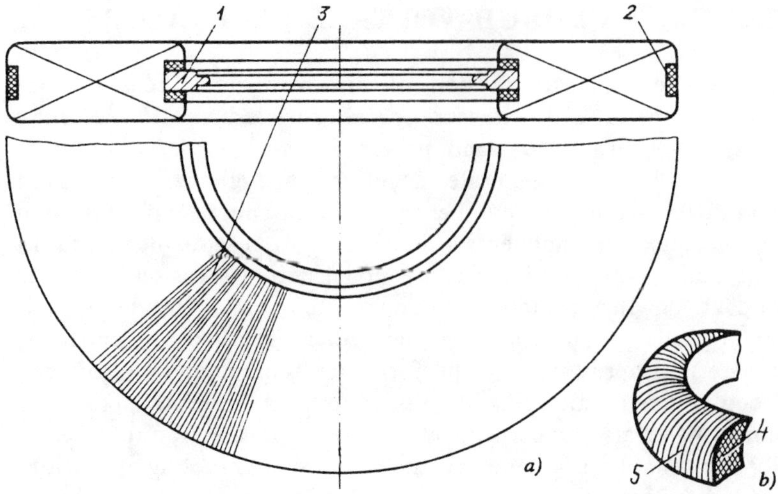


Figure 6.37 *a)* Disk of sectionalized magnetic core of ELT-15 accelerator. *b)* Part of high-voltage electrode of air-core transformer transparent for variable magnetic flux. *1*—Solid copper ring; *2*—glass tape soaked with epoxy compound; *3*—plates of transformer iron; *4*—ring of insulating material (textolite, plexiglas, etc.); *5*—wire.

Two principal modifications of magnetic circuits are employed in transformers, viz., solid (closed) and divided (sectionalized). In the first case, the magnetic circuit is at ground potential (see Fig. 6.32). Sectionalized magnetic circuits are generally used in axial-balanced versions of transformers. Disks of the central cores of such magnetic circuits are insulated from one another and are energized with different voltages: from zero to full operating voltage (Fig. 6.18*b* and *c*). The main objective pursued in sectionalizing magnetic circuit cores is removal of high-voltage insulation between sections of secondary winding and magnetic circuit core. However, it should be noted that such a magnetic circuit is much more expensive than a solid one. That is why in some single-phase systems the central core is not used at all (Fig. 6.18*d*) regardless of increase of number of dispersion fluxes, deterioration of coupling coefficient, and efficiency of accelerators. To improve flux linkage between windings in some accelerating installations, the primary winding is wound on a copper shield (see, for example, 25 in Fig. 6.13) or is fabricated in the shape of a cone (Fig. 6.18*a* and *d*). Insulation between elements of sectionalized magnetic circuit is effected by the use of gas-filled gaps (Fig. 6.18*b*) or solid dielectrics (Fig. 6.18*c*). construction of a radial-laminated disk of single-phase transformer central core is exemplified in Fig. 6.37*a*.

When a single-phase transformer is not provided with a ferromagnetic circuit for improving the coupling coefficient, minimum possible resistance to magnetic flux must be provided. With this objective, part of the high-voltage electrode crowning the secondary winding, the protecting screen located adjacent to the edge of the primary winding (4 and 3 in Fig. 6.11), and some other parts of the circuitry are fabricated from materials transparent for variable magnetic flux. Insulating materials coated with a thin layer of metal or wrapped up with wire (see Fig. 6.37*b*), or a perforated thin metal material offering high ohmic resistance, is used for this purpose.

One of the most vital component parts of transformer-type voltage sources is transformer high-voltage winding. The main problem arising in designing the winding is meeting the stringent requirements on dielectric strength in aging or emergency high-voltage breakdowns. At the moment of breakdown, the voltage along the secondary winding is distributed in compliance with the action of a capacitive divider formed by the longitudinal capacitance of the winding and the transverse capacitance of the winding to the ground. The situation for a single-phase system is illustrated in Fig. 6.38. If the main operating voltage $U_2$ is distributed along the

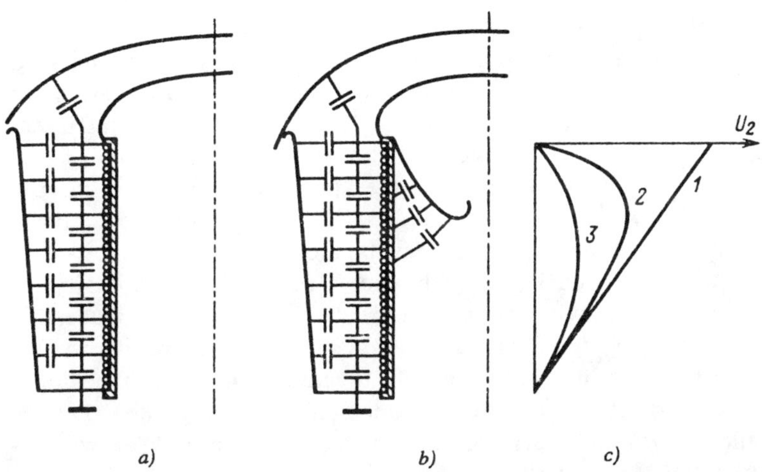


**Figure 6.38** Schematic equivalent diagrams *a)* and *b)* of capacitive voltage divider composed of longitudinal and transverse capacitance of winding. In *b)*, secondary winding contains protecting shield. *c)* Curves of voltage distribution over secondary winding. *1*—operating voltage; *2* and *3*—voltage at moment of breakdown from winding high-voltage end to ground, respectively, for *a)* and *b)*.

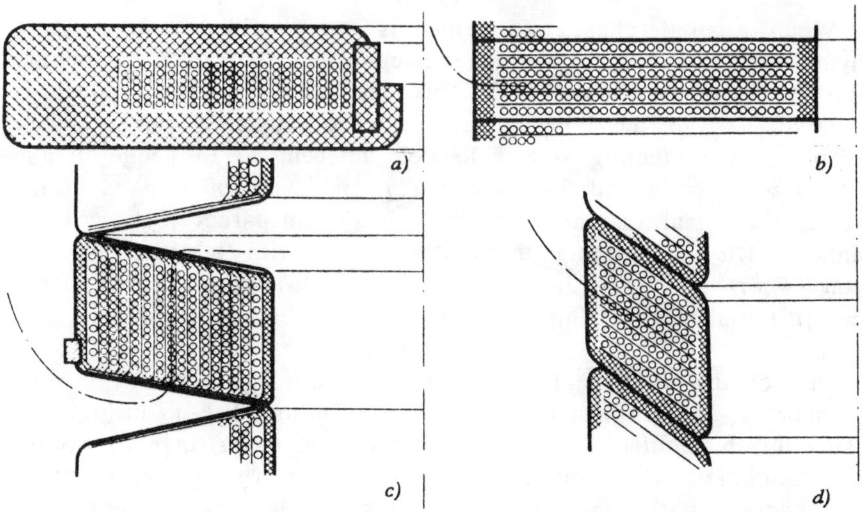


**Figure 6.39** Sections of multiturn secondary windings. *a)* and *b)* Cylindrical-layer winding. *c)* Flat-layer winding; *d)* Tapered-layer winding. Dash and dot line indicates approximate direction of electric field equipotentials.

winding approximately linearly* (Fig. 6.38*c* curve 1), then at the first moment after the high-voltage lead is shortened to ground the voltage distribution lengthwise, the secondary winding will be of the nature roughly shown in Fig. 6.38 (curve 2).

The full-discharge time of the high-voltage electrode amounts to fractions of a microsecond, which is much less than the period of secondary winding natural oscillations. This time is not sufficient for appreciable shifting of charges in the secondary winding of the transformer, as a result of which the potential gradient in the upper part of the winding may increase by 10, 15, or more times. The overvoltage wave thus formed travels along the winding. Decaying oscillations arise, during which the electric field intensity at different points of the winding may exceed the tolerable values. To protect the secondary winding, a shield is placed adjacent to the upper end of the winding. The shield, together with the upper part of the winding, forms an additional capacitance (Fig. 6.38*b*). Due to this fact, the overvoltage arising as a result of the breakdown decreases (curve 3 in Fig. 6.38*c*).

In installations with gas or vacuum-type insulation, it is common practice to carry out preliminary aging of high-voltage gaps by weak-current breakdowns, starting with voltage equal to 30 to 50

* Due to the presence of dispersion fluxes, the winding turns must be laid with variable pitch to obtain a permanent gradient of elecric field potential.

percent of the operating voltage. To protect the generator against breakdowns, the load is provided with an additional resistor connected in series with the secondary winding. High-voltage breakdown may be removed completely, on condition that twofold or threefold margin in dielectric strength is provided for in all high-voltage gaps.

Figure 6.39 illustrates diagrams of several typical sections of multiturn secondary windings. In Fig. 6.39*a*, the winding is composed of cylindrical layers of turns, a solid dielectric provides wide insulating spaces, and the copper-filling factor of the winding section is relatively small. Such sections are used, in particular, in insulating core transformers produced by the High Voltage Engineering Corp. Figure 6.39*b* and *c* show two versions of windings of an electron transformer (ELT). In Fig. 6.39*b*, turns are laid in cylindrical layers with paper insulation made in the shape of a trough so as to hold the turns in position during winding process. To enhance resistance to aging breakdowns, the winding sections are fitted with metal coating, and interturn insulation of final layers is reinforced. The tapered shape ensures a variable gas gap between turns of the adjacent sections in full conformity with potential difference between them (see also Fig. 6.13). The sections are electrically connected at the points where they contact one another.

Sections fabricated in compliance with the version *b* in Fig 6.39 (and also with versions *c* and *d* discussed below) are impregnated with heat-cured epoxy compound. Vacuum drying is first performed, and then impregnation is accomplished under a pressure of up to 2 to 4 MPa. The frameless solid coils obtained this way exhibit perfect electrical and mechanical properties.

In the construction shown in Fig. 6.39*c*, the winding layers are essentially flat disks with helical-wound turns. The arrangement of the winding layers approximates the equipotential directions [shown approximately by a dot-dash line (—·—)] close to the sectionalized magnetic circuit (see also Fig. 6.40). In regions of small radius of curvature of equipotentials, the gradient of the electric field potential is substantially higher. As Fig. 6.40 shows, a sec-

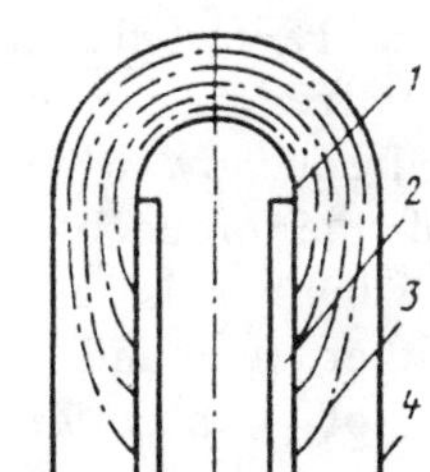


**Figure 6.40** Equipotentials in axial-balance high-voltage generator. *1*—High-voltage electrode; *2*—secondary winding; *3*—equipotential; *4*—vessel.

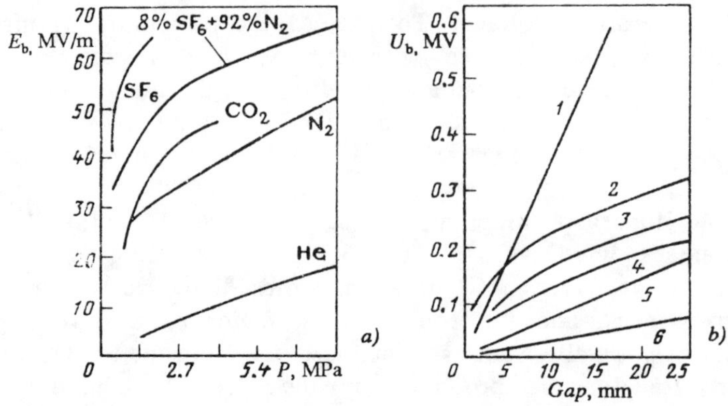


**Figure 6.41** *a)* Dependence of dielectric strength of dielectric on pressure (direct voltage). *b)* Breakdown voltage of some dielectrics for homogeneous field and direct voltage. *1*—$SF_6$ gas at pressure of 0.7 MPa; *2*—vacuum; *3*—transformer oil; *1*—porcelain; *5*—$SF_6$ gas at atmospheric pressure; *6*—air at atmospheric pressure.

tion assembled of flat layers ensures a smoother variation of equipotentials and, correspondingly, a lesser field intensity and higher dielectric strength. The copper-filling factor of the secondary winding in such a construction is also made much better. The sections may be placed directly one on top of one another without leaving intermediate gaps. But the process of laying flat layers of windings and subsequent series connection of the layers require a number of labor-intensive technological operations.

The next achievement in the development of frameless coils can be exemplified by the construction illustrated in Fig. 6.39*d*. Arrangement of the layers over the tapered surfaces enables their production and interconnection to be materially simplified [87]. Tapered coils are also impregnated with epoxy compound following the technology described above. The gradient of the potential along the windings constituted of sections shown in Fig. 6.39*b*, through *d* reaches 1 to 2 MV/m.

High-voltage windings are assembled of separate sections when the winding contains tens and hundreds of thousand turns. The sectionalization method facilitates the production of secondary windings and permits winding components to be readily tested and replaced. However, it is found unfeasible to sectionalize the secondary winding in generators with higher operating frequency where the secondary winding has much less turns, i.e., of the order of a hundred or a thousand. The cost of such a few turn windings is, as a rule, not too high and, if need be, it can be either rewound or replaced entirely. Two versions of constructions of nonsectionalized windings are shown in Figs. 6.11 and 6.12.

Insulating materials used in high-voltage generators determine, to a large measure, the dimensions and reliability of the generators. The technical characteristics of some dielectrics typical for the accelerators are presented in Fig. 6.41. (The depicted curves are taken from comparatively small specimens and may be considered as fairly relative characteristics. Voltage gradients that may be tolerated in real conditions are appreciably less.) As was noted above, basic high-voltage gaps in generators of industrial electron accelerators are insulated by the use of compressed gas. The most wide-spread gas insulation is a mixture of $SF_6$ gas and N at a pressure of 1 to 1.5 MPa. Although a higher dielectric strength is gotten with 100 percent $SF_6$ gas, its cost is rather high, so that the insulating gas is composed of 5 to 8 percent $SF_6$ in N. The dielectric strength of this mixture is greater than that of N by approximately 50 percent. A dielectric strength approximating that $SF_6$ gas is possessed by Freon-12 ($CCl_2F_2$); however, at room temperatures, Freon-12 may be utilized only at a pressure not exceeding 0.5 to 0.6 MPa (see Table 6.1) [88].

The perfect insulating properties of $SF_6$, $CCl_2F_2$, $CO_2$, and a number of other gases can be accounted for by the ability of the molecules of these gases to actively annex free electrons and to form negative ions, which substantially impedes the development of the avalanche processes causing breakdowns. That is why the common name of these dielectrics is negative electrogas. Compared to liquid and solid insulating materials, gas insulation allows for much better recovery of dielectric strength after breakdowns.

Transformer and capacitor oils are used as liquid insulation. Both the gas and the oil utilized in high-voltage generators are subject to dewatering, so that the content of moisture does not exceed $1.5 \times 10^{-3}$ percent. As is evident from the description of some

**Table 6.1 Characteristics of some gases used in high-voltage generators**

| Characteristic | $SF_6$ | Freon ($CCl_2F_2$) | $N_2$ | $CO_2$ |
|---|---|---|---|---|
| Relative dielectric strength at $P$ = 0.1 MPa, $E_{max}/E_{air\ max}$ | 2.5 | 2.5 | 1 | 0.9 |
| Boiling (sublimation) temperature at $P$ = 0.1 MPa, °C | –62 | –28 | –196 | –78.5 |
| Liquefaction pressure at temperature of 20°C, MPa | 2.2 | 0.6 | – | 5.6 |
| Density at atmospheric pressure $P$ = 0.1 MPa, and $T$ = 0°C, kg/m³ | 6.98 | 1.32 | 1.25 | 1.98 |

generators, oil in combination with paper or some other laminated dielectric is also used. However, in generators rated at voltages of the order of 1 MV and provided with high-voltage gaps of complicated configuration, liquid insulation is inferior to compressed gas.

Solid insulation has not yet found application in insulating gaps of generators with direct voltage of the order of 1 MV. In principle, solid insulation can afford comparatively high gradients, particularly when applying a method of alternation of thin layers of insulation and conductive medium (as well as in case of forced spread of potential) or in designing a dielectric with predetermined bulk conductivity.

For insulation of generators, a fine vacuum may be used. In this case, particles are subjected to acceleration directly in the high-voltage gap of the generator, which obviates difficulties associated with the transfer of high voltage and creation of dielectric partitions for separation of different insulating media. At present, the circuitry of this type has been implemented only in some installations with a high-frequency accelerating field. This is related primarily to the fact that the gradient of the electric field potential obtained in a vacuum is much lower than in other media. Additionally, generator cooling conditions are impaired, and it becomes impossible to separate the voltage source from the accelerating device and to arrange it beyond the radiation shield. At constant voltages exceeding 200 to 300 kV, the admissible gradient of the electric field potential in a vacuum cannot be above 20 to 30 kV/cm as compared to the value of ~100 kV/cm obtained in compressed gas. The indispensable increase of high-voltage gaps can bring about a drastic deterioration of performance characteristics of generators, e.g., decrease of coupling coefficient in transformers. At the same time, a gradient of several hundred kilovolts per centimeter is quite tolerable for a high-frequency electric field in vacuum, which makes it practicable to design accelerating circutis with such fields (see Section 6.8).

The characteristics of direct-voltage cascade generators are dictated, to a great extent, by parameters of their rectifying devices. In contemporary generators, semiconductor silicon rectifiers are used, as a rule. Relative energy losses in rectifying elements, which are determined by the ratio of direct voltage drop to inverse voltage, are much less in semiconductor diodes than in kenotrons and selenium rectifiers. Voltage of tens of kilovolts usually generated through one stage is rectified by the use of a rectifying stack composed of series-connected rectifiers. Found most suitable for this purpose are avalanche diodes possessing an extremely nonlinear reverse branch of the current-voltage characteristic, which allows for connecting them in series without involving balancing dividers. With many rectifiers connected in series, the voltage under fault conditions may be distributed nonuniformly due to availability of spurious capacitances (analogous to distribution of voltage along high-voltage windings of transformers, see above). In such a case, a dangerous situation may arise not only by the amplitude of the arising voltage but also by the rate of voltage rise. In case of trouble, some rectifiers are more likely to be damaged by a short-circuit effect, but a series of rectifiers can remain serviceable if relatively few rectifiers become nonfunctional. In order to have spare capacity in the stack of series-connected rectifiers, a number of additional rectifiers are introduced in this circuit.

The other elements governing operation of cascade generators are capacitors. Except for the dynamitron, where the purpose of coupling capacitors is served by construction capacitance between electrodes, all other types of generators employ commercial capacitors of various kinds: ceramic, oiled paper, etc. In capacitance-coupled cascade generators, capacitors are placed one atop another or assembled into piles and confined in insulated housing and serve, as a rule, as a supporting structure, since they mount rectifiers, high-voltage electrodes, and some other elements of the circuitry.

Casings of oil-filled capacitors placed in vessels filled with compressed gas may be ruined by gas bubbles in capacitors. To prevent this elastic expanders are sometimes used for changing the volume of capacitors.

The elements, which are also considered important for pulsed generators of many types, are spark gaps. They are used for cutting capacitors in circuit, for connecting them to loads, and for other functions. Quick-acting switch gaps are also widely used for reducing pulse duration, for crowbar in emergencies in relevant facilities, including continuous-action installations. Generally, spark gaps should feature minimum inductance and resistance, high rate of current rise, and large amplitude of current, and should provide maximum possible number of on/off operations. Usually, compressed gas serves as a working medium for spark gaps in industrial accelerators. In some cases, standard thyristors or hydrogen thyristors are used as keys. Accuracy of switching moment, rate of current rise, and some other parameters of spark gaps are materially dependent on technique of firing.

Concluding this review of high-voltage generators used in direct-action accelerators, it should be noted that at the present time, accelerating voltages in most industrial installations do not exceed 0.5 to 0.7 MV. The electron energy generated by these installations suffices to accomplish one- or two-sided irradiation of most articles subjected to radiation processing. Sources of bremsstrahlung with the indicated accelerating voltage provide radioscopy of many articles subject to nondestructive inspection. Characteristics of some installations intended for processing metals by electron-beam emission approach the parameters of industrial accelerators. Power supply sources in some of them are rated at voltages in the range of 0.15 to 0.2 MV and a power of tens of kilowatts. Power supply sources of these facilities are step-up transformers with solid or oil insulation.

The possibility of connecting industrial accelerators directly to high-voltage transmission lines has been considered [89]. A line voltage of 0.3 MV should apparently provide a power supply to a considerable part of industrial accelerators, but the point is that in such installations a need arises to involve fairly intricate high-voltage contrivances to regulate accelerating voltage, to protect supply network in emergency situations, and so on. At present, no information has been obtained on direct supply of industrial accelerators from high-voltage transmission lines.

## 6.10 ACCELERATING DEVICES

At present, direct current of hundreds of milliamperes has been reached in direct-action accelerators. The amplitude of the current in pulsed radiation installations with hot cathodes amounts to the value of the order of 100 A, while that in installations with cold cathodes is some $10^5$ A.*

As distinct from various methods of voltage generation, the operating principle of accelerating devices with potential electric fields is comparatively simple (see Section 6.1). One of the basic differences between accelerating devices lies in the quantity of electrodes or grids interposed between cathode and anode. The simplest type uses diode systems (i.e., no grid is placed between cathode and anode), and current is controlled by accelerating voltage or by cathode heater. In a triode (Fig. 6.1), the electron beam current is under grid control, and therefore a greater density of current can be achieved. The presence of many intermediate electrodes with preset potentials between cathode and anode makes it possible to increase the gradient of an accelerating field potential and to increase current density. Intermediate electrodes may be

*The data hold for a single accelerating device. Aggregate values of summarized currents will be much higher if accelerator employs several accelerating devices operating at a time.

applicable for accommodating focusing lenses and for shielding insulators as well as for some other purposes.

Some differences among accelerating devices is in construction of insulators, cathode units, and other components. Particular electron-optical systems and components of accelerating devices are selected with regard for parameters of electron beam, such as energy, current, cross-section, and pulse duration.

A great variety of layout of accelerating devices with respect to generators can be found. Thus, in accelerators rated at energy less than 0.5–0.7 MeV, accelerating systems are often arranged at a distance of several meters from the generator to which it is connected with a cable. This enables generators to be located out of the radiation hazard zone, thereby materially reducing volume of objects under protection and shielding mass. Moreover, a single generator can supply a number of acclerating devices. Where high voltages are used, it is good practice to have the accelerating device built into the high-voltage generator or placed together with the generator in a common gas-filled space. This will eliminate the need for high-voltage cables and, consequently, to the necessity of using coaxial gas-filled cables, which present difficulties and entail large expenses. Since construction of high-voltage generators are closely interrelated, information on accelerating devices can be found in descriptions of generators.

All generators rated at voltage of the order of 1 MV, (except ILU generators) are provided with gas or oil insulation. Because of this, the accelerator must be equipped with a dielectric barrier for separating the vacuum, where the acceleration process takes place, from the external medium (e.g., gas, oil, air), provided that full operating voltage is maintained along this barrier. Figure 6.42 shows diagrams of some typical versions of particle acceleration equipment. In Fig. 6.42*a*, voltage is fed to the vacuum space through solid insulator (3); in Fig. 6.42*b*, the insulator is made in the shape of a pipe with rings (8) intended for uniform distribution of voltage along the pipe. Internal and external rings (8) are connected electrically in pairs through vacuum-tight conductors (9). Potentials on rings are preset by a voltage divider (here and in subsequent figures, the voltage divider is not shown). In the version shown in Fig. 6.42*c*, the insulating tube is located on the side of the vacuum space with high-quality insulation (1) (for example, compressed gas or oil). In this type, it is much easier to provide adequate mechanical strength of the insulator because it experiences only pressing forces. However, there is an additional high-voltage vacuum space between rings (8) and rod (1). In the diagram shown in Fig. 6.42*d*, the circuit employs a high-voltage flat insulator (11).

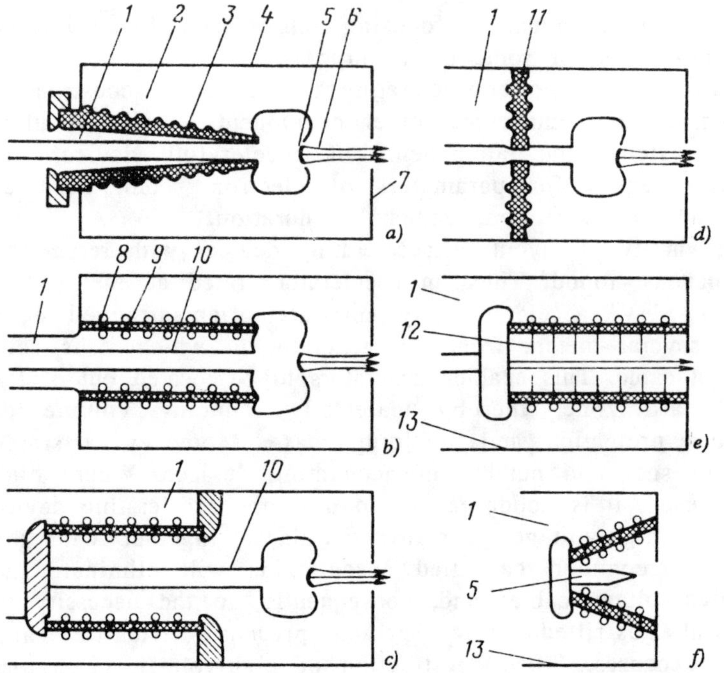


**Figure 6.42** Accelerating devices. *a)* Injection of voltage into vacuum through monolithic insulator. *b)* and *c)* Lead-in insulators with voltage distribution *d)* Flat insulator separating vacuum and outside insulating medium. *e)* Accelerating tube. f) Tapered sectionalized insulator. *1*—External insulating medium (gas. oil, etc.); *2*—vacuum; *3*—insulator; *4*—vacuum chamber; *5*—cathode; *6*—beam of accelerated particles; *7*—anode; *8*—voltage distribution rings (equipotential rings); *9*—conductors, interconnecting inner and outer rings; *10*—rod; *11*—flat insulator; *12*— electrodes of accelerating tube; *13*—tank.

Figure 6.42*e* shows one of the most widespread circuits, i.e., a sectionalized accelerating tube (see also Figs. 6.3*a* and 6.14). Ceramic, porcelain, glass, and some organic materials are used to insulate the accelerating tube. Electrodes are fabricated from titanium, stainless steel, or aluminum alloys. In some acceleration tubes, molybdenum rings are placed in the center of the electrodes (close to accelerating channel). Vacuum connection of nonorganic insulators with electrodes is effected by the use of special glues (polyvinyl acetate, coil-cure compounds, etc.) or by means of ceramic-metal welding. An important role is played by the geometry of joints of metal and insulation. Traces of metal on insulator surface (particularly from the side of vacuum) may result in significant deterioration of dielectric strength. Dielectric strength of insulators may be somewhat increased at the expense of extending its vacuum surface. When high-quality insulation is used as an external medium, the external surface of the accelerating tube features higher dielectric strength than its vacuum surface. Solid insulating pipes used with voltage spread under the action of metal rings similar to the insulators are shown in Fig. 6.42*b* and *c*. With increase of accelerating tube length, it becomes necessary to increase the diameter of the tank (13) to ensure a more uniform distribution of voltage along the tube. Although a divider is connected to electrodes (12) for ensuring a constant electric field potential gradient, it will consume considerable power in case of disadvantageous ratio of geometric characteristics. At a constant accelerating voltage of 0.2 to 0.3 MV and a pulsed voltage of 0.4 to 0.5 MV (and sometimes higher), accelerating tubes often employ solid insulators affording no voltage spread. The length of such an insulator shall be increased as compared with a sectionalized tube in which potential distribution is preassigned.

Even though insulators made of organic materials find little use in accelerating tubes of industrial electron accelerators, they may be preferred in continuous-pumping experimental installations. Production of insulators from Plexiglas and other organic materials does not require complex technological operations, and vacuum-tight rubber may be used for sealing joints of insulator with electrodes. Small $\varepsilon$, optimal geometry, and proper resistance to breakdowns allow for high intensity of electric field and long service life. Such tubes were used in Elita installations (Fig. 6.11) and showed high dielectric resistance and reliability during approximately a 20-year period of operation.

Fig. 6.42*f* shows a sectionalized tapered insulator. The illustrated version allows for a more reasonable application of cross section of the installation, since larger insulating gaps correspond to higher voltages (both in vacuum and in gaseous or other media outside the insulator). Such insulators are widely used on pulsed accelerators with cold cathodes. When operating in a short-pulse emission mode, the distribution of voltage among the electrodes is substantially dependent on their crosswise capacitance to the tank walls. Optimum geometry is usually found by the use of an electrolytic bath or by calculations taking account of distributed capacitance, $\varepsilon$ of dielectric of the tube, environment, etc.

Dielectric strength of vacuum gap depends on the materials of the electrodes, conditions of origin, and propagation of secondary particles, as well as many other factors. Whenever a flux of accelerating particles exists in the vacuum gap, breakdown can occur more readily due to ionization of residual gas, dispersion and multiplication of particles on intermediate electrodes, flow of secondary particles from anode or target positioned beyond the vacuum space. It has not been yet possible to provide a quantitative estimation of these processes, and therefore designing new accelerating devices is mainly based on extrapolation of the known empirical regularities.

The radical factor having an effect on operation of accelerating devices is duration of acceleration process. On the basis of this parameter, the existing installations may be subdivided into two groups:

1. For accelerating electron beams within a duration range of $10^{-5}$ s to the stable-state conditions.
2. For accelerating electron beams impulses lasting for less than $10^{-6}$ s (in some cases in the range of $10^{-6}$ to $10^{-5}$ s).

In the first case, at an absolute difference of potentials of 0.5 MV and more, the gradient of the accelerating field potential amounts to 2 to 5 MV/m.; while at a voltage of 0.4 to 0.5 MV, it is 8 to 10 MV/m. In the second case the potential gradient reaches 50 MV/m which is accounted for by the fact that the time required for acceleration of electrons is less than the duration of development of electrical breakdown. In the first case, the density of the accelerating electron beam respectively lies within 10 A/cm$^2$, in the second case, it reaches 10 kA/cm$^2$ and higher. In long-pulse and stationary accelerators use is made of hot cathodes: $LaB_6$ (working temperature 1550°C), Ta (1900°C), W (2200°C). In evacuated systems these are oxide-coated cathodes (800°C) and impregnated (1100°C).* The service life of cathode units in modern industrial electron accelerators runs into several thousand hours. In short-pulse electron accelerators use is made of cold cathodes.

Typical electron-optical systems of long-pulse and stationary electron accelerators are presented in Fig. 6.43. Three versions of electron-optical system are used in accelerating tubes: flat–parallel accelerating field (Fig. 6.43*a*, *b*); field with variable tilt of equipotentials (Fig. 6.43*c*); and magnetofocusing channel (Fig. 6.43*d*). With flat-parallel electrodes (Fig. 6.43*a*) a ratio of $b \geq 3a$ is generally maintained, which prevents the electron beam from being affected by random charges arising on the inner wall of the insulator and also does not allow stray electrons of the beam to strike this wall. Another version of shielding is represented by the tapered electrodes (Fig. 6.43*b*). In some cases, the electrodes are drawn together at the tube axis for increasing perveance and de-

* The value of the working temperature is tentative, since it depends on specified service life, vacuum, etc.

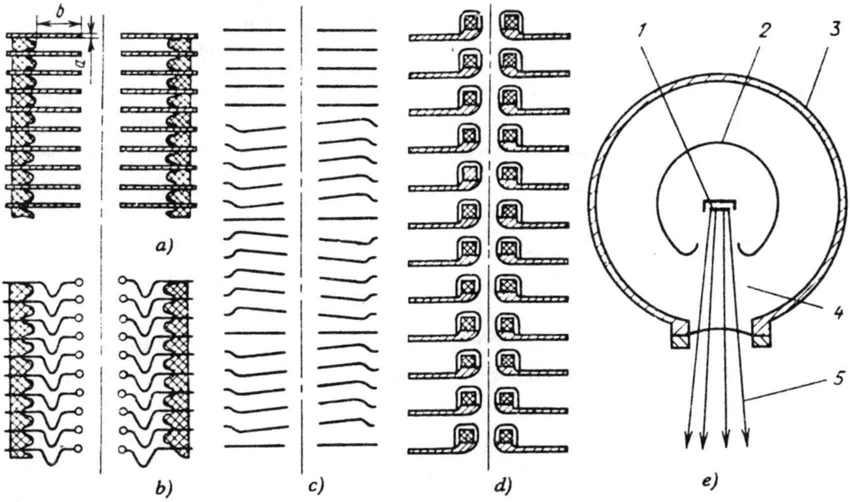


**Figure 6.43** Electron-optical systems of electron accelerators. *a)* and *b)* Accelerating tube with plane-parallel accelerating field. *c)* Accelerating tube with inclined electrodes. *d)* Accelerating tube with focusing magnets. *e)* Single-gap accelerating system. *1*—Cathode; *2*—cathode-adjacent electrode; *3*—vacuum chamber; *4*—accelerating gap; *5*—electron beam.

creasing the gradient of the electric field potential along the insulator (see, for example, Fig. 6 .11). In accelerating tubes with inclined electrodes, the accelerating electron beam passes quite freely, whereas secondary particles are restricted in their travel. (Fig. 6.43*c*).

Due to the high intensity of electron beams in industrial electron accelerators, the accelerating tube aperture must substantially exceed the electron beam diameter, since ingress of even a small portion of electric current may cause heating of electrodes, liberation of gas, and other undesirable aftereffects. In accelerating tubes of industrial accelerators, often at currents of 10 to 100 mA, the aperture of the accelerating channel is usually at least 2 to 3 cm in diameter. Arranging permanent magnetic lenses adjacent to the electron beam axis will materially improve beam focusing and, at the same time, hamper free passage of secondary particles. (Fig. 6.43*d*) [65, 90]. For instance, magnetic lenses providing a magnetic field of 0.06 T make it possible to obtain a beam current density of 1 to 10 A/cm$^2$.

It should be emphasized that certain studies have been carried out to find a possible use for transverse magnetic fields and to devise some other methods for decreasing the flux of secondary particles accelerated along accelerating tubes. However, no essential increase of electric field voltage gradient along the accelerating tube has been reached. For protecting the tube against full-voltage breakdowns that can occur because of the eventual breakdown of some insulating rings, provision is made for protective spark gaps outside the tube between the rings (8). As a rule, there are two or three rows of spark gaps (arranged through 180° and 120°, respectively) provided to minimize inductive impedance from the point of breakdown to the nearest spark gap. Intensive studies of vacuum breakdown aspects, development of new materials and elaboration of technology of joining the rings have not yet resulted in notable growth of potential gradient that could be withstood by accelerating tubes at constant voltages. During this same 25- to 30-year period, the magnitude of current of particles accelerated in the tube has risen by a factor of several orders.

One way that might lead to a considerable increase of dielectric strength of insulators is sectionalizing into smaller divisions or utilizing materials with homogeneous bulk conductivity [91]. In the first case, the thickness of the insulator of each tube section might be reduced to tens of micrometers as against the adopted height of the rings measuring 1 to 3 cm. It is well known that vacuum and other varieties of insulation in micrometer-wide gaps are capable of standing electric field voltage gradient of 500 to 1000 MV/m. Should a gradient in such multilayer insulation of only 10 to 20 MV/m be obtained, the current density in acceleration channels could have been increased to 100 A/cm$^2$ and more. Materials with bulk conductivity should additionally possess proper insulating properties. Conductance uniformity should not be worse than 1 to 5 percent and should remain unchanged in the presence of temperature gradients. At the same time, apart from insulating properties of sectionalized insulation of accelerating tubes, there still exists a problem of dielectric strength of vacuum channels wherein acceleration procedure takes place. It is not inconceivable that the gradient of accelerating voltage in these channels could be materially increased by implementing many parallel channels of smaller diameters.

Figure 6.43*e* shows a cross sectional view of a device intended for accelerating a long sheet beam of electrons in one gap. The potential difference in this device does not exceed 0.3 to 0.4 MV. The cathode (1) and surrounding electrode (2) are fixed on an insulator on the end of a vacuum chamber (3). The electrodes are so shaped as to obtain a beam divergence angle corresponding to the crosswise dimension of the outlet window. The characteristic dimensions of the beam passing through the foil window are width, 4 to 6 cm and length, 100 to 150 cm. Such a device is used in Elektrocurtain accelerators manufactured by Energy Science (U.S.).

A version of a device with several parallel sheet beams is illustrated in Fig. 9.5*d*.

Industrial electron accelerators employ both continuous-pumping and evacuated accelerating devices. Selection of a particular version depends on such factors as the voltage and power of the electron beam, mode of use of electrons (extraction to the outside or bombardment of internal target), material of component parts, and service life. An evacuated accelerating system may incorporate a small pump to maintain the required vacuum. Evacuated accelerating tubes are often used in bremsstrahlung generators. Figure 6.44 illustrates a number of versions of such devices. All the tubes use glass flasks (1) as insulators. In the tube of accelerator RUP-400 (Fig. 6.44*a* and *b*) accelerating electrodes (5) are shaped like cylinders, thereby ensuring shielding of insulation against ingress of accelerating particles. The electrostatic field existing between each pair of electrodes accelerates and focuses electron beams. Such electrodes were widely used in early direct-action electron accelerators. Their shortcoming lies in the fact that the accelerating field is concentrated at some points of the trajectory, as result of which the average gradient of potential tends to decrease and the length of the accelerating tube increases. In this case the perveance declines and, accordingly, the density of the accelerating electron beam drops. The potential gradient in such a tube amounts to ~0.5 MV/m, the current magnitude being a few milliamperes. The accelerating tubes of resonance transformers produced by General Electric (Section 6.4) has the same construction. For accelerating electron beams with currents of tens of milliamperes and more in accelerating tubes, use is made of electron-optical systems described above (Fig. 6.43). Figure 6.44*c* shows single-gap accelerating device for pulse accelerator Raps-1 for 0.6 MV (Section 6.8). The device depicted in Fig. 6.44*d* is rated at direct and sinusoidal voltage of 0.25 MV, is provided with a forced-cooling anode (4), and is rated at an average beam power of a few kilowatts. An evacuated tube with a more intensive electron beam and a rotary anode is described in section 8.2.

Several versions of evacuated tubes with cold cathodes are shown in Fig. 6.45. Tubes IMA 2-150 D and IMA 5-320 D (Fig. 6.45*a* and *b*) find application in commercial apparatuses produced by the NPO Burevestnik (Section 6.4, Fig. 6.12). The basic characteristics of the tubes are given in Table 6.2 [61]. Here the pressure is $10^{-4}$ to $10^{-5}$ Pa. Molybdenum glass serves as the insulator; the cathodes are made of tungsten. The cathode (3) in the IMA 2-150 D tube is a cylinder of 2 mm in diameter, made of tungsten foil of 20 μm thick. The purpose of cathode 3 in tube IMA-5-320D is served

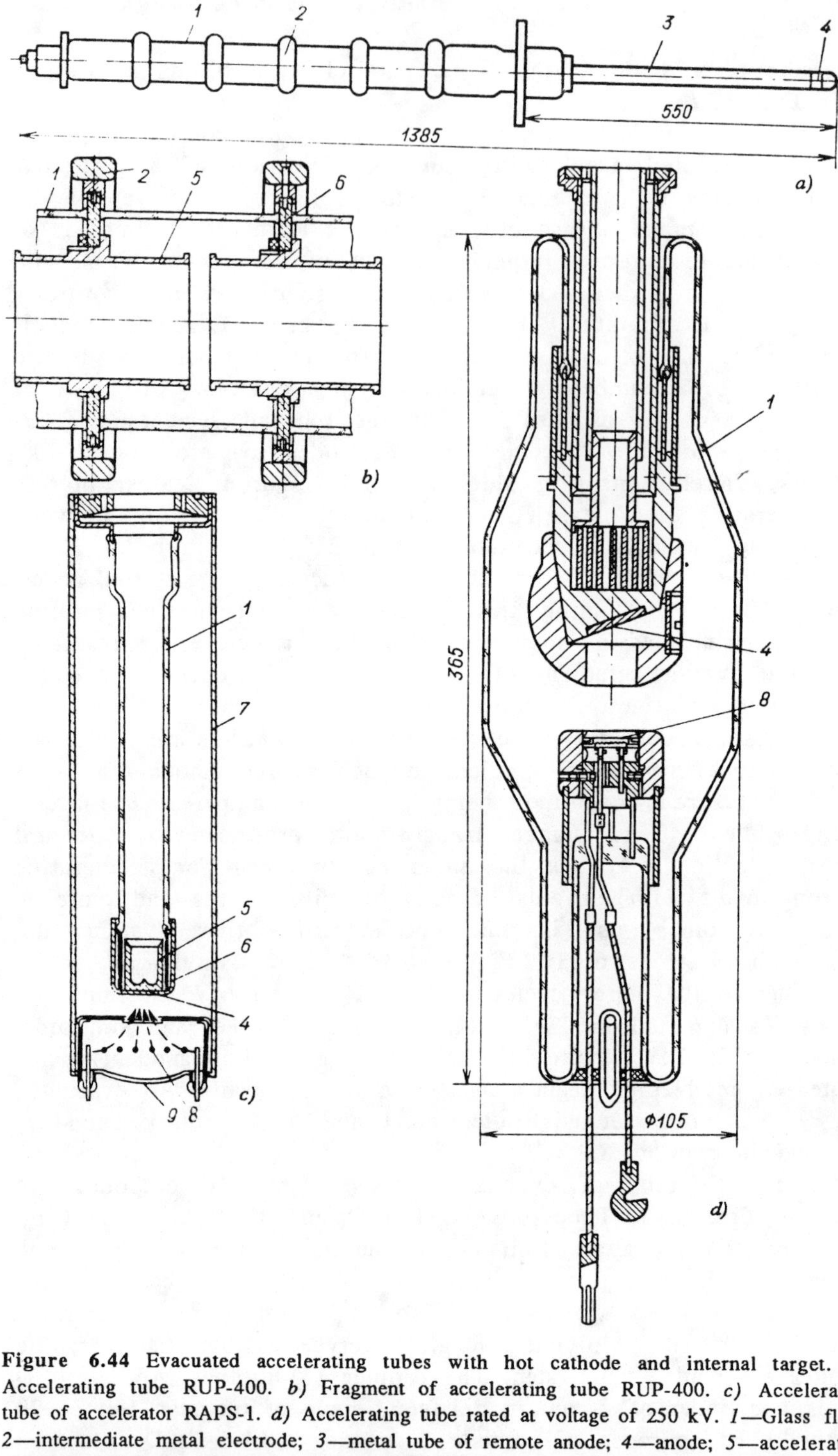


**Figure 6.44** Evacuated accelerating tubes with hot cathode and internal target. *a)* Accelerating tube RUP-400. *b)* Fragment of accelerating tube RUP-400. *c)* Accelerating tube of accelerator RAPS-1. *d)* Accelerating tube rated at voltage of 250 kV. *1*—Glass flask; *2*—intermediate metal electrode; *3*—metal tube of remote anode; *4*—anode; *5*—accelerating electrodes; *6*—cover rings; *7*—metal casing; *8*—cathode; *9*—foil window. Dimensions are expressed in millimeters.

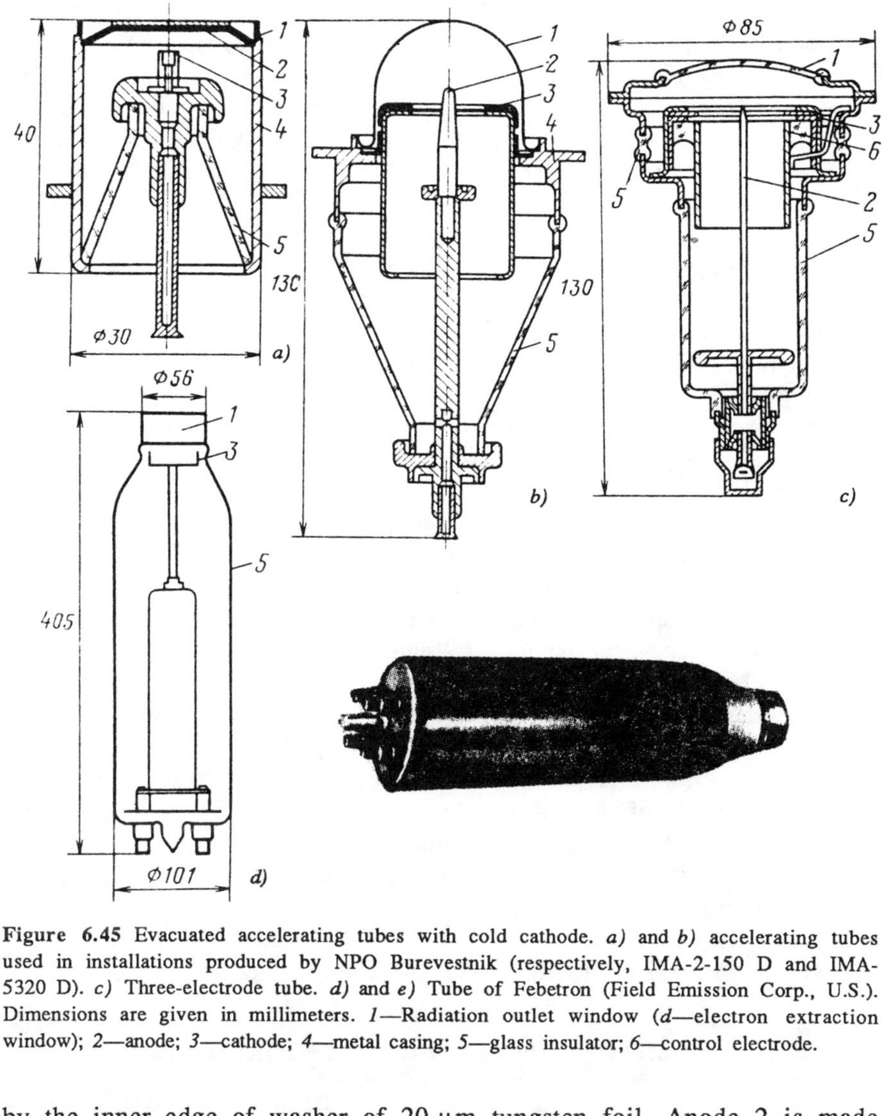


**Figure 6.45** Evacuated accelerating tubes with cold cathode. *a)* and *b)* accelerating tubes used in installations produced by NPO Burevestnik (respectively, IMA-2-150 D and IMA-5320 D). *c)* Three-electrode tube. *d)* and *e)* Tube of Febetron (Field Emission Corp., U.S.). Dimensions are given in millimeters. *1*—Radiation outlet window (*d*—electron extraction window); *2*—anode; *3*—cathode; *4*—metal casing; *5*—glass insulator; *6*—control electrode.

by the inner edge of washer of 20-μm tungsten foil. Anode 2 is made of a tapered tungsten bar 4 mm in diameter, with a 14° apex cone angle. The tube is used in apparatus Mira-3D and is found suitable for x-ray examination of hollow objects from the inside. As the illustrations show, provision is made for parts to prevent direct ingress of beam-sprayed anode material to the insulator. The service life of such tubes is ~$10^6$ pulses.

The tube shown in Fig. 6.45*c* employs a control electrode (6). It undergoes an effect of a pulse of voltage applied from an additional source. As a result, an auxiliary discharge which arises between electrode and cathode presets the instant of electron acceleration and time of origin of bremsstrahlung pulse. This allows the working pulse to be synchronized with the process under study with the aid of an additional electrode. As a result, stability of parameters of pulses in intermittent operation mode is improved, and it also becomes possible to regulate the current in the accelerating tube.

Figure 6.45*c* and *e* demonstrates the accelerating tube of Febetron-705 (Section 6.8). It is designed for acceleration of electron beams of 2 MeV in energy and 5 kA in current magnitude. Pulse duration lies within ~30ns. The electron beam is extracted outward through the outlet window (1). In the case of generation of bremsstrahlung, the target is installed outside the vacuum chamber.

The beam energy in a pulse is about 300 J. At a distance of 300 mm from the target the pulse dose amounts to 2.6 mCal/kg. The tube is designed for single pulses.

Some versions of electron-optical systems of accelerating devices of bremsstrahlung generators are shown in Fig. 6.46 [92]. Multipoint or multiblade cold cathodes (2) are used in these systems. Tungsten tapered anodes (3) are 3 to 7 mm in diameter at the base, the apex angle is 10 to 30°, and the curvature radius is 0.5 to 1.5 mm. The voltage in such accelerating devices amounts to 0.2 to 2 MV, and the pulse length is about $10^{-7}$ s. Despite some advantages of direct-action accelerators (simple construction of accelerating mechanism, high efficiency, etc.), their practical use is limited whenever energy of particles is increased. The need for an accelerator part being under a potential equal numerically to the final energy of particles results in increase of high-voltage gaps and three-dimensional characteristics of the installation. Due to this fact, the mass and cost of production of the accelerator tend to rise almost proportionally to the rise of electron energy, i.e., to the

**Table 6.2 Parameters of tubes produced by NPO Burevestnik**

| Characteristic | IMA-2-150 D | IMA5-320 D |
|---|---|---|
| Voltage amplitude, kV | 150–200 | 320 |
| Diameter of effective focal spot, mm | 2.3–3.0 | 2.3–3.0 |
| Highest pulse repetition frequency, $s^{-1}$ | 50 | 15 |
| Length, mm | 40 | 120 |
| Diameter, mm | 30 | 62 |
| Mass, g | 60 | 200 |
| Energy stored in output circuit, J | 0.6 | 3 |
| Exposure dose per pulse at distance of 50 mm, μC/kg | 7.7 | 13 |

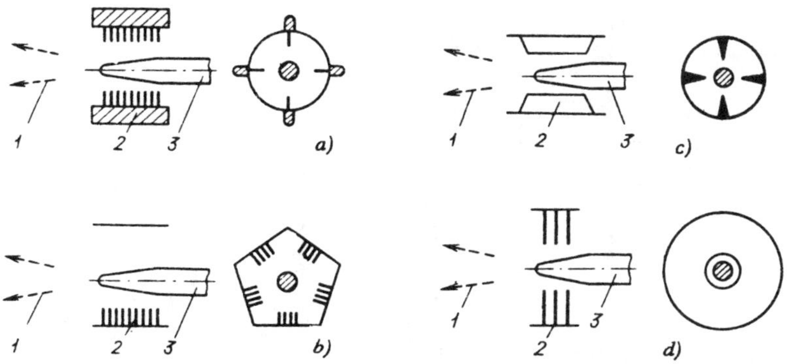


**Figure 6.46** Electron-optical systems of accelerating devices of generators and bremsstrahlung pulsed sets (in two projections) *a)* Hewlett-Packard (U.S.). *b)* CSF (France). *c)* CGR (France). *d)* NPO Burevestnik (USSR). *1*—Bremsstrahlung; *2*—cathode; *3*—anode.

power of 2.5 to 3. (It should be noted that some elements of accelerators, for example, injector and control system, remain practically unchanged in design. On the other hand, the length of the accelerating tube and thickness of some types of insulation tend to increase more than linearly with an increase in energy of accelerating particles)

At the same time, electron beams with energies of tens of megaelectronvolts with a relatively small power of 10 to $10^4$ W are required for solving many applied problems. Techniques and associated facilities for obtaining such beams with the use of electric fields of comparatively small voltage is described in Chapter 7.

## MATERIAL ADDED IN PROOF

Cockroft-Walton type of cascade generators are most widely used in Japan. More than one hundred cascade generators of different types are now used in industry to process electric wire, plastic tube and sheet, automobile tires, etc. [182]. The design of the latest development of the Cockroft-Walton type of electron accelerator (Nissin High Voltage Co., Ltd., and others) rated at 5 MeV and a power of 300 kW [182] is shown in Figure 6.24*b*. Silicon rectifiers and oil filled capacitors are used in the generator. The generator and the accelerating tube are installed in a vessel filled with $SF_6$. The horizontally accelerated electron beam is used to generate the bremsstrahlung radiation on the surface of the tantalum target arranged in the atmosphere. In order to obtain the data on the accelerator design, experiments have been carried out at the JAERI TAKASAKI 2 MeV, 60 kW running plant.

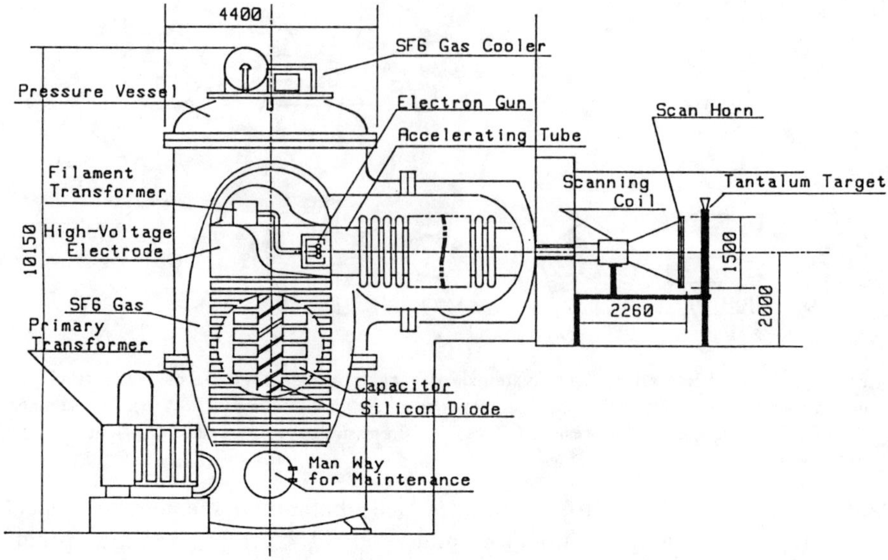


**Fig. 6.24*b*** 5 MeV 300 kW Cockroft-Walton type of electron accelerator.

CHAPTER

# SEVEN

# INDUCTION AND RESONANCE ELECTRON ACCELERATORS

## 7.1 ACCELERATION OF ELECTRONS IN VORTEX ELECTRIC FIELDS

When a particle is moving in a potential electric field (used in direct-action electron accelerators), the increment of energy is determined unambiguously by difference of potentials $\Delta U$ at points at the start and end of the trajectory. For a closed trajectory (particle returns back to initial point), particle energy remains unchanged.

Change of energy of particle traveling in a vortex (induction) field is dependent on trajectory. Acceleration is also feasible whenever a particle moves along closed trajectory, provided this trajectory covers magnetic flux changing with time.*

In a circuit covering a variable magnetic flux $\Psi$, the following electromotive force (EMF) is induced:

$$\oint E dl = -\frac{1}{c}\int \frac{\partial B}{\partial t}\, dS \quad \text{or} \quad U_{\text{loop}} = -\frac{1}{c}\frac{d\Psi}{dt}$$

where $U_{\text{loop}}$ is the loop voltage.

*Vortex electric field arising due to change of magnetic flux is described with equation

$$\text{rot}\,\mathbf{E} = -\frac{1}{c}\frac{\partial \mathbf{B}}{\partial t}$$

It will be recalled, that intensity $E$ of a potential electric field is related to potential $U$ through ratio expressed as $E = -\,\text{grad}\,U$.

Acceleration of electrons by vortex electric field is diagrammatically shown in Fig. 7.1*a*. Magnetic flux $\Psi = B \cdot S$ is concentrated in a ferromagnetic circuit (1) with cross section $S$. Current flowing in the primary winding (4) generates a magnetic flux. Electrons travel along a rectilinear trajectory (2) through an aperture in the magnetic circuit or orbit around the magnetic circuit in an additional (guiding) field (not shown in the figure) along a trajectory (3).

Figure 7.1*a* corresponds to movement of particles in linear induction electron accelerators. The second version pertains to motion of electrons in a circular orbit of approximately constant radius and is characteristic of betatrons. As a result of one pass through the magnetic circuit in linear accelerators or one revolution in circular orbit in betatrons, the electrons increase in energy by an increment

$$\Delta \varepsilon = eU_{\text{loop}} = -\frac{e}{c}\frac{d\Psi}{dt}$$

Electrons may also travel along a helix or flat spiral (Fig. 7.1*b* and *c*).

The operating principle of induction electron accelerators is similar to that of a transformer with the difference that here the role of the secondary winding is played by an electron beam. Linear induction electron accelerators usually incorporate several magnetic circuits (Fig. 7.2*a*) with parallel-connected primary windings. "Secondary turns" of some inductors can be considered connected in series. Electrons in betatrons orbit many times around the magnetic circuit and their final energy is higher (numerically)

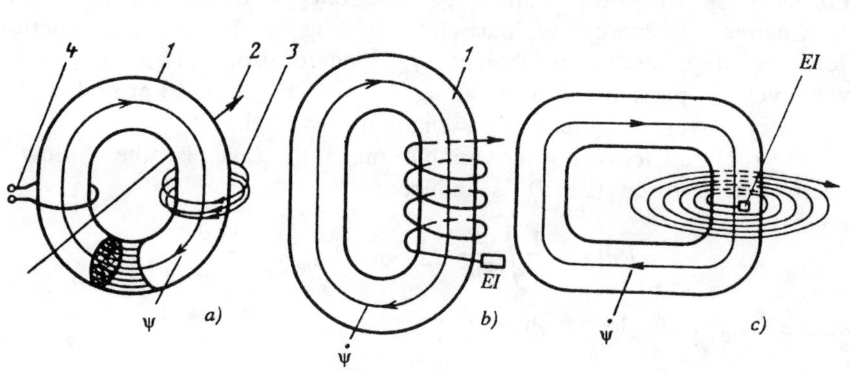


**Figure 7.1** Acceleration of electrons by vortex electric field. *a)* Acceleration of electrons traveling along rectilinear and approximately circuit trajectories. *b)* and *c)* Acceleration of electrons traveling along helix and flat spiral. *1*—Magnetic circuit; *2*—rectilinear trajectory of electrons; *3*—trajectrory of electrons traveling around magnetic circuit; *4*—primary winding; *EI*—electron injector; ψ—magnetic flux.

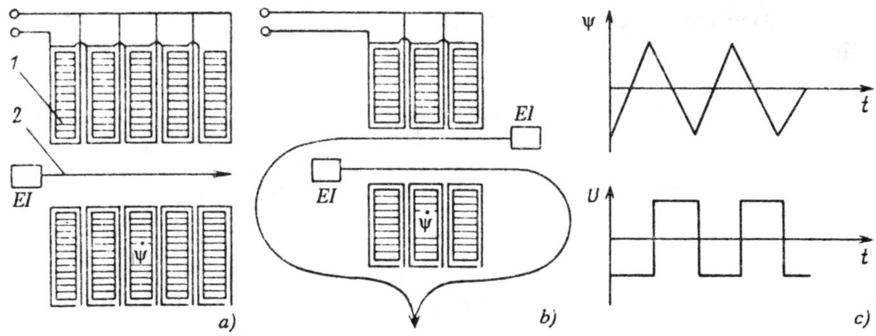


**Figure 7.2** Acceleration of electrons in linear electron accelerators. Designations are as for Figure 7.1. *a)* Pulse linear electron accelerator. *b)* Linear electron accelerator generating continuous electron beam—trajectrory is curved by use of additional magnets (not shown). *c)* Curves demonstrating operation of linear electron accelerator generating continuous electron beam. *U*—accelerating voltage.

by many orders than the value of primary winding voltage. The time of acceleration of electrons in induction electron accelerators is determined by time of rise of magnetic flux up to the maximum value. Then the magnetic flux drops and the vortex electric field reverses in sense. At this time, electrons can be accelerated in the opposite direction. There is, in principle, the possibility of designing linear induction electron accelerators capable of generating a continuous flow of constant-energy accelerated electrons (Fig.7.2*b* and *c*) [93]. In this case, it is essential to linearly change magnetic flux in time and to converge electrons in both directions. Acceleration of electrons in both directions performed in betatrons sometimes finds practical application.

### 7.1.1 Betatron Electron Accelerators

Let us consider in more detail the peculiarities of acceleration of electrons in betatrons. One of the principal conditions for operating betatrons is maintaining a permanent radius of orbit of electrons during the entire acceleration process, since this makes it possible to build up the electron energy during many revolutions, and keep the vacuum chamber and guiding magnetic field comparatively small in volume. Curvature radius $R$ of a particle trajectory in a magnetic field is related to particle pulse $p$ by the ratio (see Section 1.3)

$$p = \frac{e}{c} BR$$

The average intensity $E$ of the vortex electric field on orbit of radius $R$ is

$$E = \frac{U_{\text{loop}}}{2\pi R} = -\frac{1}{2\pi Rc}\frac{d\Psi}{dt}$$

Let us assume that $R$ = const, then from equation $\partial p/\partial t = -eE$ (taking $\Psi = \overline{B}\,\pi R^2$), we find that

$$\frac{\partial B}{\partial t} = \frac{1}{2}\frac{\partial \overline{B}}{\partial t}$$

Consequently, regardless of its energy, a particle will revolve in a constant radius orbit if guiding magnetic field $B$ on this orbit tends to rise at a rate that is half that of average field $\overline{B}$ embraced by the orbit. In this case, there is nothing to restrict the nature of change of $B$ and $\overline{B}$ in time (except that they should vary synchronously). An orbit that satisfies this condition is referred to as equilibrium or betatron orbit, and its radius $R_e$ is termed equilibrium radius. The derived ratio is called betatron condition (also 2:1 condition or Wideroe flux condition). Electrons that appear to be not exactly on equilibrium orbit but nearby at the outset of acceleration, will also be subjected to acceleration, thereby approaching an equilibrium orbit asymptotically.

It might be well to point out that, from the above calculations, it follows that the electron-pulse increment is directly proportional to change of magnetic flux enveloped with electron traveling orbit, i.e.,

$$p(t) - p(0) = \frac{e}{2\pi R_{\text{loop}}c}\left[\Psi(t) - \Psi(0)\right]$$

Apparently, a common way of synchronous changing of guiding magnetic flux (passing in gap between poles 3, Fig. 7.3*a*) and magnetic flux enclosed in equilibrium orbit lies in excitation of these fluxes by one and the same field winding to obtain different magnetic resistances to these fluxes for complying with the condition 2:1.

The value of magnetic flux with preassigned number of ampere turns $Iw$ can be derived from the equation

$$\Psi = \mu_0 \frac{Iw}{\Sigma l/\mu S}$$

where $I$ is expressed in amperes, and $\Psi$ in webers (Wb). $\Sigma l/\mu S$ stands for total resistance of all sections of magnetic circuit, where $l$ is expressed in meters, $S$ in square meters, and $\mu_0$ = 1,257 × $10^{-6}$ Wb/A · m.

Magnetic fluxes pass through the ferromagnetic circuit (1) (closing yokes are not shown in Fig. 7.3*a*, see Fig. 7.7). As seen from the illustration, the guiding magnetic flux has to overcome a larger air gap than that passed by the flux in the central part of the accelerator (in closing yokes such air gaps are not provided, as a rule). Configuration of magnetic poles (3) dictates the shape of magnetic lines of guiding field $B$ needed for focusing electrons (see Fig. 7.6). The vacuum chamber (4) is located in the gap between the poles. An electron gun (6) is placed close to the equilibrium orbit

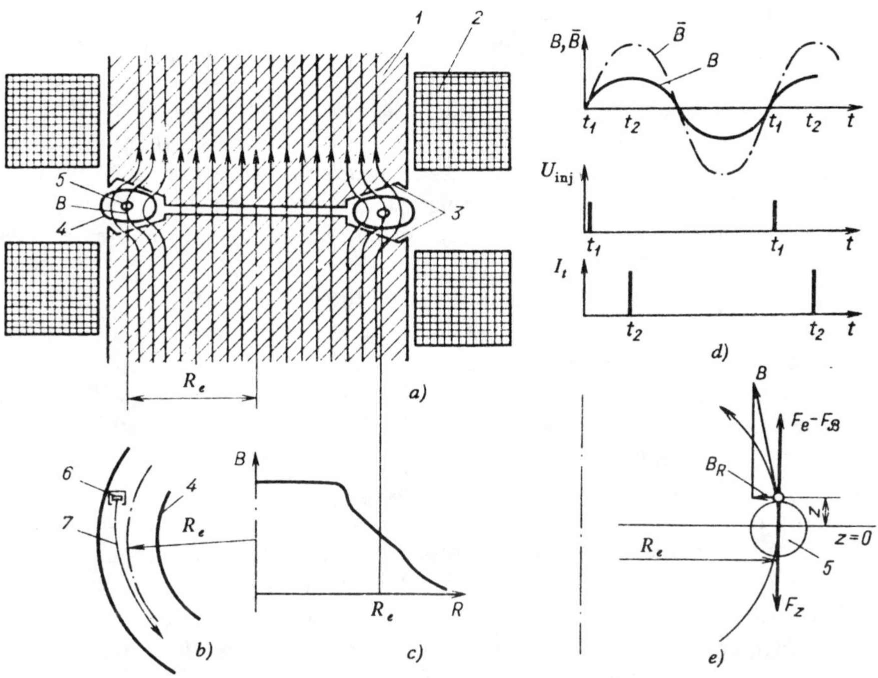


**Figure 7.3** Diagrams and curves illustrating operation of betatron. *a)* View of betatron fragment in which magnetic flux and flux inside orbit are excited by one and the same winding. *b)* Arrangement of electron gun in vacuum chamber (plan view). *c)* Magnetic induction versus radius. *d)* Change of $B$ and $\overline{B}$ with time, and moments of injection of electrons and deflection of beam to target. *e)* Forces acting on electron being at distance $z$ from median plane. *1*—Magnetic circuit; *2*—field winding for exciting guiding and central magnetic fluxes; *3*—poles presetting shape of magnetic lines of guiding magnetic field; *4*—vacuum chamber; *5*—electron beam on equilibrium orbit; *6*—electron gun; *7*—electron beam at injection moment; $R_e$—radius of equilibrium orbit, $B$—guiding magnetic field; $\overline{B}$—intermediate magnetic field inside orbit; $B_R$—radial component of guiding magnetic field; $t_1$—injection moment; $t_2$—moment of ceasing of acceleration process deflection of beam to target); $U_{inj}$—injector voltage; $I_t$—current of accelerated electrons to target; $z$—vertical coordinate of electron extremely shifted from median plane; $F_z$—focusing force of external (guiding) magnetic field acting on electron in vertical direction; $F_e - F_B$—resultant force of natural fields of electron beam.

and injects electrons (7) in the initial stage of acceleration approximately tangentially to the equilibrium orbit (Fig. 7.3*b*). Generally, windings, (2) in a betatron are supplied with power from a common power supply network, and magnetic fields change sinusoidally (Fig. 7.3*d*). When accelerating voltage $U_{\text{inj}}$ is applied to the electron gun and injection of electrons is effected, the time is selected so that the field $B$ on the equilibrium orbit at this instant approximates the value derived from the equation (see Section 1.3)

$$U_{\text{inj}} = e^{-1}\sqrt{9B^2R^2 + 0.511^2} - 0.511$$

In this case the injected electrons will orbit along a trajectory approximating an equilibrium orbit, and thereby a part of them will be involved in the acceleration process. Then the magnetic field on and inside the orbit increases in compliance with the 2:1 condition, and electrons are accelerated until time $t_2$ when the magnetic field closely approximates its maximum. The final energy $W \cong 3\,BR$ (see Section 1.3). For example, at $B = 0.4T$ and $R_e = 15$ cm, $W = 18$MeV. Accelerated electrons are either extracted into the atmosphere or deflected to the target located inside the vacuum chamber with a view for obtaining a bremsstrahlung pulse. The particles are extracted by the use of pulsed magnetic or electrostatic deflectors. For deflection of particles to the target, it is sufficient to upset the betatron condition. The radius of the beam trajectory will either rise or decrease depending on which field is increased at a higher rate than that required by the 2:1 condition (i.e., average magnitude of magnetic field enclosed in orbit or guiding field).

**Electron focusing in betatrons.** For retaining electrons in the process of acceleration the same orbit in addition to satisfying the 2:1 condition, the charged particles must be focused in the median and vertical planes. Forces required for returning the electrons to the equilibrium orbit are produced with the help of the guiding magnetic field. Bending of lines of force of guiding magnetic field (as indicated in Fig. 7.3*a* and *e*) gives rise to a radial component of magnetic field $B_R$ (all over the effective region of the guiding magnetic field except the median plane). This component is dependent on vertical coordinate $z$ in conformity with the law

$$B_R \cong z\left(\frac{\partial B_z}{\partial R}\right)_{z=0}$$

As the result of the interaction between electrons with circular velocity $V$ and deflection at distance $z$ from the median plane and the component of magnetic field $B_R$ a Lorentz force directed vertically arises:

$$F_z = \frac{e}{c} VB_R$$

Force $F_z$ is directed toward the median plane (causing the charged particle to return to the equilibrium orbit) if $\partial B_z/\partial R < 0$, i.e., the magnetic field tends to decrease with an increase in radius (Fig. 7.3*b*). Now we can express the law of change of magnetic field with radius as $B \sim R^{-n}$. The index of magnetic field decay

$$n = -\frac{R}{B}\left(\frac{\partial B_z}{\partial R}\right) z = 0$$

and correspondingly

$$B_R = -\frac{zn}{R} B_z$$

Motion of particles along coordinate $z$ is found from the equation

$$m\ddot{z} = F_z = -\frac{e}{c} vB_z \ (n/R)$$

Since

$$\frac{e}{c} vB_z = \frac{mv^2}{R}$$

$v/R = \omega_0$ where $\omega_0$ is the angular rotation frequency of a particle. Therefore, the equation describing motion of electrons along coordinate $z$ takes the form

$$\ddot{z} + n\omega_0^2 z = 0$$

In solution of equation $z = z_m \sin \sqrt{n}\, \omega_0 t$, it should be taken into account that $\sqrt{n}\, \omega_0 = \omega_z$ stands for angular frequency of vertical oscillations.

The same method is applied in determining the restoring force in case of minor deflections of electrons in the radial direction, with radial oscillation frequency $\omega_R = \sqrt{1 - n}\, \omega_0$. To make both the types of oscillation stable, the index of decay of magnetic field must lie within $0 < n < 1$. More rigorous restrictions are imposed on $n$ by the necessity to prevent resonance between betatron oscillation frequency and rate of revolution (orbiting), as a result of decay of guiding magnetic field usually complies with the condition 0.25 <

$n < 0.75$, for at $n = 0.25$, $\omega_z = 0.5\ \omega_0$, while at $n = 0.75$, $\omega_R = 0.5\ \omega_0$. In this case, the possibility still exists of producing resonance between vertical and radiation oscillations at $n = 0.5$, which, as a rule, is not dangerous. (In determining the resonance values of $n$, its value was considered to be constant throughout the region of particle motion.)

Deflections of electrons from the equilibrium orbit are observed both in the initial stage of acceleration (i.e., at injection moment $t_1$) due to discrepancy between $U_{\text{inj}}$ and $B$ as well as vertical displacement of electrons and in the process of acceleration due to effects in inherent space charge, heterogeneity of magnetic field, etc. It is easy to demonstrate that amplitudes of betatron oscillations tend to decrease with the rise of guiding magnetic field as ~ $B^{-0.5}$. Therefore, the size of the effective region of the guiding magnetic field and the cross section of the vacuum chamber are determined primarily by the cross section of electron beam in the beginning of the acceleration process. The radial size of the vacuum chamber usually exceeds the vertical dimension because of necessity to accomodate the injector, internal target, and additionally some variation of radius of equilibrium orbit in the process of acceleration due to any possible error in following the 2:1 condition.

Beam intensity in betatron is a major parameter of the facility. Experiments suggests that in the process of acceleration electrons are practically not lost and electron beam current is determined by the number of particles involved in acceleration procedure at the injection moment. Collective interaction of electrons in the process of injection has not yet been calculated precisely despite numerous studies. However, it is feasible to make an estimate of the betatron limiting current obtained when optimal conditions of injections are satisfied.

Figure 7.3*e* depicts the cross section of an electron beam on equilibrium orbit. The electron with coordinate $z$ is affected by vertical focusing force

$$F_z = \frac{mv^2}{R}\frac{nz}{R} = \gamma\beta^2\frac{m_0c^2}{R}\frac{nz}{R}$$

and in the opposite direction, by the force of inherent space charge

$$F_e - F_B = 2\pi e^2 N\gamma^{-2} z$$

(see Section 1.3). Thus, the resultant force exerted on electron, namely $F = (\gamma\beta^2 m_0c^2 R^{-2}n - 2\pi e^2\gamma^{-2}N)z$. Obviously the density of charged particles should not exceed the value N found for the case $F = 0$. Taking into consideration the probable resonance that may arise between the frequency of vertical oscillations and rate of orbiting, restriction of $N$ is determined on the basis of more stringent conditions:

$$(\gamma\beta^2 m_0 c^2 R^{-2} n - 2\pi e^2 \gamma^{-2} N_m) z = \gamma\beta^2 m_0 c^2 R^{-2} n_{\mathrm{res}}\, z$$

where $n_{\mathrm{res}} = 0.25$ [105]. From this, it is inferred that

$$N_m = \frac{n - n_{\mathrm{res}}}{2\pi} \frac{m_0 c^2}{e^2} \frac{\gamma^3\beta^2}{R^2}$$

and the maximum current of particles circulating in betatron for round beam with cross section $\pi z^2$:

$$I_m = \frac{Q}{T} = \pi z^2 \, 2\pi R_e e N_m \frac{\beta c}{2\pi R_e} = 0.5\,(n - n_{\mathrm{res}}) \frac{m_0 c^2}{e^2} \gamma^3\beta^3 \left(\frac{z}{R_e}\right)^2$$

The same result is obtained from analysis of radial focusing. Consequently, keeping in mind that $m_0 c^3/e = 17{,}000$ A and $\gamma$ and $\beta$ are minimal at the injection moment, and substituting typical value $a$ equal to one half the vertical or radial aperture of the chamber for value $z$ we shall obtain

$$I_m = 10^4 \left(n - n_{\mathrm{res}}\right) \gamma_0^3 \beta_0^3 \left(\frac{a}{R}\right)^2$$

where $I_m$ is expressed in amperes, $\gamma_0$ and $\beta_0$ are, respectively, the relativistic factor and the relative velocity of electrons at the moment of injection. A resonance value of magnetic field decay index closest to the value $n$ is substituted for $n_{\mathrm{res}}$. It has been apparent from the obtained formula that betatron current may be increased with the increase of injection energy and extension of relative aperture in acceleration area.

The injection energy in industrial betatrons commonly amounts to 30 to 70 keV, provided the injector is located inside the vacuum chamber (Fig. 7.3*b*). $I_m$ in this case will not be above 1 A. The number of electrons involved in acceleration process is $\sim 10^{10}$. Experimental installations sometimes use external injectors rated at higher voltages. For example, in [106] an installation is described in which at injection voltage $U_{\mathrm{inj}} = 0.5$ MeV, $I_m \cong 100$ A has been obtained (approximately $2 \times 10^{12}$ electrons per acceleration cycle). The final energy is 100 MeV. The estimates of $I_m$ derived from the above-mentioned formula are in good agreement with the experiments.

The average intensity of an electron beam (as well as bremsstrahlung) generated by betatron can be raised at the expense of increase of repetition rate of operating cycles. Intensity of bremsstrahlung is doubled in acceleration of electrons in both directions.

### 7.1.2 Comparison of Betatrons and Linear Induction Electron Accelerators

A primary difference between linear induction electron accelerators and betatrons relates to accelerating conditions. Injection in linear accelerators can be carried continuously as long as the magnetic flux is changed, while in betatrons injections are accomplished in extremely short lapses of time, usually for duration that is less than 0.1 percent of total duration of acceleration process. Another difference is that in linear induction electron accelerators provision is usually made for a more intensive focusing of the electron beam by the use of magnetic lenses or a longitudinal magnetic field, which enables the beam current to be increased substantially. Injection and extraction processes in linear accelerators are much easier. Injection energy here, as a rule, exceeds 1 MeV, pulse current reaches hundreds and thousands of amperes. As a result, the average intensity of electron beam in a linear induction electron accelerator may exceed that of betatron beam by a factor of many orders. At high beam currents a linear accelerator operates as a loaded transformer, and its efficiency may reach tens of percent. These properties of linear accelerators attract widespread attention since it is believed to be a rather promising source of powerful electron beams which are desired for industrial production processes.

Betatrons operate practically at no load, and efficiency of betatrons in industrial use is much less than 1 percent. The cardinal advantage of betatrons over linear induction electron accelerators is their capability of obtaining high energy in a comparatively small-sized, portable apparatus. That is why betatrons are used where high energy is needed but electron beam is a minor importance.

### 7.1.3 Nonlinear Induction Electron Accelerators

In induction electron accelerator with helical trajectory of electrons (Fig. 7.1*b*) each helix turn may possess an individual track with a time-constant guiding magnetic field varying from turn to turn as the energy level of electrons increases. Such accelerators may preserve many advantages that are characteristic of linear accelerators (heavy current, ease of injection and extraction of electrons) and at the same time ensure higher energy capacity of one magnetic circuit. Energy imparted to electrons is directly proportional to the number of revolutions around the magnetic circuit (flux change being linear in time). However, the maximum length of pulse of accelerated particles drops in the same proportions as with linear induction electron accelerators.

In the case of a flat spiral (Fig. 7.1*c*) the guiding magnetic field is also permanent and current loops are adjacent to one another. With this type of accelerator, a vacuum chamber with a large radial aperture can be used [94, 96]. Such a version of the circuit ensures prolonged injection of particles and correspondingly higher average current as opposed to the betatron. Where there are a great number of loops, the time required for an electron to pass throughout the spiral path may be comparable to the rise time of magnetic flux in the central core. Thus the injection must be discontinued before the flux assumes its maximum value, so that all the particles under injection can complete their spiral traveling to the extreme radius. The circuit illustrated in Fig. 7.1*c* uses internal injector. Arrangement of spiral turns closely to one another in one chamber results in interaction of charges of turns and leads to limitation of maximum current involved in acceleration (similar to that discussed above).

In some induction electron accelerators, in particular, installations characterized by helical trajectory of electrons, there is the problem of shielding the region of electron passage to guard against effects of stray magnetic fields of accelerating flux that are close to electron beam (for example, see [96]).

## 7.2 CONSTRUCTION AND POWER SUPPLY CIRCUITS OF INDUCTION ELECTRON ACCELERATORS

### 7.2.1 Linear Induction Electron Accelerators

Linear induction electron accelerators have not yet found wide application in spite of their simple acceleration circuit and the ease of obtaining an average power of hundreds of kilowatts, efficiency of 70 to 80 percent, and accelerated particle energy of tens of megaelectronvolts [97]. At present, there are approximately 15 linear induction accelerators used in research. The construction diagram of a component unit of the first accelerator of this type (Astron), is given in Fig. 7.4 [98]. The unit consists of 20 cores (1). The accelerating field of each core provides for an increment of energy levels of electrons up to 12.5 keV. The cores are essentially toroidal magnetic circuits wound of a tape of alloy (50 percent Fe and 50 percent Ni), 25 μm thick. Each core carries a single-turn winding to which a coaxial cable is discharged across a pulse thyratron. Placed inside the cores adjacent to the unit axis is a

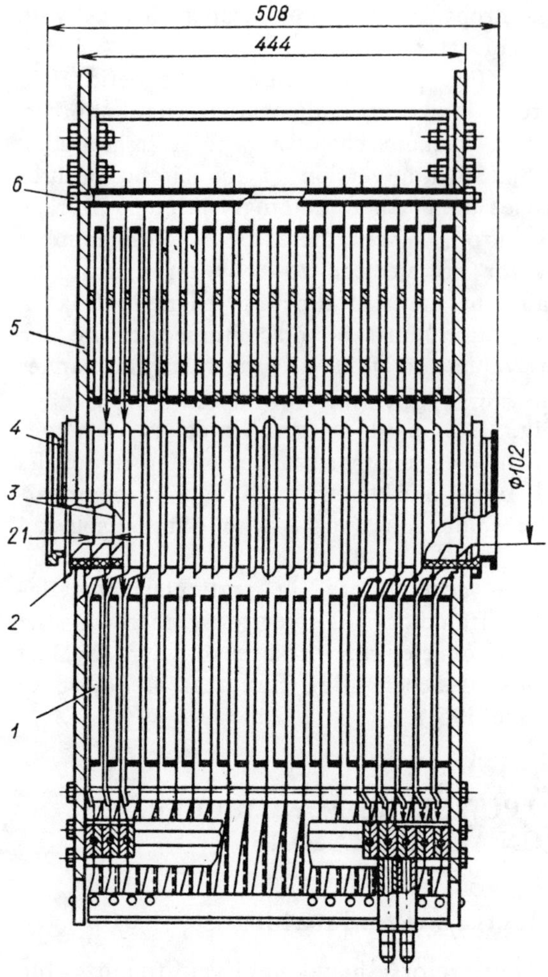


**Figure 7.4** Linear induction electron accelerator Astron. *1*—Core; *2* and *3*—insulator and electrode of accelerating tube; *4*—flange of accelerating tube; *5*—construction plate; *6*—coupling bolt.

length of a sectionalized accelerating tube. The electrodes of the tube (3) are spaced apart at a distance of 2.1 cm (the same spacing used for the cores). The tube aperture is 10.2 cm in diameter and overall length of the unit is about 0.5 m. The accelerator comprises 17 units, plus 2 units with a larger aperture (20.3 cm diameter), located at the forward end (close to the injector). The beam current at the output of the injector units is 1200 A, with an energy rating of 550 keV. The installation employs a hot cathode 25 cm in diameter. Between every two units there is a space for accommodating focusing magnetic lenses, vacuum pumps and diagnostic apparatus. The total length of the accelerator is 30 m. In its initial version in the 1960s, Astron produced an electron beam of 4.2 MeV, 800 A, with a pulse duration of 0.3 μs. The frequency repetition rate amounted to 60 Hz. The subsequent modification of the installation allowed the energy of accelerated particle to be increased to 6 MeV.

The principle difficulties in designing linear induction electron accelerators are presented by the complex technology of fabrication of large-diameter ferromagnetic cores, development of high-power quick-action keys with high repetition rate, and the problem of transport of intensive electron beams. For accelerating heavy currents it is necessary to strive for increasing intensity $E$ of the accelerating electric field.

$$E = R\,\frac{\Delta B}{\Delta t}$$

where $R$ is the outer radius of the cores. Therefore, at the preset induction change $\Delta B$, the value of $E$ can be increased at the expense of increase of $R$ or decrease of induction change time $\Delta t$. Increase of $R$ will bring about a prompt increase of the accelerator mass (as $\sim R^2$), whereas decrease of $\Delta t$ necessitates employment of a more expensive ferromagnet and leads to more intricate construction of keys and to shortening of acceleration cycles. The parameters selected for the foregoing installations with ferromagnetic cores approximate the optimum for the modern standard of technology.

A new successful trend in further modification of linear induction electron accelerators may be exemplified by a version suggested in 1968 (independently and simultaneously in [99] and [100]), in which use was made of radial lines (Fig. 7.5*a* and *b*) instead of ferromagnetic cores. A single section in such accelerators represents torroidal casing (1) open along an inner perimeter (2). Located in the median plane of the casing is a circular electrode (3) to which a voltage is impulsively applied relative to the casing, after which a multichannel discharger (4) is shorted to one of the ends of the casing (1). An electric field arises along the axis of the section. The life of this field depends on the time required for the electromagnetic wave to cover the distance from discharger to open end of the radial line (5), i.e., equal to $\sim 2\,R/v$ where $R$ is the outer radius of the electrode and $v$ is the velocity of the electromagnetic wave in the line. The electromagnetic wave reflects from the open end of the radial line and moves back. The electric field on the axis in this case reverses in sign and its lifetime at the inversed polarity is twice the lifetime of the initial pulse. In such installations it is found practicable not to use a magnetic circuit made from

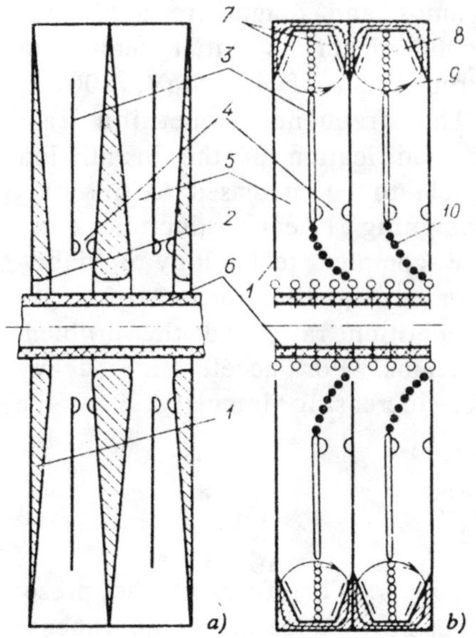


**Figure 7.5** Two versions of linear electron accelerators with radial lines. *a)* Circuit with charging from external high-voltage source (Marx generator). *b)* Circuit with generation of high voltage by transformer placed inside line. *1*—Toroidal casing; *2*—internal perimeter; *3*—circular electrode; *4*—multichannel discharger; *5*—open end of radial line; *6*—vacuum accelerating tube; *7* and *8*—primary and secondary windings of charging transformer; *9*—magnetic flux of transformer; *10*—rings for potential distribution.

expensive and scarce materials and to obtain rather easily a small value of $\Delta t$. At the same time, accelerators with radial lines employ a more complicated keys (spark gaps, 4). The said installations may function normally only in each case across six to eight channels at a synchronization accuracy of some 1 ns (with section diameter equal to ~1 m). Breakdowns of the spark gaps in the adjacent sections are shifted in time with allowance for electron transit time.

In the circuit shown in Fig. 7.5*a* the lines are charged by means of the Marx generator (supply feeders are omitted in the illustration), and deionized water is used as the dielectric [99]. In the version pictured in Fig. 7.5*b* [100], each section incorporates a shock-excited step-up pulse transformer for charging the line (see Section 6.4). The primary winding (7) of the transformer has a conical shape. The secondary winding (8) is flat, enabling uniform distribution of voltage at the external edge of the electrode (3). Figure 7.5*b* also shows a variant of voltage distribution near the inner end of the electrode with the aid of many rings (10) interconnected through an inductive (or combined) divider.

Using the LIU-10 accelerator at a current of 50 kA, an energy level of 13.5 MeV has been achieved, with a pulse duration ranging from 20 to 40 ns [101]. The charging voltage in each line is 0.5 MeV. The length of the accelerator is 7 m, its width is 4 m, and its height is 2.2 m. About 400 synchronized gas-filled trigatron-type spark gaps have been used as keys. The electron beam has a tubular structure with a maximum diameter 180 mm, and thickness of 5 to 20 mm. In the initial region (i.e. the injector), the electron energy reaches 2.5 MeV. The beam is focused by a longitudinal magnetic field of 0.5 T produced by solenoids arranged near the internal surface of a polyethylene vacuum tube (6).

At the present time, linear induction electron accelerators are fairly complex physical devices and are difficult to use for industrial purposes. Nevertheless progress in development of their components, particularly keys, is expected to make these accelerators applicable for technological purposes. Thus, for example, some information on development of a serial module of industrial linear accelerator has been obtained [102]. The increment of the module energy is 0.25 MeV. It is expected that sequential introduction of a variety of the modules depending on the required parameters of electron beams will occur. Some data are presented in [103] concerning the development of modules of linear induction electron accelerators rated at an energy of 0.5 MeV. Consideration is being given to possibility of production of industrial linear accelerators on the basis of these modules, for operation at energies of 2.5 and 10 MeV.

### 7.2.2 Betatrons

The main components of betatrons are shown in Fig. 7.3*a* and *b*. The magnetic circuit (1) is composed of transformer iron plates. The magnetic flux closes two symmetrical inverse yokes. In some betatrons, the position of the equilibrium orbit is adjusted by varying the size of the air gap in the magnetic circuit central part. The winding (2) is supplied with power from industrial power supply mains of 50 to 60 Hz or from a power supply source generating a voltage of higher frequency. To improve cos $\varphi$, a capacitor unit is connected in parallel with the winding. Generally, $U_{loop}$ amounts to a few tens of volts. Thus, for gaining an energy of about several tens of megaelectronvolts the electrons have to complete some 1 million revolutions.

The required configuration of the magnetic field for focusing of electrons, i.e., $B \sim R^{-n}$, is provided by the shape of the magnetic poles. The magnetic field between the poles infinitely stretching in radial directions varies in inverse proportion to the distance between poles: $B \sim h^{-1}$, (Fig. 7.6*a*). Consequently, for obtaining the prescribed value of $n$, it is essential to ensure $h \sim R^{n}$. However, because of the final radial dimensions of the magnetic poles, the edge effects play an important role in shaping the field. In order for a quick variation of the magnetic field near edges 2 and 3 to be somewhat retarded, the gap between the poles is made somewhat smaller

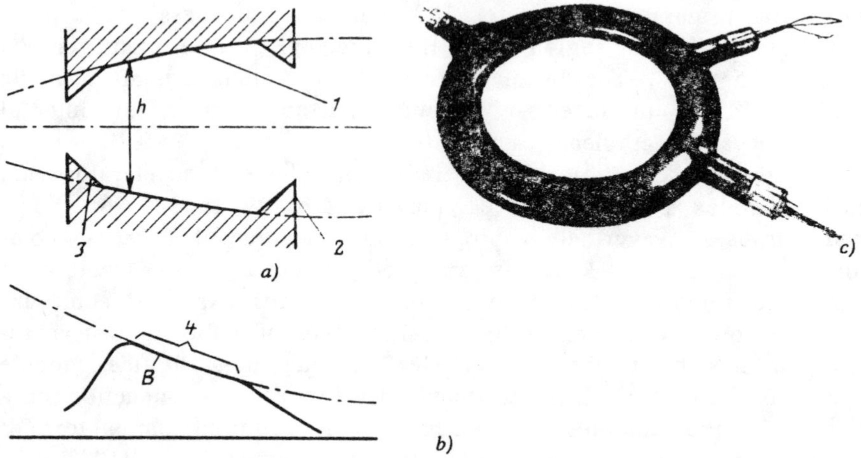


Figure 7.6 Components of betatron. *a)* Shape of poles shaping guiding magnetic field. *b)* Distribution of guiding field *B* over radius (qualitatively). *c)* Vacuum chamber. *1*—Magnet poles; 2 and *3*—edges of poles; *4*—working section of guiding field. Dotted line shows infinite poles.

than that specified by the requirement $h \sim R^n$, (Fig. 7.6*b*). The effective section (4) of the guiding field where $n$ lies within the predetermined limits is always less than the radial size of the poles.

An exception occurs for experimental installations in which the windings producing magnetic field are arranged directly adjacent to the interpole gap both inside and outside and are similar with the gap in height. In this case, the required value of $n$ [104] can be maintained throughout the region bounded by the poles and windings.

The vacuum chamber in which electron acceleration is performed is fabricated of glass, ceramic, or some other insulating materials and is lined with a thin layer of conductive material for leakage of random charges. The chamber is provided with lead-ins for feeding the injector and deflector which deflects the particles at the end of the acceleration cycle (Fig. 7.6*c*).

A magnet use in another version of the betatron developed by the Nuclear Physics Institute in Tomsk, is illustrated in Fig. 7.7 [105].

The gap between the betatron poles periodically varies in azimuth, as a result of which azimuth components of the guiding magnetic field arise along with additional focusing forces. This makes it practicable to increase the quantity of particles involved in acceleration process, and correspondingly to enhance the intensity of bremsstrahlung generated by the betatron. The magnet poles accomodate six ridges (4) for changing the gap and magnetic field in azimuth at a pitch of 60° (Fig. 7.8*c*). The direction of the transformer steel plates of which the ridges are assembled is shown in the illustration. The betatron magnetic circuit (1) incorporates six closing yokes (2). The average radius of the equilibrium orbit is 6 cm. The index of decay of an average magnetic field on the orbit $n$ = 0.6. The interpole gap on the equilibrium orbit radius is about 4 cm. The radius of location of the injector unit is 7.6 to 7.8 cm. The injection energy is 25 keV. The energy of accelerated particles amounts to 6 MeV. Installations of this type are in serial production.

The most popular energy of betatrons used in applied purposes is $W$ 5 to 35 MeV, the average current of accelerated electrons is $10^{-2}$ to $10^{-1}$ μA, and the x-ray radiation intensity is $10^{-3}$ to 1 C/kg min. The cost of betatrons increases rapidly with an increase in the

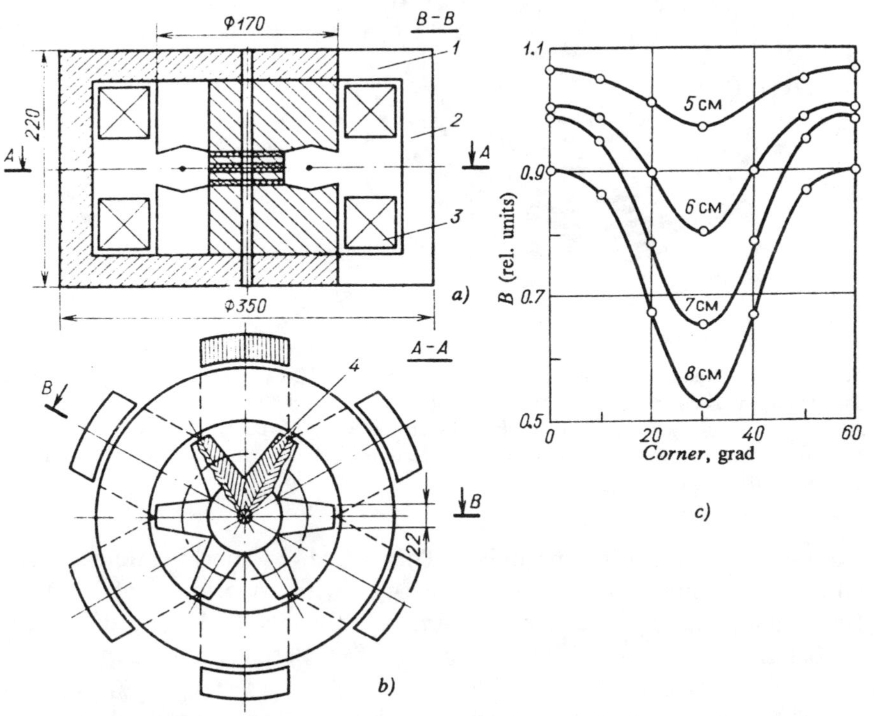


**Figure 7.7** Magnet of betatron with azimuth variation of field. *a)* Section through *BB*. *b)* Section through *AA*. *c)* Azimuth distribution of guiding magnetic field *B* in median plane over one variation cycle. *1*—Magnetic circuit; *2*—closing yoke of magnetic circuit; *3*—winding; *4*—comb; (I, II, III, IV)—distribution of field *B*, respectively, for radii measuring 5, 6, 7, and 8 cm.

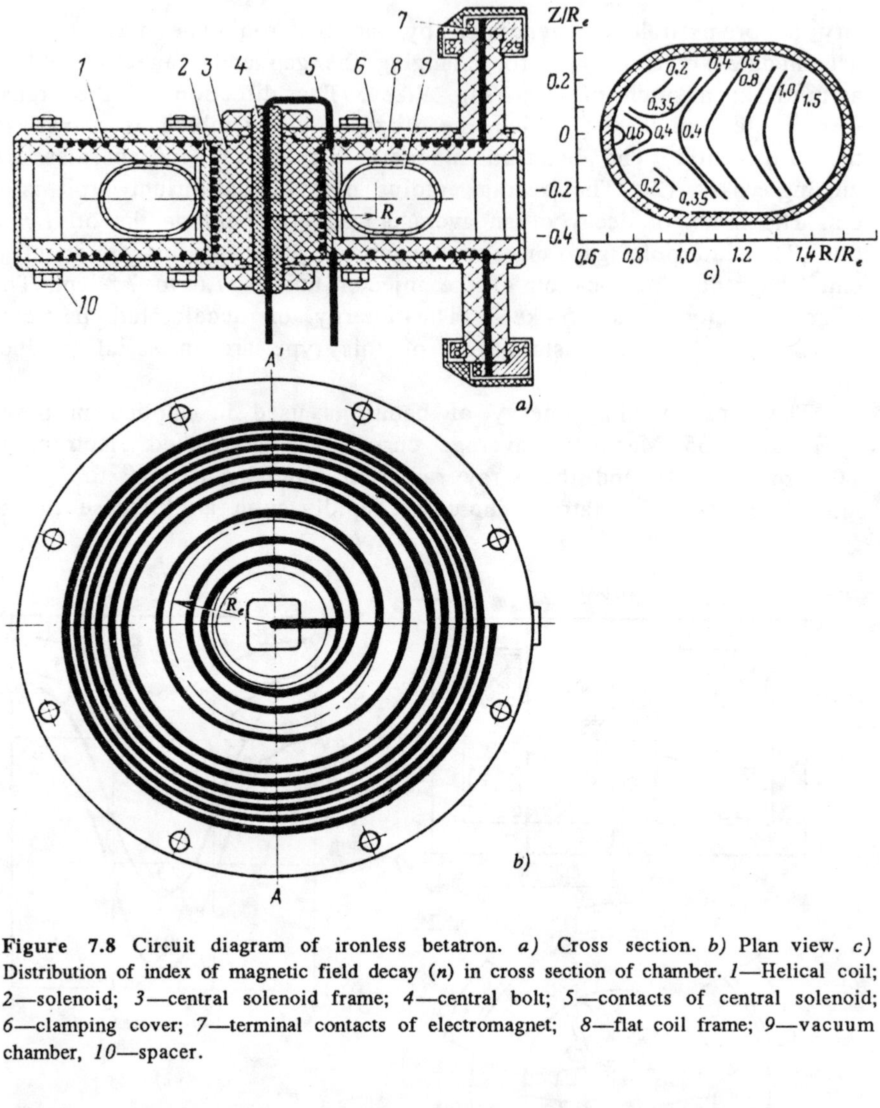


**Figure 7.8** Circuit diagram of ironless betatron. *a)* Cross section. *b)* Plan view. *c)* Distribution of index of magnetic field decay (*n*) in cross section of chamber. *1*—Helical coil; *2*—solenoid; *3*—central solenoid frame; *4*—central bolt; *5*—contacts of central solenoid; *6*—clamping cover; *7*—terminal contacts of electromagnet; *8*—flat coil frame; *9*—vacuum chamber, *10*—spacer.

energy of accelerating particles. This trend in modification of construction in increasing particle energy (with amplitude of magnetic field and relative aperture remaining unchanged) leads to increase of betatron mass as $W^{-3}$. The mass of a serial production betatron rated at 25 MeV amounts to 4 to 5t.

The other reason for restricting the limiting energy of electrons in betatron is losses of energy for synchrotron radiation. After each revolution, the particle loses $\Delta\varepsilon = 10^{-5} W^4/R_e$ (see Section 2.9). For instance, at $W$ = 50 MeV and $R_e$ = 25 cm, the value of $\Delta\varepsilon$ = 2.5 eV, which may be comparable with the increase of energy within one revolution. (It will be recalled that due to the sinusoidal change of field at the end of the acceleration, $U_{loop}$ tends to drop.)

Upset of the 2:1 condition due to synchrotron radiation losses results in decrease of the beam orbit radius and in ceasing further acceleration of electrons. Correction of rate of rise of $B$ or $B$ required for satisfiying the betatron condition seems to be expedient only when losses $\Delta\mathcal{E}$ are materially less than $eU_{loop}$. Noncomformance with the betatron 2:1 condition is substantially less in installations with short-time acceleration and higher $U_{loop}$.

Betatrons with ironless magnets have been developed for decreasing their size and reducing acceleration time. Maximum magnetic field $B_m$ in such betatrons is several times the field existing in accelerators with ferromagnetic circuit. The volume of the accelerator decreases as $B^{-3}{}_m$. The construction of an ironless betatron is shown in Figure 7.8*a* and *b* [106]. The position of the equilibrium orbit is preassigned by the number of turns and magnitudes of current in the coils (1) and solenoid (2). Arrangement of the turns dictates the configuration of the magnetic field. Figure 7.8*c* shows the distribution of index $n$ characterizing the decay of the magnetic field in cross section of chamber 9. Azimuth heterogeneity of the magnetic field does not exceed 0.5 percent and mainly results from effects of the power supply busbars. Mechanical strength is provided by parts 3, 4, 6, 8, and 10, most of which are made of textolite. Electromagnets of this type were used in betatrons with $R_e$ = 3.9, 7.8, 11.7, and 23.4 cm. The electromagnets were energized from gang capacitors. The minimum accelerating time was about $10^{-5}$ s. The maximum intensity of the magnetic field depended to a large extent on mechanical strength and was equal to 2 to 4 T on an equilibrium orbit and 10 to 20 T in the central region. (Recall that forces produced by magnetic field of 20 T corresponds to pressure of 160 MPa.)

At short-time modes of acceleration, the ohmic losses in windings (despite high current magnitudes) amounted to a value that was less than in betatron with iron magnet rated at the similar energy. An accelerator with $R_e$ = 3.9 cm (acceleration time 13us) was used to operate at a frequency of 25 Hz. For increasing the intensity, an external injector with a voltage of 0.5 MV was used (duration of injection pulse being $10^{-7}$ to $10^{-8}$ s). Electrons entered the orbit through the channel shielded at the injection moment against adverse effects of external magnetic fields. The maximum energy obtained in betatrons of this type was 100 MeV, with a circulating current of ~100 A.

Even though the described construction has not found wide application, it has been used as a basis for developing portable accelerators that have been successfully employed for various applied purposes.*

## 7.3 RESONANCE ACCELERATION OF ELECTRONS

The principle of resonance acceleration of charged particles underlies operation of every contemporary high-energy electron accelerator. In installations designed for nuclear research, this principle has made it possible to achieve an energy of the order of $10^6$ MeV. As far as industrial electron accelerators are concerned, the resonance acceleration method is applied starting with the energy of 4 to 5 MeV, i.e., when dimensions and cost of direct-action accelerators tend to grow substantially.

In resonance acceleration method a particle repeatedly crosses the regions of variable electric field, each time getting in it in the same voltage phase $\varphi_e$ at which a certain increment of particle energy occurs, viz.: $\Delta\varepsilon = eU_m \cos\varphi_e$

Let us consider the principle of operation of resonance accelerators on one of early versions proposed in the 1920s (Fig. 7.9*a*). A number of drift tubes (2) of variable length is connected to a source of sinusoidal voltage *u*. Negative-charged particles (3) can be injected during a half-cycle of voltage *u* when polarity corresponds to that shown in Fig. 7.9*a* (minus on negative-charged particle source 1 and plus on the first drift tube). Particles whose transit time through the first drift tube is $T/2$ (i.e., half-cycle of voltage *u*) will be affected by change of voltage polarity when they reach the subsequent gap (between the first and second tubes). As a result, these particles will be accelerated. For particle acceleration in the next gaps, the length of tubes (2) must be gradually increased because of increased velocity (and also due to constancy of supply voltage cycle *T*). The main condition for arrival of charged particle at each gap in the same phase of accelerating voltage $\varphi_e$ (Fig. 7.9*b*) is constancy of time of transit through every drift tube: $l_i/\beta_i c = T/2$,** where $l_i$ is the length of the *i*th tube, and $\beta_i$ is the relative velocity of a charged particle inside the tube. Velocity $\beta_i$ can be derived from the equation

$$\beta_i = \sqrt{1 - \left(\frac{\varepsilon_0}{\varepsilon_0 + i\Delta\varepsilon}\right)^2}$$

*See new material added in proof on page 191.

**Since the time of transit through the gap comes to a few degrees of cycle of voltage *u*, the time of arrival of electron in phase $\varphi_e$ should be related to passage across a definite plane, for example, the center of the gap.

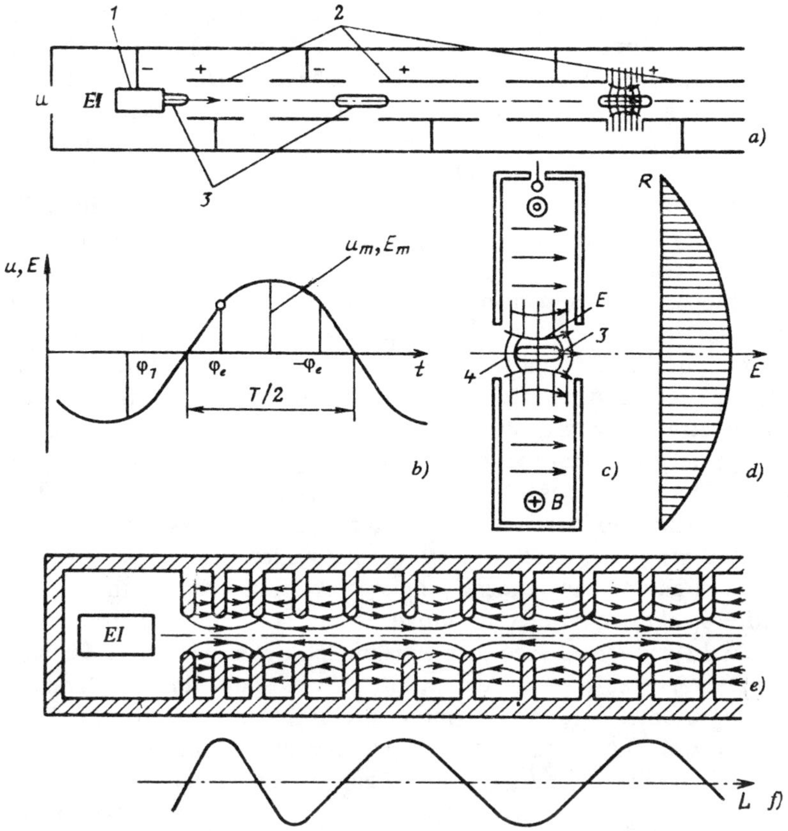


**Figure 7.9** Resonance acceleration circuit. *a)* Accelerator with drift tubes. *b)* Temporal variation of difference of potentials and voltage gradient of accelerating electric field. *c)* Cylindrical resonator (direction of electric and magnetic fields in excited resonator is shown). *d)* Distribution of electric field intensity. *e)* Wave-guide linear accelerator. *f)* Instantaneous distribution of voltage gradient of electric field over wave-guide axis. *1*—source of negatively charged particles; *2*—drift tubes; *3*—accelerating particles; *4*–equipotentials of accelerating electric field; *EI*—electron injector; *u*—accelerating voltage; *E*—gradient of voltage of accelerating electric field; *B*—direction of magnetic field in excited resonator; $\varphi_e$—equilibrium phase of accelerating electric field; *L*—distance coverd by electrons.

where $i$ is the number of the passed accelerating gaps (see Section 1.3). The said condition can be complied with only for a particular group of particles permanently held in phase $\varphi_e$. These particles are referred to as equilibrium particles, and phase $\varphi_e$ is called an equilibrium phase. The absolute value of the final particle energy is many times the value of $u_m$ applied between the drift tubes.

Large-scale development of electron accelerators based on the resonance acceleration method was stimulated by the discovery made by Veksler in 1944. This is the so-called autophasing phenomenon in which a stable state of voltage is possible not only for individual equilibrium particles but also for particles getting in accelerating fields within a wide range of phases—from $-\varphi_e$ to $\varphi_1$, as illustrated in Fig. 7.9*b* (for a definition of $\varphi_1$, see below). It has been noted that particles with phase $\varphi_1 < \varphi < -\varphi_e$ are also capable of undergoing an acceleration effect for an extended time with performing stable fluctuations and gaining, on average, approximately the same energy (during multiple passes of accelerating field) as equilibrium particles.

The particles that happen to be to the right of the equilibrium phase (between $\varphi_e$ and $-\varphi_e$) will assume energy and velocity exceeding those of equilibrium particles. As a result, they arrive at the region of the accelerating field in a time interval less than $T/2$, and, consequently, they are shifted in phase to the left. This occurs until they appear to be to the left of $\varphi_e$. Then the increase in their energy level becomes less than that of equilibrium particles, and their phase shifting to the left starts decelerating. However, due to the fact that during the time of stay in the region $\varphi_e < \varphi < -\varphi_e$, the particles have gained an excess energy as compared with equilibrium particles, they continue to move to the left until their energy becomes equal to that of the equilibrium particles. At that instant, the time of their transit between two ingresses into the accelerating gap (into its middle part, to be more precise) will amount to $T/2$. The particles are then found to be somewhat to the left of $\varphi_e$; therefore, their gain in energy in the gap is less than that of the equilibrium particles (they even lose some energy when they get in the retarding field). The energy and velocity of the particles then become less than those of the equilibrium particles, and, consequently, their transit time exceeds $T/2$. As a result, the particles start moving in phase to the right. The point at which phase motion velocity inverses in sense is denoted as $\varphi_1$. In this way, the particles maintain stable phase fluctuation relative to $\varphi_e$ still being included in the cluster of accelerating particles. The length of clusters of accelerating particles (3) are shown in Fig. 7.9*a* fictitiously. Actually, the clusters at the initial stage of acceleration are much longer, and their length corresponds to the duration exceeding a half-cycle of electric field variation from $\varphi_1$ to $-\varphi_e$ (Fig. 7.9*b*).

For finding an equation of phase fluctuation of electrons use is made (as initial data) of expressions for values of energy variation along electron motion direction ($x$ coordinate) for equilibrium and non-equilibrium particles when they cross the accelerating electric field:

$$\frac{d\varepsilon_e}{dx} = eU_m \cos \varphi_e \quad \text{and} \quad \frac{d\varepsilon}{dx} = eU_m \cos \varphi$$

where $\varepsilon_e$ is the energy of an equilibrium particle. Departure of the particle energy from equilibrium $\Delta\varepsilon = \varepsilon - \varepsilon_e$ obeys the equation

$$\frac{d\,\Delta\varepsilon}{dx} = eU_m\,(\cos\varphi - \cos\varphi_e)$$

After a number of transformations (see, for example, [107]), the equation expressing electron phase fluctuation will take another qualitative form:

$$\ddot{\varphi} = A\,(\cos\varphi - \cos\varphi_e)$$

where $A$ is a slowly varying function. It will conform to oscillations of a pendulum to which a constant (slowly varying, to be more correct) torque is imparted. Analysis of the equation suggests that a phase fluctuation cycle exceeds fluctuation cycle of an accelerating electric field by tens of times. The process of acceleration is accompanied with attenuation of fluctuation and decrease of phase spectrum of accelerating particle cluster. The value of $\varphi_1$ can be found from the same equation.

In the modern resonance electron accelerators, use is made of a high-frequency electric field, as a rule, within a range of $10^2$ to $10^4$ MHz (wavelength 3 m to 3 cm). Electromagnetic oscillations are excited in cavities of different geometry made of metal or lined with any conductive material. One of the most widespread versions of an accelerating system is a cylindrical cavity resonator shown in Fig. 7.9*c*. Magnetic lines of electric field $E$ are directed approximately parallel to the resonator axis. Magnetic field $B$ is axially symmetric. Radial distribution of $E$ is illustrated in Fig. 7.9*d*. Shown in the top part of the illustration is a coupling loop for feeding the power to the resonator. Accelerating particles (3) travel along the resonator axis. As the full energy is gained by particles in one resonator, they are injected when the electric field intensity approximates $E_m$. Although in principle particle acceleration is practicable all over the voltage half-cycle of the particular polarity, the pulse duration is selected in compliance with allowable spread of energy of accelerating particles. For increasing particle energies, it is necessary to install in series several resonators whose high-frequency oscillations are phase-shifted with regard to transit time of electrons. Use of considerable number of resonators gives rise to particle oscillations with respect to equilibrium phase, as described above.

### 7.3.1 Waveguide Linear Accelerator

In most industrial accelerators, a system of many separate resonators is substituted by a cylindrical waveguide to propagate a traveling wave with a longitudinal component of electric field (Fig. 7.9*e*). Electromagnetic waves and accelerating particles move in one and the same direction and the particles experience the action of accelerating field. The waveguide linear electron accelerators, as a rule, are used for accelerating electrons to the energy of 1 to 2 MeV and above. To ensure synchronous motion of particles and electromagnetic waves, the wave guide geometry is selected with regard to rise of electron velocity. Special diaphragms are introduced for decelerating wave propagation to the value equal to velocity of electrons. The electron injector EI generates electrons with an energy of 30 to 100 keV. The spacing of the diaphragms varies at the initial section. Starting from an energy of 2 to 3 MeV, the electron velocity practically remains unchanged (see Fig. 1.3), and the wave-guide structure provides the wave propagation velocity similar to velocity of light. At the initial stage of the acceleration process, there exists an equilibrium phase of the accelerating voltage, and the particles in this phase move steadily at the wave velocity:

$$\beta = \sqrt{1 - \left(\frac{\varepsilon_0}{\varepsilon_0 + eE_m L \cos \varphi_e}\right)^2}$$

where $L$ is the distance covered by electrons. For a maximum number of injected particles to be involved in the acceleration process and bunched near the equilibrium phase for reducing spread of electron beam energy, particles undergo prebunching in the starting part of the wave guide. One of the methods of prebunching is modulation of electrons for velocities in an alternating electric field and subsequent decrease of length of electron cluster (bunch) in electron transit space wherein fast-moving particles overtake particles traveling at lesser velocities. Use is also made of a buncher with variable parameters in which phase velocity and accelerating field amplitude can be smoothly varied over the length of the wave guide. Such a buncher provides practically full capture of particles (up to 90 to 95 percent of injected particles) and the length of the cluster after passing through the buncher is 10° to 30° in phase, the energy spread of particles being 3 to 7 percent (see, for example, [107]).

Apart from traveling-wave acceleration, an ever-increasing application of the method of standing-wave operation has been noted in the last few years. In the first case, high frequency energy is applied to one end of the wave guide, while the other end is connected to a load for saturation of residual power. In the second case, the wave guide is shorted from both sides and energy is injected into its central parts. A wave guide excited with standing wave may be considered as a set of coupled resonators containing traveling waves of different phase velocities. In this case, particles are accelerated in field of fundamental harmonic whose velocity coincides with that of particles.

In traveling-wave accelerators, sometimes use is made of the method of recuperation of high-frequency energy, i.e., residual energy at the wave guide output is applied again to the input, thereby raising efficiency of the accelerating installation.

Focusing of particles in transverse direction in linear electron accelerators is impeded by defocusing effect of accelerating electric field. For example, in an accelerator with drift tubes, the gap is essentially an electrostatic lens (see Figs. 7.9*a* and 7.10), and forces caused by transverse components of the electric field focus electrons in the first half of the lens, but defocus them in the second half. Equilibrium phase is known to be on the ascending branch of the voltage curve, therefore, particles, on the average, pass the defocusing part of the lens with a more intense field, than when they pass the focusing part. The resultant action of the electric field here is defocusing. It is easy to show that the electric field in the wave guide also defocuses charged particles. In low-energy wave guide accelerators (i.e., rated at a few megaelectron-volts) focusing is usually ensured by the use of longitudinal solenoidal field. In the case of high energies of electrons, the defocusing action of the accelerating field is weak, and therefore, focusing is effected by means of magnetic quadrupole lenses spaced apart at rather substantial distance. A cross-sectional view is shown in Fig. 7.10*b*, where electrons move at a right angle to the drawing. Each quadrupole lens focuses particles in one transverse direction and defocuses in the other. Joint action of a pair of lenses whose magnetic field directions are shifted crosswise through 90° ensures focusing of particles in both directions. Focusing solenoid and lenses are generally arranged on the outside of the wave guide.

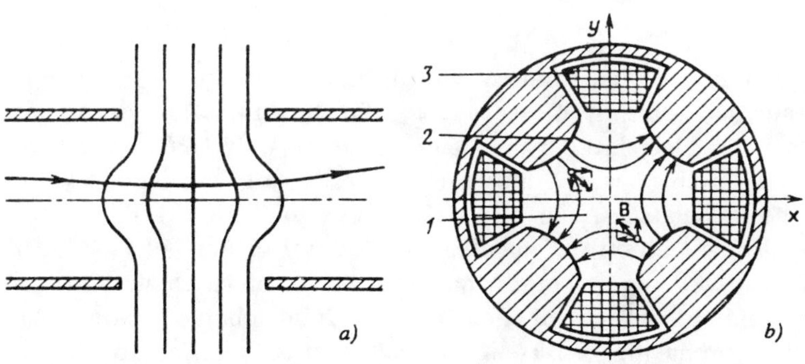


**Figure 7.10** Focusing of electrons in linear electron accelerator. *a)* Diagram of electron trajectory in accelerating gap between drift tubes. *b)* Cross section of quadrupolar magnetic lens. *1*—Region of motion of electrons; *2*—magnetic pole; *3*—winding.

Electron beam intensity of linear electron accelerators is limited not only by operating capabilities of focusing systems but also, even to greater degree, by effects of interaction of electrons with high-frequency fields excited in a resonance system both by external generator and by electron beam itself. Review of this interaction can be found in a number of the works for example, in [107]. Maximum current of electrons in industrial wave-guide accelerators is as a rule, of the order of 1A.

### 7.3.2 Microtrons

Another variety of resonance electron accelerators used for applied purposes can be exemplified by microtrons. Acceleration of electrons in microtrons is performed in high-frequency fields of a cylindrical resonator (Fig. 7.9*b*) located between the poles of a permanent magnet. Electrons move in circular orbits with a common tangency point inside the resonator (Fig. 7.11*a*). To make a particle pass through the resonator each time in equilibrium phase $\varphi_e$, two conditions should be satisfied:

1. The increment of electron energy after each pass through the resonator should be whole number times as less as energy of electron entering the first circular orbit: $\Delta\varepsilon = \varepsilon_{ent}/\alpha$ where $\varepsilon_{ent}$ stands for full energy of charged particle, (i.e., rest energy plus energy gained at injection moment), and $\alpha$ is a whole number (integer).
2. The amount of change of orbit time of electron after each pass through resonator should be equal to cycle of accelerating high-frequency (HF) field:

$$\Delta T = \frac{2\pi\Delta\varepsilon}{ecB} = T_{HF}$$

Thus, the orbiting period of the equilibrium electron ($T$) is a multiple of the accelerating field cycle ($T_{HF}$), and after each pass through resonator increases by a unit $T_{HF}$ (Fig. 7.11*b*). The foregoing formula enables the value of intensity of guiding magnetic field $B$ to be calculated. As a rule, it amounts to 0.1 (0.2) $T$.

Unlike linear electron accelerators, equilibrium phase $\varphi_e$ in microtrons (as well as in other magnetic resonance accelerators) is on the descending branch of accelerating field characteristics. The principle of autophasing works in a way that is similar to that of linear accelerators, as described above. Electrons that appear to be left of $\varphi_e$ will obtain higher energy than equilibrium particles. The cycle will be, respectively, larger than that of equilibrium elec-

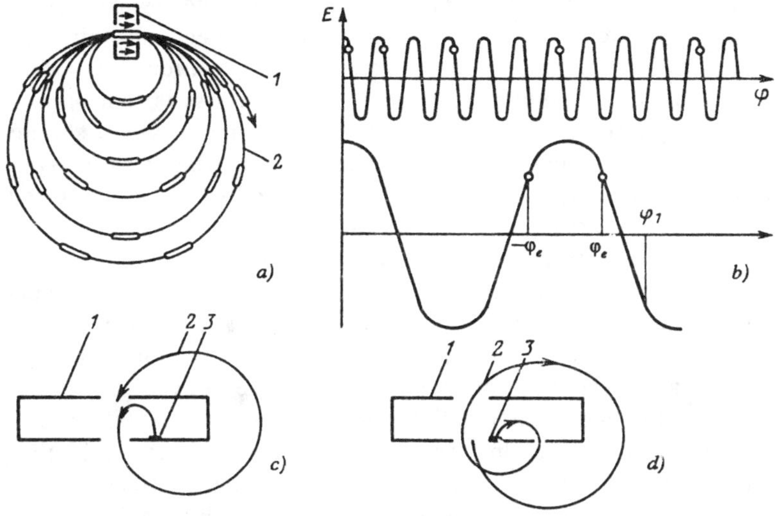


**Figure 7.11** Acceleration of electrons in microtrons. *a)* Diagram of acceleration of particle (guiding magnetic field *B* is directed perpendicular to drawing). *b)* Moments of passes of accelerating high-frequency field by equilibrium particle. *c)* and *d)* Diagrams of injection *1*—Resonator; 2—trajectory of electrons; *3*—cathode.

trons, and they will come in the resonator later,* i.e., they move in phase to the right. After getting in the equilibrium phase with an excess energy, as compared with equilibrium electrons, non-equilibrium particles continue traveling to the right until their energy is equalized (phase $\varphi_1$). Due to the subsequent decrease of energy, as compared to equilibrium electrons, the orbiting cycle of the particles will start to diminish and will move in phase to the left. As a result, nonequilibrium particles will display stable oscillations relative to $\varphi_e$ accumulating, on average, an energy similar to that of equilibrium particles. The amplitude of oscillations will permanently decrease in the process of acceleration. At the moment of transit over equilibrium phase, the energy of phase-oscillation particles will be most different from that of equilibrium particles. Correspondingly, their orbits at this moment will be of maximum or minimum radius (maximum in phase motion of particles from left to right; minimum in reverse motion). Changes of the radius caused by the oscillations in question shall be taken into account in selection of size of path of guiding magnetic field in resonance accelerator

*This is in contrast to linear accelerators, where particles with energy exceeding that of equilibrium particles reach the gap with accelerating field earlier than equilibrium ones. This is the cause of the difference in positioning of equilibrium phases, i.e., either on the ascending or the descending branch of the accelerating field voltage curve.

with a relatively small aperture. (For example, in the synchrotron). Oscillations of particles in microtrons with respect to circular orbits in radial direction are of less value. The equation for radial phase oscillations in magnetic resonance accelerators is similar to the equation of phase oscillations for linear accelerators. It should be noted that, like linear electron accelerators, the autophasing phenomenon enables most of the injected particles to be captured and reliably held in acceleration conditions.

Although the microtron was one of the early accelerators proposed after the discovery of the autophasing phenomenon, considerable difficulties were presented by development of an injector that provides substantial current and, at the same time, is not a barrier to motion of accelerating particles. Progressive development of microtrons actually began after discovery and experimental verification (in the early 1960s in the Soviet Union) of the injection method, which enabled pulse current of accelerating particles to be increased by two orders, i.e., to 20 to 50 mA [108]. Figure 7.11*c* and *d*, show this injection method. In both versions, the internal wall of the resonator (1) mounts a hot cathode (3) (usually $LaB_6$ at a temperature of 1500° to 1600°C). Under the action of variable electric and magnetic fields of the resonator and permanent magnetic field of a microtron, electrons are accelerated and move along specified trajectories. The positions of cathodes are determined by calculating optimum trajectories of electrons.

**Particle focusing.** Focusing of particles in microtrons is effected by means of the electric field of the resonator. Under the action of the high-frequency field, electrons are focused at the resonator input and defocused at its output (see equipotentials in Fig. 7.9*c*). Since the equilibrium phase is on the descending branch of the electric field, particles pass over the first part of the resonator under the effect of a more intensive field as compared to their transit across the second part of the resonator. As a result, the high-frequency field focuses both equilibrium electrons and electrons liable to phase oscillation, since the latter electrons are subject to the focusing effect of the accelerating field in most passes through the resonator. The focusing in microtrons is not a difficult problem, because particles under acceleration quickly gain adequate energy, and therefore, the force of their own space charge is considerably compensated for by magnetic convergency (see Section 1.3). The guiding magnetic field in microtrons is constant in radius to an accuracy of 0.1 and 0.5 percent and is not used for focusing the particles.

**Extraction of accelerated electrons.** Extraction of accelerated electrons is performed through a channel in a ferromagnetic tube tangent to the last orbit. As a rule, the tube is located at the site of greatest divergence of the orbits, diametrically opposite to the resonator. The channel aperture and wall thickness are increased in the direction of electron beam motion. The interior of the tube (the channel where particles move) is practically void of any magnetic field. Magnetic field disturbance on the final orbits is minimized with the aid of correcting devices. In most microtrons the extraction efficiency is nearly 100 percent.

**Electron beam intensity.** The intensity of the electron beam in the microtron is restricted chiefly by the maximum tolerable pulse current. Starting from a certain current level, the load imparted to the resonator by the electron flux (by effective and random electrons) so much exceeds ohmic losses in the resonator that the Q factor of the latter drops, and the resonator loses its resonance properties. The basic reserve for rise of electron beam power in a microtron is the increase of pulse repetition rate of accelerating voltage (until achievement of continuous operating conditions are achieved). To this end use should be made of the cooling system for the resonator and associated components. An average power of electron beams in present-day microtrons amounts, as a rule, to tens and hundreds of watts (average current reaching 20 to 25 $\mu$A).

**Limiting energy of charged particles.** The limiting energy of charged particles in microtrons is also stipulated, to a large degree, by limiting current in resonator that is due to simultaneous passes of many clusters of electrons of various energies through the resonator. With increase of energy, more stringent requirements are imposed for homogeneity of guiding magnetic field $B$. Since the field is permanent, a slight misalignment of magnetic poles or nonuniformity of their magnetic resistance may cause orbit shifting. The tolerated heterogeneity is very much dependent on the number of orbits $N$: $\Delta B/B \sim N^{-2}$. The diameter of microtron poles is usually less than 1 m. The most common energy level ranges from 5 to 30 MeV. Pulse duration of high-frequency accelerating voltage, as a rule, does not exceed a few microseconds. The pulse repetition frequency is 50 to 500 Hz, and the period-to-duration ratio is 1000. The gain of energy after one revolution in orbit is generally 0.3 to 0.5 MeV. The number of revolutions amounts to a few tens.

## 7.4 ELECTRIC CIRCUITS AND CONSTRUCTION OF RESONANCE ELECTRON ACCELERATORS

Figure 7.12 illustrates Elektronika, a typical industrial electron accelerator rated at an energy of 3 to 7 MeV and an average power in

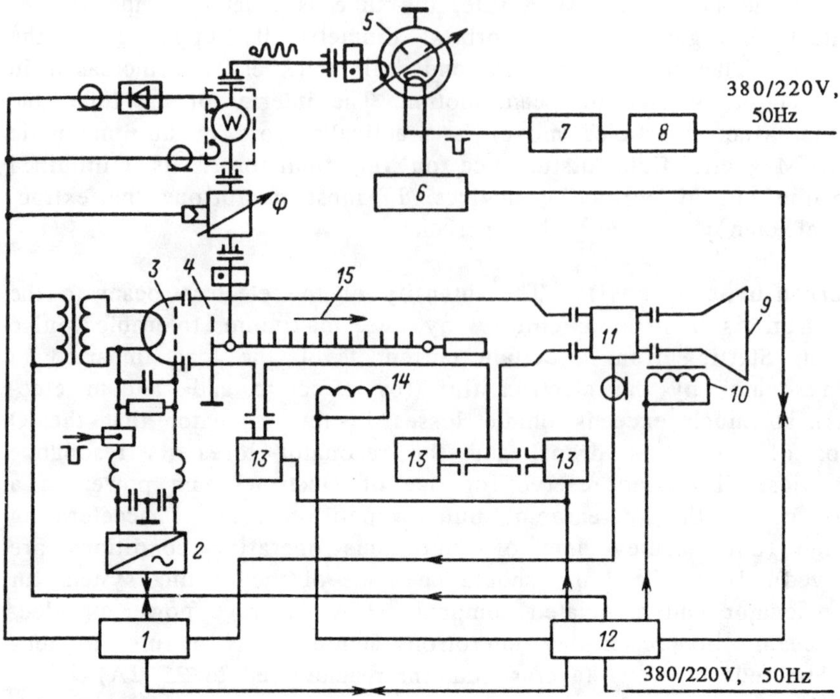


**Figure 7.12** Circuit diagram of industrial electron accelerator Elektronika. *1*—Control panel; *2*—stabilized rectifier; *3*—electron gun; *4*—wave guide; *5*—generator; *6*—generator unit; *7*—pulse modulator; *8*—high-voltage rectifier; *9*-extraction (outlet) window; *10*—sweep circuit; *11*—current pickup; *12*—power supply board; *13*—vacuum pumps; *14*—focusing solenoid; *15*—direction of travel of accelerating electrons.

the order of 5 kW. Acceleration is effected by means of traveling-wave wave guide (4). An electron gun (3) is a grid-controlled unit. Focusing is performed by the use of a longitudinal magnetic field produced by a solenoid (14). The accelerated electron beam is scanned by a magnetic field (system 10) and is extracted through an outlet foil window (9). The accelerator is available in two versions differing from each other in parameters of accelerating wave guides. Table 7.1 compares the basic specifications of the two versions of the Elektronika accelerators [109].

Power supply for both versions comes from three-phase AC mains of 380/220V, 50 Hz. Figure 7.13*a* and *b* illustrates circuit diagrams of wave guides with wavelength $\lambda$ = 10 cm of accelerating high frequency field and with wave (and particles) velocity equal to velocity of light $\beta \cong 1$. Referring to the illustration, the wave-guide diameter depends minimally on the accelerating channel aperture. Figure 7.13*c* shows the propagation velocity of a high-frequency

**Table 7.1 Comparison of two versions of Elektronika**

| Parameter | ELU-4 | ELU-6 |
|---|---|---|
| Rated energy of accelerated electrons, MeV | 4 | 6 |
| Range of energy control, MeV | 3–5 | 4–7 |
| Average rated power of electron beam, kW | 5 | 5 |
| Beam pulse current, A | Up to 1.5 | 1.5 |
| Rated pulse repetition frequency, Hz | 200 | 200 |
| Duration of current pulses, s | 4.5 | 4.5 |
| Length of sweep line, mm | 400 | 400 |
| Power consumption, kW | 75 | 75 |

wave as a function of the ratio of the aperture diameter to the wave-guide diameter $b$. Decrease of aperture diameter $a$ (for similar flux of high-frequency energy) may cause increase of electric field intensity $E$ as well as decrease of accelerator length. However, in this case, the high-frequency energy flow tends to decay faster, deteriorating the efficiency of the system. Decrease of $a$ will also intensify sensitivity to any errors in construction, as well to instability of the high-frequency and temperature fluctuations in the process of operation.

Wave-guides are made of oxygen-free copper with an inner surface finish ∇8. The Q factor of the wave-guide is generally equal to $10^4$.

In the wave guides shown in Fig. 7.13*a* and *b* (as in Fig. 7.9*e*) four diaphragms are uniformly arranged within one wavelength. Versions also exist in which distances between diaphragms vary periodically. Thicknesses of diaphragms are selected with regard to

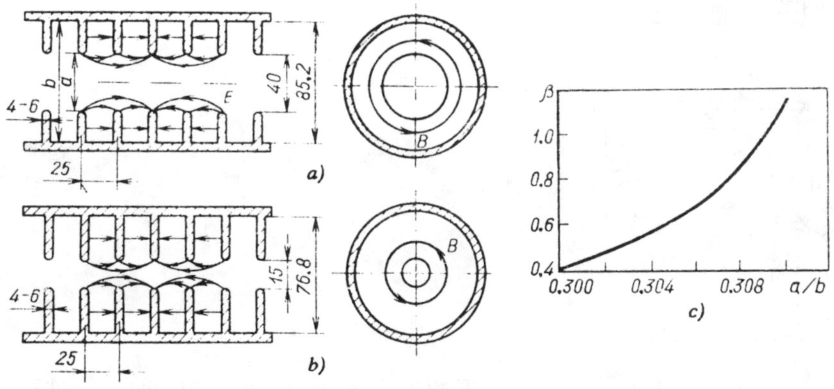


**Figure 7.13** Examples of dimensions of wave guide for *a)* and *b)* $\lambda$ = 10 cm and $\beta$ = 1, and *c)* dependence of high-frequency wave propagation velocity to ratio of diameter of aperture $a$ to diameter of aperture $b$ for $\lambda$ = 10 cm. $B$ are lines of intensity of magnetic field.

technological considerations and usually amounts to a few millimeters. The diaphragms are arranged with an accuracy of several microns. Figure 7.13*a* and *b* also shows the direction of magnetic field lines. The maximum axial gradient of potential for different accelerating structures lies in the range of 10 to 20 MV/m.

Construction and parameters of linear electron accelerators are materially determined by incorporated high-frequency generators. Use is usually made of commercially manufactured magnetrons or klystrons.

Figure 7.14 illustrates the industrial accelerator Cassitron [110] recently developed by the firm CGR MeV (France). The accelerator contains three resonators (2, 6, and 8) in which the electron beam is accelerated in series. Magnetic lenses (1, 3, and 7) are mounted adjacent to an injector (4) and between the resonators. The working frequency of the high-frequency field amounts to 200 MHz. Excitation is effected by the generator (5) employing a triode with anode voltage of some 40 kV and peak anode current of 1000 A. The control grid voltage is 1.5 kV, the anode-dissipated power is about 65 kW, and peak power of the triode is 5 MW.

Some parameters of electron beam generated by Cassitron are energy, 8 to 10 MeV, and average power 20 to 25 kW. Figure 7.14 shows a system for the conversion of electron beam into bremsstrahlung with the following components: scanning device (10), extraction (outlet) window (11), and target (12).

Figure 7.15 illustrates a microtron [111] installed at the Lebedev Physics Institute of the USSR Academy of Sciences. The energy

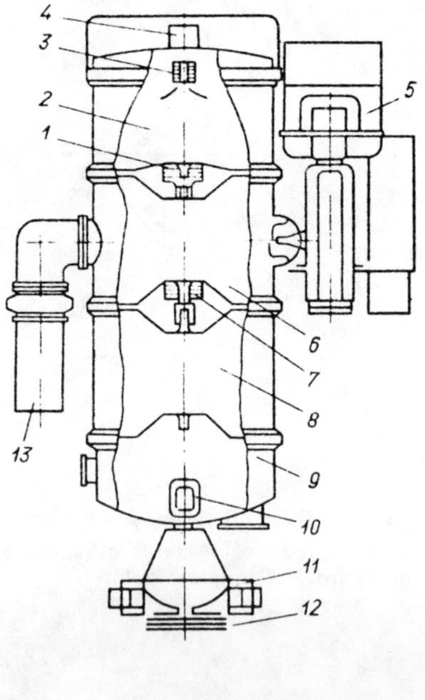


**Figure 7.14** Kassitron accelerator. *1* and *7*—Magnetic lenses between resonators; *2*, *6*, and *8*—resonators; *3*—magnetic lens near injector; *4*—injector; *5*—high-frequency generator; *9*—vacuum space; *10*—scanning device; *11*—extraction window; *12*—target; *13*—pump.

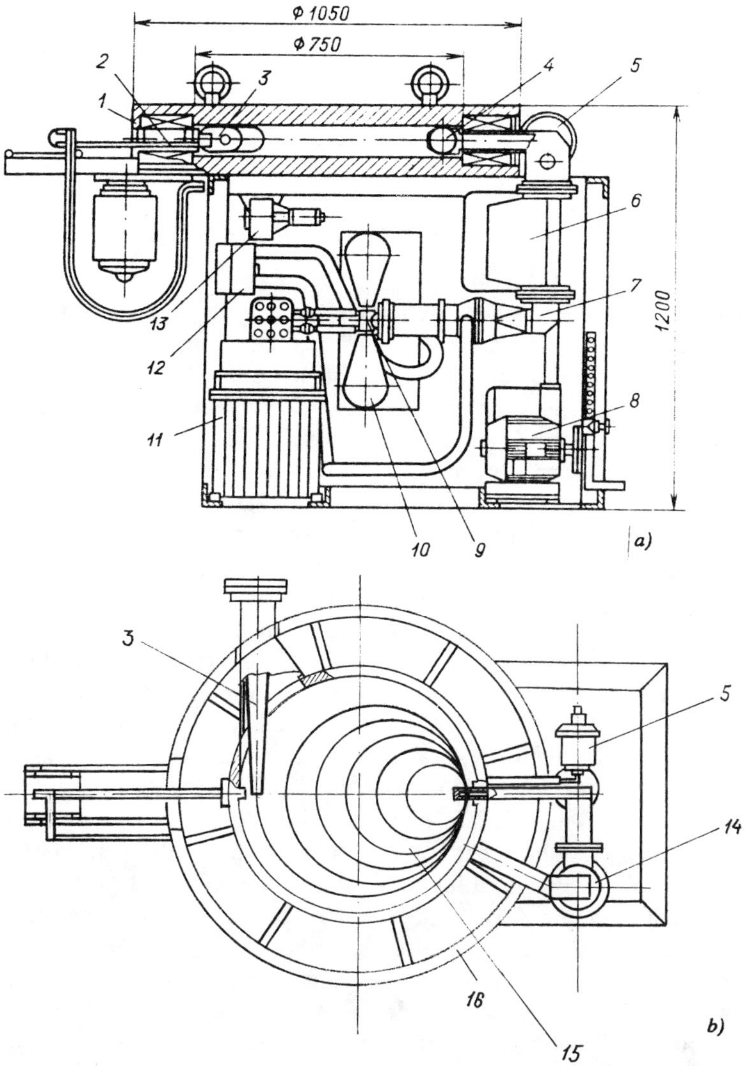


**Figure 7.15** Construction of 15-MeV microtron. *a)* Cross section. *b)* Plan view. *c)* Enlarged fragmentary view of magnet and vacuum chamber. *1*—Magnet pole; *2*—extractor shifting mechanism; *3*—electron extraction channel; *4*—resonator; *5*—electromagnet of ferrite rectifier; *6*—high-vacuum pump; *7*—wave guide; *8*—preevacuation pump; *9*—magnetron; *10*—magnet of magnetron; *11*—pulsed transformer; *12*—cooling air blower; *13*—fan; *14*—resonator readjustment mechanism; *15*—orbits of electrons; *16*—inverse magnetic circuit; *17*—electromagnet winding; *18*—cooling pipes; *9*—wall of vacuum chamber; *20*—vacuum-tight sealing; *21*—vacuum chamber.

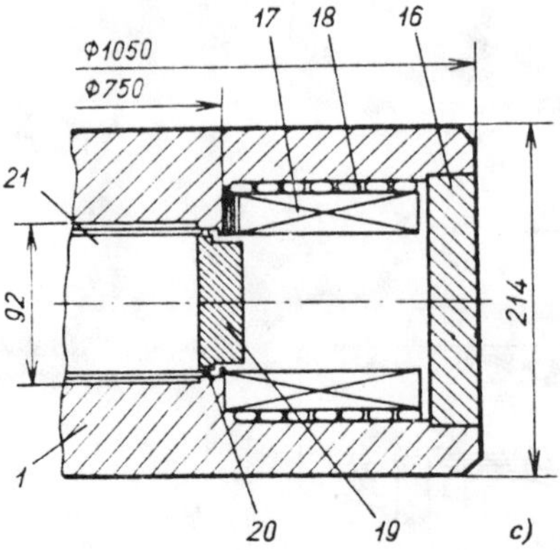


**Figure 7.15** (*Continued*) (*c*)

of the accelerated electrons is 15 MeV. The poles of the magnet (1) are interconnected with a circular yoke (16). The vacuum chamber (21) is formed by the poles of a permanent magnet (1) and circular steel wall (19) sealed by vacuum-tight packing (20). Series-connected magnet coils (17) are wound of a copper band 25 mm wide. At the tightness points on the magnet poles there are projections making for decrease of magnetic field decay towards the edges (see edges on betatron poles in Fig. 7.6). Tolerance for nonparallelism of magnet poles is 0.03 mm. Resonator (4) is coupled to a 2 MW magnetron (9) through a wave guide (7). The resonator frequency may be varied within a minor range by deforming one of the resonator walls made in the form of a membrane. The gradient of the electric field potential in the resonator is approximately 0.5 MV/cm. One of the resonator walls bears (see Fig. 7.11*c* and *d*) a cathode of $LaB_6$ in the form of a cube with a side of 1.5 mm. The cathode is soldered to a tantalum plate which serves as a preheater. The position of the channel (3) for extraction of the electrons is regulated by channel shifting mechanism (2). The channel aperture at the inlet for particles is 8 mm, and the wall thickness dimensions increased. High-vacuum oil-diffusion pump (6) with a capacity of 500 *l/s* and pre-evacuation pump (8) provide in the chamber (21) a vacuum of approximately $10^{-4}$ Pa. Air blowers (12) and a fan (13) are used for cooling. The high-frequency generator modulator circuit and emitter are power-supplied from a motor generator set type VPL-30. The dimensions of the microtrons are given in the illustration.

A microtron modification to obtain a stereogram generating two beams of bremsstrahlung can be used for flaw detection. Two thermionic cathodes are placed on opposite walls of the microtron's resonator. Two targets are mounted in the area of the final orbit. First, the electrons emitted by one cathode are accelerated and then decelerated on one of the targets. Then the direction of the guiding magnetic field is reversed by switching the power supply of the magnet windings. Thereupon the electrons emitted by the second cathode are accelerated, and the particles now travel in the reverse direction and are decelerated on the second target. By remote control of both target positions, the intersection point of bremsstrahlung beams can be changed.

It might be well to point out that microtrons, along with betatrons are more compact and and less expensive than linear electron accelerators rated at the same energies. However, the power of microtrons, as a rule, is one order less than that of industrial linear accelerators. Microtrons operate mainly in the range of 10-cm wavelengths, though there are some experimental versions operating on 3-cm wavelengths. The problem that is being solved now is potentialities of use of such power supplies as magnetron amplifiers (platinotrons) whose efficiency materially exceeds that of magnetrons (i.e., 60 to 70 percent as compared to 40 to 50 percent). Microtrons find the same applications as betatrons and linear electron accelerators, such as, for flaw detection, radiotherapy, and activation analysis.

## MATERIAL ADDED IN PROOF

According to recent reports [183], the pilot scale linear induction accelerator (LIA) with an electron energy of 1.6 MeV and an average power of 20 kW is to start operating in 1987. It is planned that, on the basis of this plant, LIA for commercial use will be built later with an electron energy up to 20 MeV and power of 50–500 kW. It was noted in the quoted reports that the bremsstrahlung radiation obtained at the plant with an energy of 5 MeV and power of 500 kW is approximately equivalent to that of the 4 MCi cobalt source. The cost of such an accelerator would be close to that of a cobalt source. However, the accelerator can be turned off, its energy and power are controlled, and a higher dose rate could have a decisive meaning for some applications.

CHAPTER

# EIGHT

# COMPONENTS OF RADIATION INSTALLATIONS

## 8.1 DEVICES FOR EXTRACTION OF ACCELERATED ELECTRONS FROM VACUUM

As a rule, accelerated electrons or electron-generated bremsstrahlung are used in industrial installations for irradiation of objects arranged out of vacuum. All the extracting devices are based on two basic methods of extraction of particles: (1) through a thin foil made of light material which separates regions of different pressure; and (2) along a channel communicating regions of different pressure. In the first method, the foil must have adequate mechanical strength for withstanding the pressure drop. The losses of the electron energy in the foil should not cause heating of the foil to a temperature that would deteriorate the mechanical strength. In the second method, the channel or duct is provided with several vacuum pumps to evacuate the gas which tends to flow toward the low-pressure region. The length and diameter of the channel, as well as the number and output capacity of the pumps, are selected with regard to maintaining the preset pressure drop.

### 8.1.1 Extraction of Electrons through Foil

The foil used in the extracting windows is fabricated of light, strong, and heat-resistant materials. The thickness of the foil ranges from 15 to 100 μm. The minimum thickness of the foil is dependent on the technological abilities of fabricating vacuum-tight foils. The foil must stand up to a high dose of radiation without deterioration of its mechanical strength. The dose for the extraction windows of the accelerators in industrial use is at least $10^6$ MGy; the lifetime of the foil is thousands of hours. The most commonly used materials for the extracting windows are titanium and aluminum. For extracting low-current beams ($j = 10^{-6}$ to $10^{-7}$ A/cm$^2$) and in pulsed radiation installations, use may also be made of polymeric materials coated on both sides with a thin layer of metal, for example, a polyamide film 30 to 50 μm thick with an aluminum or beryllium coating of ~1 μm.

The specific losses of the electron energy ($\Delta E$ MeV/cm) depend minimally on the energy, beginning at 0.5 MeV (see Section 2.1). The basic properties of the foils used in the extracting windows are given in Table 8.1. A disadvantage of beryllium is the toxicity of its vapors.

When passing through the foil the electrons are deflected from their initial line of travel. The thicker the foil and the less the electron energy, the greater is their scattering (see Table 2.1). The energy loss of the electrons in the foil usually amounts to 1 to 2 percent for an energy of ~1MeV and 10 to 20 percent for 0.15 to 0.2 MeV. Heating of the foil is partially intensified by the effect of the secondary electrons generated by the primary beam in the material under irradiation and in the atmosphere and received by the foil from the outward side. The foil is cooled by blowing out with a jet of air or other gas, and also at the expense of heat transfer to the parts of the extracting window. A small portion of the energy is taken away by thermal radiation. When operation is performed by sending single pulsed beams, the foil undergoes adiabatic heating and cools down the time intervals between successive pulses.

Shown in Fig. 8.1*a* and *b* is the diagram of the outlet window meant for extracting an electron flux with an average density of current of $1 \times 10^{-4}$ to $2 \times 10^{-4}$ A/cm$^2$. Figure 8.1*e* illustrates a

**Table 8.1 Main performance characteristics of foils used in beam-extracting windows**

| Characteristics | Beryllium | Aluminum | Aluminum alloys | Titanium |
|---|---|---|---|---|
| Specific loss of electron energy, MeV/cm | 3.5 | 4.5 | 4.5 | 6.4 |
| Ultimate strength, kg/mm$^2$ | 25 | 7–9 | 20 | 57–60 |
| Coefficient of thermal conductivity, W/cm °C | 1.8 | 2 | 1.3 | 0.15 |
| Heat capacity, J/g °C | 2.1 | 0.93 | 0.95 | 0.54 |
| Density, g/cm | 1.85 | 2.7 | 2.8 | 4.5 |
| Tolerable working temperature, °C | 450 | 100 | 200 | 450 |

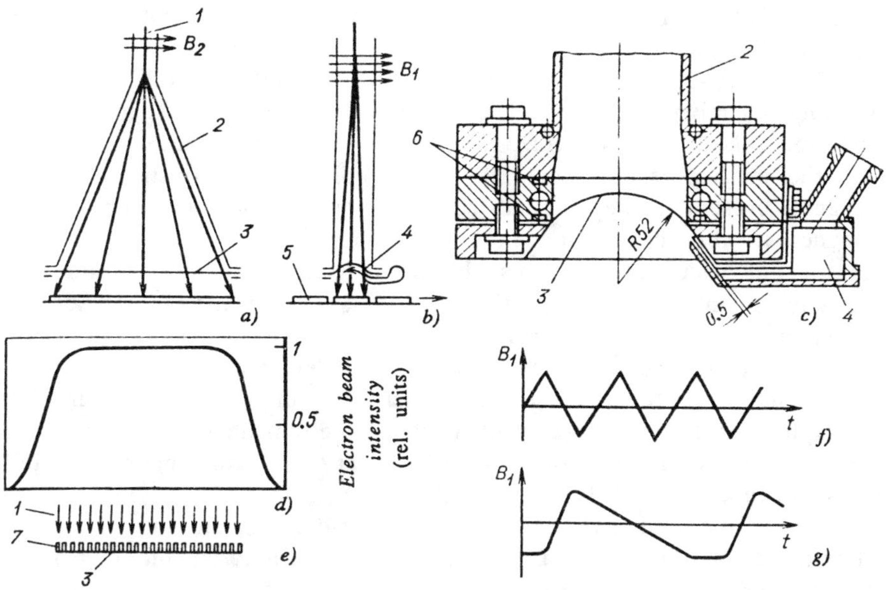

**Figure 8.1** Extraction of stationary electron beam through foil window. Diagram of extraction window with scanning in two directions, as viewed *a)* from longer side and *b)* shorter side of window. *c)* Distribution of intensity of beam extracted along longer side of extracting window. *d)* Window with supporting ribs. *e)* Typical construction of extraction window. *f)* and *g)* Time variation of scanning magnetic field $B_1$ for accelerators *f)* ELV and *g)* ILU-6. *1*—Electron beam; *2*—vacuum chamber (horn); *3*—foil; *4*—air cooling; *5*—conveyor with articles under irradiation; *6*—seal; *7*—supporting ribs; $B_1$—induction of scanning magnetic field along longer side of extraction window; $B_2$—induction of scanning magnetic field along shorter side of extraction window.

standard construction of the window. The beam with a small cross section at its origin is scanned in two directions by means of transverse magnetic fields $B_1$ and $B_2$. Requirements for the scanning along the longer side of the window are dictated by the necessary uniformity of the radiation dose on the object under irradiation and even heating of the foil. If the articles being irradiated move in the direction of the shorter side of the window, scanning along that side is required only for reducing the heat load experienced by the window. To ensure scanning along the longer side, use is usually made of the magnetic field $B_1$ with a period of 50 to 5000 Hz varying in adherence to the sawtooth principle. The typical curves of variations of magnetic field $B_1$ in time are given in Fig. 8.1*f* and *g*.The coils generating the magnetic field $B_2$ are generally fed with a sine voltage of 50 to 60 Hz. Under the action of the scanned beam each region of the extracting window is periodically heated and expanded which may result in thermal fatigue failure. The radiation effect of the electrons causes abnormal oxidation and embrittlement of the foil.

The beam-scanning magnets may be accomodated both inside the vacuum chamber and beyond it. In the first case, the gap between the poles is less, and, correspondingly, the power required to supply the magnet is less too. However, cooling and adjustment of the magnet present some difficulties, and the component parts of the magnet in this connection must be fabricated of materials free of gassing in vacuum conditions. If the magnet is located outside the vacuum chamber, the magnetic field passing the metal wall of the chamber tends to get distributed. The scanning magnet may be installed outside the vacuum chamber provided that the thickness of the chamber wall is much less than that of the skin layer. For example, for the stainless steel wall of the chamber (~1 mm thick) in the interpolar gap, the rate of variation of the magnetic field must not exceed 2 to 3 kHz.

The typical distribution of the extracted beam in the air is illustrated in Fig. 8.1*c*. Decrease of the intensity at the edges can be attributed to nonuniformity distribution of the current in the initial beam and the scattering of the electrons in the foil and in the air. As the object under irradiation recedes from the foil, the edge effects are enhanced and the flat section of the distribution gets shorter. It is common practice to arrange the object to be treated at a distance not exceeding 5 to 10 cm from the beam extracting window. The characteristic transversal dimensions of the window for the case depicted in Fig. 8.1*a*, *b*, and *e* amount to 40 to 80 mm. The dimension along the longer side of the window is, as a rule, 0.5 to 1.8 m. To decrease mechanical stresses in the foil the latter is fabricated in the shape of a cylinder. The deflection equals about one-third of the window width.

When a need arises to have a larger width of the extracting window, for example, 10 to 20 cm (or more), or to inject the electrons into a space with high pressure, crosswise ribs are used to support the foil (Figs. 8.1*g* and 8.2). Such windows are used, in particular, in electronic ion lasers (Section 2.8) where the beam is extracted from the vacuum and transferred into a space featuring a pressure of several atmospheres. The energy liberated in the foil is taken off primarily at the expense of the heat transfer through the foil to support ribs and then to the window circumference which is cooled with running water. The ribs are usually a few millimeters thick and they capture 30 to 40 percent of the beam. Calculations of the optimal geometry of such windows are presented in [112]. In order to minimize heating of the ribs caused by direct hits of electrons, provision may be made for additional crosspieces.

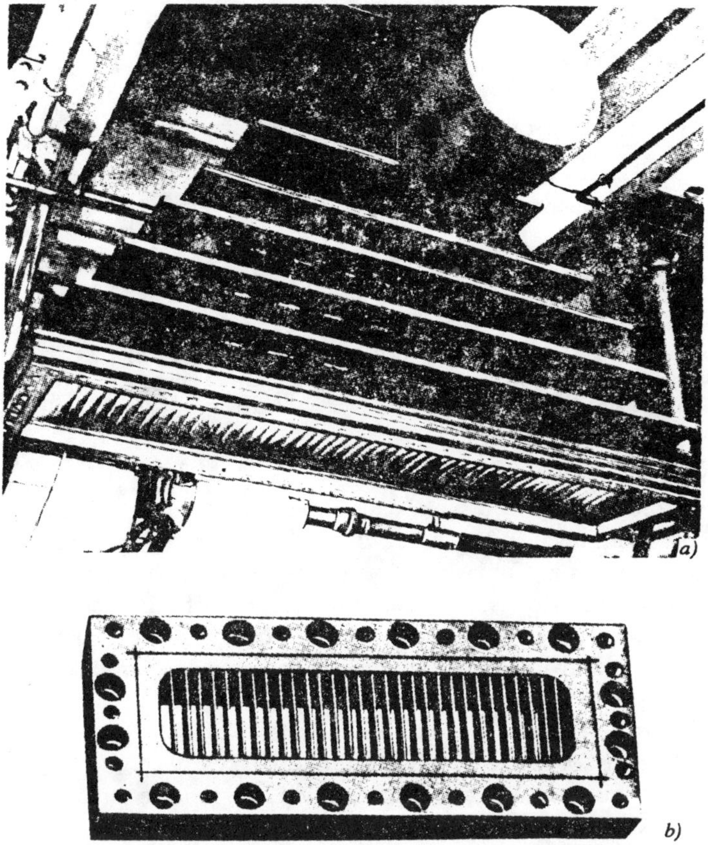

Figure 8.2 Extraction windows with supporting ribs. *a)* General view of extractor and outlet window with area of 125 × 25 cm (firm HVEC) [141]. *b)* Flange of extraction window with supporting ribs [140].

When heat is removed because of heat conduction of the foil, the temperature is only slightly dependent on the thickness, since both the amount of the evolved energy and the value of the heat flow are directly proportional to the foil thickness. In case of a round window, the maximum current flowing across it is practically independent of the window diameter and for example, for aluminum amounts to $\sim 10^{-2}$ A. In the case of a window made in the form of a slot the current per unit length of the slot tends to grow with decrease of its width. Prior to extraction, the beams accelerated in short pulses (0.1 to 20 $\mu$s) at a high repetition frequency are increased in cross section, as a rule, by the use of a quadruple lens, for it is technologically very difficult to provide a uniform scanning of the beam during the life of a short pulse. In accelerators

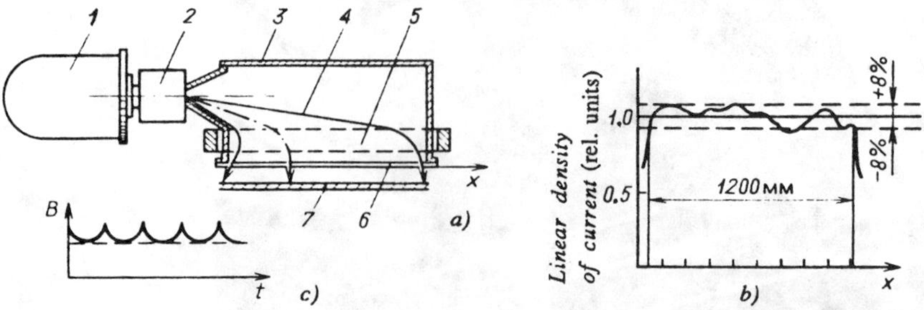


**Figure 8.3** Beam extractor with reduced-size horn. *a)* Circuit diagram of extractor. *b)* Distribution of linear density of current along window (found experimentally). *c)* Time variation of magnetic field *B* in scanning device. *1*—Accelerator; 2—scanning device; *3*—vacuum chamber; *4*—accelerated electrons; *5*—region of direct magnetic field; *6*—extraction window; *7*—object under radiation.

rated at energies of 150 to 400 keV, an extended cathode is often used. In this case, the need may arise to increase merely the cross section of the electron flux, for which purpose use may be made of scanning in an electrostatic field. Yet the forces of the electric field acting on the relativistic electrons are comparatively small (see Section 1.3); therefore, for obtaining substantial angles of deflection, provision should be made for long plates (in the direction of the traveling electrons) to be held under variable potential. In many cases, the use of an extended cathode makes it possible to select a shape of electrode in the acceleration region so that the beam is provided with the required divergence angle and possesses the assigned cross-section in the window area even without scanning.

The most widely used industrial accelerator is the device pictured in Fig 8.1*a* and *b*. Because of the need to obtain an approximately permanent distribution of the radiation dose along the extracting window, the scanning angle governed by the magnetic field $B_1$ usually does not exceed 25° to 30°, which leads to provision of an essentially large vacuum chamber horn, in Fig. 8.1*a* in the direction of the traveling beam and, respectively, to a considerable increase of the volume of the room to be under radiation protection. In the last few years with the object of decreasing the dimensions of the beam extractor, use has been made of the circuit demonstrated in Fig. 8.3*a* [113, 114]. Electrons (4) from the accelerator (1) are applied to a scanner (2) whose magnetic field *B* varies in compliance with the law depicted in Fig. 8.3*c*. The scanned beam is shifted through some angle and then comes to region 5 of the permanent magnetic field, which is perpendicular to the drawing, as is magnetic field *B*. Electrons (4) turn in region 5 through an angle approximating 90° and then pass through the foil of the extraction window (6) and lastly reach object 7 to be irradiated. As can be seen from Fig. 8.3*c*, magnetic field $B$ has a direct component and changes at a variable speed to provide a uniform density of the electron current along the window. The electronic systems for ensuring the required law of variation of the current in magnet 2 have been described in [115]. The experimentally derived distribution of the linear density of the electric current along the extraction window is exhibited in Fig. 8.3*b*. The nonuniformity of the current lies within ±8 percent.

Other versions of devices designed for extracting electrons through foil windows are illustrated in Fig. 8.4. Figure 8.4*a* shows a

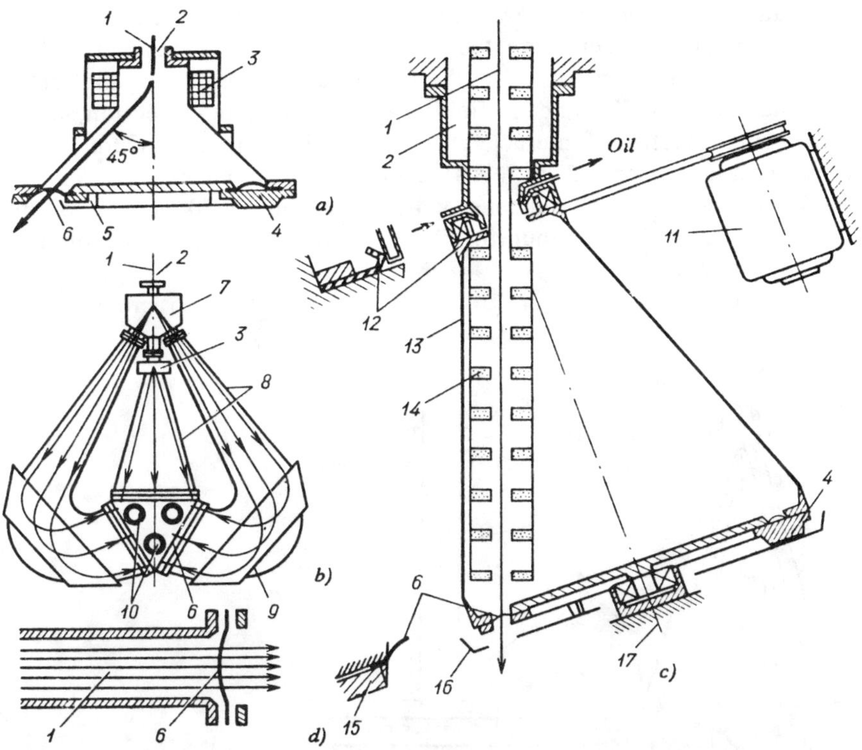


**Figure 8.4** Versions of device for extraction of electrons through foil windows. *a)* Extraction of circular beam. *b)* Three-sided irradiation of industrial articles. *c)* Extraction of concentrated beam through rotary foil. *d)* Extraction of single pulses of intense electron beam. *1*—Electron beam; *2*—vacuum chamber; *3*—scanning magnet; *4*–rib; *5*—delivery of cooling air; 6—extraction window; *7*—beam distributor; *8*—horns; *9*—deflection or turning magnet; *10*—articles under irradiation; *11*—motor rotating cone with foil; *12*—vacuum seal of rotary cone; *13*—rotating cone; *14*—fixed focusing magnets; *15*—lead seal; *16*—fixed shield; *17*—rotation axis of cone and foil.

circular extraction window (6). The circular scanning of the beam (1) is effected by the use of a stator (3) of a standard induction motor producing a magnetic field of 350 Gs rotating at a frequency of 1000 Hz. This device has been tested in an experimental installation intended for treatment of grain [116]. To ensure three-sided irradiation of the articles to be treated in the industrial installations the beam is separated into three parts that are delivered alternately to three extraction windows arranged at an angle of 120° with respect to one another (Fig 8.4*b*) [117]. The electron beams come to lateral windows after passing across the beam-bending magnets. If the electrons are accelerated in the pulsed manner, then the process of changing the beam to different trajectories is very simple. In the case of a continuous beam, this change-over entrains considerable losses of electrons when the beam is transferred from one channel to another. For uniform irradiation of articles from different sides, magnets can be installed in the atmosphere (Fig. 8.5). The electrons passing by the cable or scattered by the cable are turned by the magnetic field to the opposite direction, after which some electrons interact with the article under treatment. In the installation described in [118] at an electron energy of some 1 MeV an article 6 cm in diameter will receive up to 70 percent of the beam current, with a nonuniformity of the radiation dose of less than 10 percent.

The device illustrated in Fig. 8.4*c* is meant for extracting a con-

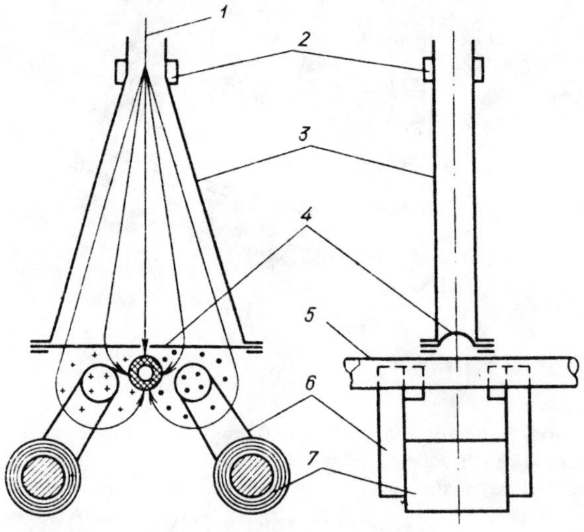


**Figure 8.5** Use of magnetic fields in the atmosphere for uniform irradiation of cylindrical articles. *1*—beam; *2*—poles of scanning magnets; *3*—horn; *4*—foil; *5*—object under radiation; *6*—magnetic circuit; *7*—winding.

centrated beam with a mean specific intensity of some 100 $kW/cm^2$ from a vacuum chamber. Beam 1 passes through fast-rotating foil (6). The vacuum seal (12) is much less in radius than the rotary foil; the sealing conditions are thus simplified. This system was used for extracting a continuous beam with an average power of 20 kW, energy of 2.2 MeV and foil diameter of ~5mm [119].

Extraction windows of 3 to 5 cm in diameter (Fig. 8.4*d*) are used for extracting single pulses of relativistic electrons with currents of tens of kiloamperes at a pulse length of the order of $10^{-7}$ s. Such windows, for example, find application in the installations discussed in Section 6.8. The ultimate density of the current and the beam duration are limited in conformity with adiabatic heating of the foil.

## 8.1.2 Extraction of Concentrated Electron Beams

When the beam density is so high that removal of heat from the foil is impractical, a channel with differential evacuation is then used. The key diagram of the device designed for extraction of concentrated electron beams is shown in Fig. 8.6*a*. The contrivance in question comprises several chambers separated by holed diagrams and evacuated by the use of individual pumps. The illustration shows an approximate distribution of pressure along the length of the device. The diaphragms of the chambers are conical in shape, as viewed from the high-pressure side. Aerodynamical effects cause the jet of gas entering the expanding horn to spread over its inner side resulting in a lesser density of the gas over the axis of the horn. The system with a differential evacuation (with two stages) was first used for extraction of x-irradiation as early as about 1920 [120]. A multistate gas-dynamic window for extraction of electrons was designed in the early 1930s [121].

For focusing the beam in the channel, provision is made for a set of magnetic lenses which ensure a minimum diameter of the beam where diaphragms separate the evacuation stages (Fig. 8.6*b*). This device provides a pressure drop from the atmospheric pressure up to $10^{-4}$ Pa with a channel of 1 to 3 mm in diameter and a few tens of centimeters in length. In the extracting device of the ELT-1.5 accelerator shown in Fig. 8.6*b*, the diameter of the outlet is about 3 mm, the electron energy amounts to 1.2 MeV, and the mean power reaches 15 kW. The efficiency of the extracting device is considerably dependent on the current consumed by the diaphragms. In the device being discussed, this amounts to a few percent of the total current of the electrons. The service life of the diaphragm may be extended, by being fabricated of refractory materials and undergoing intense cooling. In the system under consider-

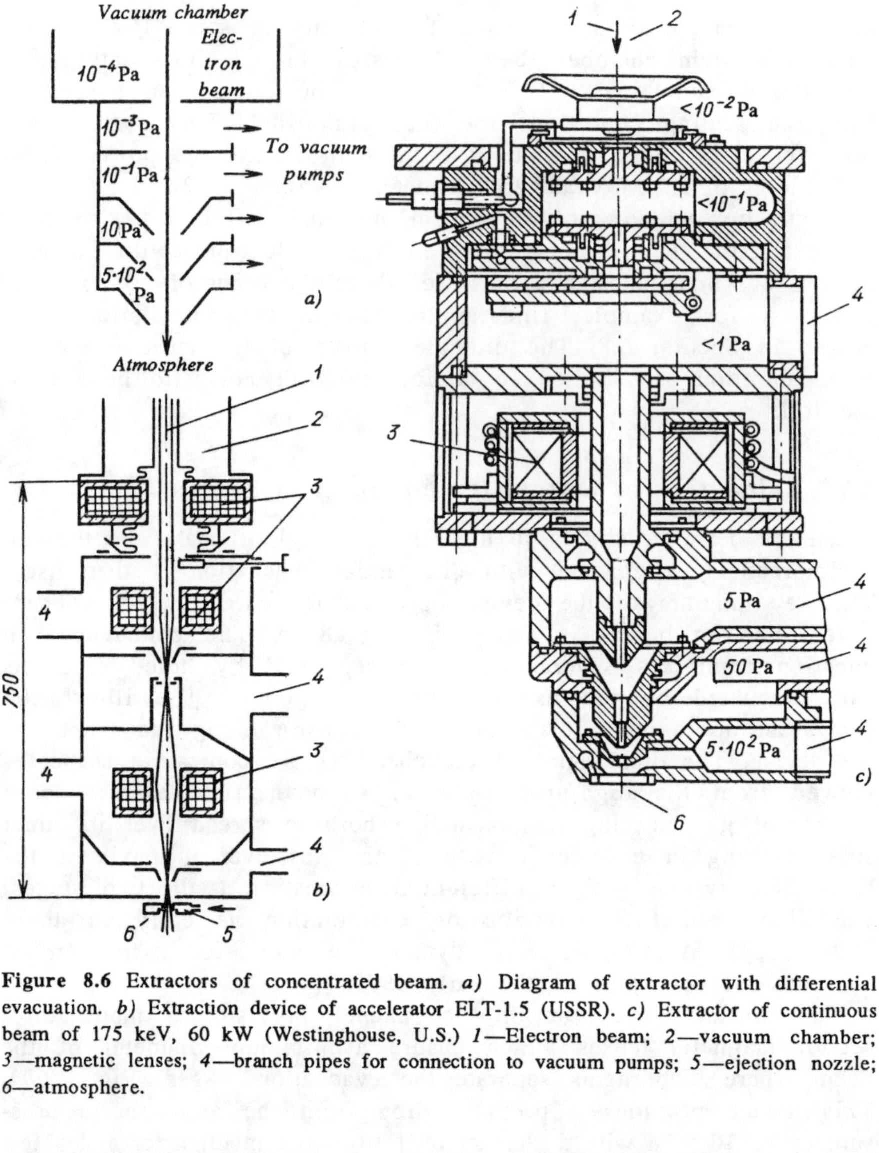


**Figure 8.6** Extractors of concentrated beam. *a)* Diagram of extractor with differential evacuation. *b)* Extraction device of accelerator ELT-1.5 (USSR). *c)* Extractor of continuous beam of 175 keV, 60 kW (Westinghouse, U.S.) *1*—Electron beam; *2*—vacuum chamber; *3*—magnetic lenses; *4*—branch pipes for connection to vacuum pumps; *5*—ejection nozzle; *6*—atmosphere.

ation, the diaphragms are replaced approximately after every 10 hours of operation. The operation of the extracting device and the accelerator as a whole may be improved if respective steps are taken to attenuate the flux of particles traveling in the direction opposite to that of the beam, particularly to provide parallel transfer of the extracted beam. The more elaborate versions of the ELV installation has permitted the extraction of a beam with an energy of 0.8 to 1.6 MeV and a power up to 40 to 60 kW [122].

Figure 8.6 indicates a device extracting beams of 175 keV, with a constant power of 60 kW, and a diameter of about 1 mm. The aperture of the outlet is of the order of 2 mm [123]. The operation of the extraction system may be somewhat improved by involving an ejection nozzle (5 in Fig. 8.6*b*), which decreases the pressure outside the first diaphragm. To lessen the scattering of the extracted electrons outside the channel, helium or a mixture of hydrogen and oxygen can be blown along the trajectory of the electron beam (the mixture will burn out at the end of the beam trajectory). One of the methods of maintaining a high density of the beam is the creation of a direct-current arc along the beam trajectory between the beam extractor and the article to be treated [124]. Coincidence of the directions of the currents in the beam and in the arc will make not only for heating the medium but also for additional focusing of the electrons.

In case of extraction of a pulsed, periodically repetitive concentrated beam, the channel should be opened only at the moments of transit of the beam. This will prevent excess inflow of gas and allow a corresponding decrease in the dimensions of the evacuation pumps. This problem can be solved by making use of the devices pictured in Fig. 8.7*a* and *b*. Figure 8.7*d* [125] shows, at the outlet of the extraction channel, a rotary drum (10) with holes (5) arranged

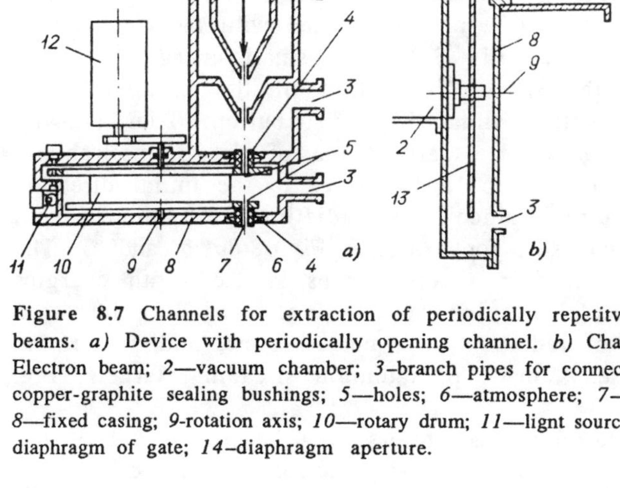


**Figure 8.7** Channels for extraction of periodically repetitve and single concentrated beams. *a)* Device with periodically opening channel. *b)* Channel with pulsed gate. *1*—Electron beam; *2*—vacuum chamber; *3*-branch pipes for connection to vacuum pumps; *4*—copper-graphite sealing bushings; *5*—holes; *6*—atmosphere; *7*—axis of extraction channel; *8*—fixed casing; *9*-rotation axis; *10*—rotary drum; *11*—lignt source; *12*—motor; *13*—movable diaphragm of gate; *14*-diaphragm aperture.

circumferentially. Two brushings (4) mounted in the casing are tightly pressed against the drum, thereby preventing ingress of air into the channel unless the axes of the holes in the disk and in the channel are brought in coincidence. The moment of the coincidence of the holes is synchronized with an electron accelerating pulse by means of a light source (11). This system has made it possible to reduce the gas inflow by the factor of one or two orders as compared with systems where the extracting channel is constantly held open.

For extracting single pulses of the beams, a pulse gate is also used (Fig. 8.7*b*). In the normal position, the hole (14) of a rotary diaphragm (13) is shifted from the axis of the extracting channel (7). By rotating the diaphragm (provided with a mechanical or electromagnetic drive), the axis of the hole is brought in line with the axis of the channel. At the moment of coincidence of the axes the extraction of the beam (1) is effected. In its further motion, the diaphragm closes up the extracting channel again, as a result of which a relatively small portion of the gas can be admitted to the vacuum chamber (2). The characteristic parameters of the beam extracted by use of this device incorporated in an industrial installation are as follows: energy, 300 keV; power, $3 \times 10^4$ kW (current of some 100 A); pulse length, $\sim 10^{-4}$ s; and extraction window diameter, 10 mm [2].

## 8.2 TARGETS FOR GENERATING BREMSSTRAHLUNG

As noted in Section 2.2, the output of bremsstrahlung tends to rise with an increase in the atomic number of the target material. It is common practice to fabricate the target from tungsten and other heavy metals. With increase of electron energy the bremsstrahlung is concentrated in a cone whose axis is in line with the direction of the traveling electrons (see Fig 2.9). For small energies (of the order of 100 keV), the space distribution of the bremsstrahlung approximates a spherically symmetrical distribution. That is why in this energy region, use is usually made of the bremsstrahlung generated at an angle of 90° to 180° relative to the initial direction of the traveling electron, since this radiation is immune from absorption in the target (see, for example, Fig. 8.8*a*, *b*, and *c*). The circuit shown in Fig. 8.8*c* in specific cases is used with energies not exceeding 1 to 2 MeV, as indicated in Fig. 6.46. With increased energy, concentration of radiation in a narrow cone, and penetrating power of photons, use is made of radiation streaming targets (Fig. 8.8*d*).

The target in high-voltage generators may be placed out of the vacuum chamber and bombarded with electrons extracted through

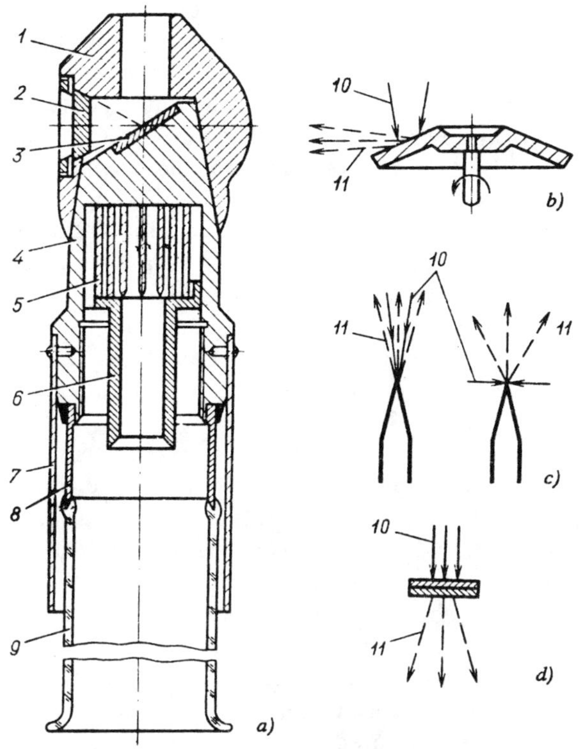


**Figure 8.8** Targets for generation of bremsstrahlung. *a)* Construction of target with running oil cooling. *b)* Rotary target. *c)* Needle-shaped target. *d)* Penetrable target. *1*—Shield; *2*—beryllium window; *3*—tungsten target; *4*—copper casing; *5*—bifilar helix; *6*—branch pipe; *7*—shield; *8*—fernico ring; *9*—anode lug (glass); *10*—accelerated electrons; *11*—bremsstrahlung radiation.

the foil window. This considerably facilitates the work of the equipment, because with operation in vacuum it is necessary to take safety precautions to eliminate gassing of the target due to heating and to protect the parts against contamination with dusty particles of the target material.

To reduce the heat load exerted on the extracting foil at the moment of beam transit through the window, the beam may first be of comparatively large cross section and then focused with the aid of the magnetic lenses in the atmosphere between the window and the target.

One of the main difficulties in developing the targets in intense-beam accelerators is the cooling of the targets. There are several ways of solving this problem: placing a metal with good heat conduction close to the heavy target and providing forced cooling of the metal, fast motion of the target, etc.

The target illustrated in Fig. 8.8*a* is used at an electron energy of 200 to 300 keV. The target is arranged inside a vacuum chamber. Tungsten target (2) and copper casing (4) are interconnected in vacuum by casting copper on a tungsten plate. The liquid coolant (for example; transformer oil) flows over a branch pipe (6) and then, via a bifilar copper helix (5), is delivered to the cylindrical part of the casing (4). The approximate value of the specific intensity in the focal spot of the target in the continuous operation mode amounts to 40 to 70 W/mm$^2$, or to 150 to 200 W/mm$^2$ in a 1-s operation. A massive shield (1) absorbs excess bremsstrahlung radiation. For extraction of the bremsstrahlung from the vacuum a beryllium window (2) of some 1 mm thickness is used. Fernico, stainless steel, glass, or other materials can be used for the extraction windows.

The beam energy in a rotary target (Fig. 8.8*b*) is distributed over a spacious circular path. The typical dimensions of the focal spot are 8 × 2 and 2 × 2 mm. As a rule, the target with this type of equipment operates in the mode of instantaneous repeated loading and is cooled by radiation. Tungsten is heated to 2200° to 2500°C, and in the course of time tends to partially evaporate. The specific intensity reaches a few kilowatts per square millimeter. After a few thousand switching operations, the radiation intensity markedly diminishes due to craters formed on the target surface, because the radiation generated by electrons captured by the craters is substantially weakened. The thermal stability of the target increases materially if tungsten is underlayed with a substrate of molybdenum. Another way of increasing the heat resistance and service life consists in providing the tungsten with an additive of rhenium (usually ~10 percent of rhenium). The target is driven to rotate by the induction motor rotor which is placed in vacuum. The rotor is essentially a massive copper cylinder (3, in Fig. 8.9*a*). The motor stator is installed outside the vacuum, in the atmosphere. The rotation speed of the rotor is of the order of $10^4$ rpm. Figure 8.9*b* shows an evacuated X-ray tube with a rotating anode, rated at 150 kW (made by Rohrenwerk GDR) [126]. In operation during 0.1 s with a focal spot measuring 2 × 2 mm the beam power will amount to 50 kW. In some modifications of the tubes featuring the same pulse duration the beam power reaches 200 kW [127].

The needle-shaped target (Fig. 8.8*c*) finds application for generating short pulses of bremsstrahlung (see also Figs. 6.45*b*, *c* and 6.46). The merit of this target is a minor size of the focal spot. Its shortcoming consists in poor removal of heat. The density of the surface energy of these targets may be as great as 50 to 200 J/cm$^2$ per pulse. The temperature may rise to 20,000 K, which causes

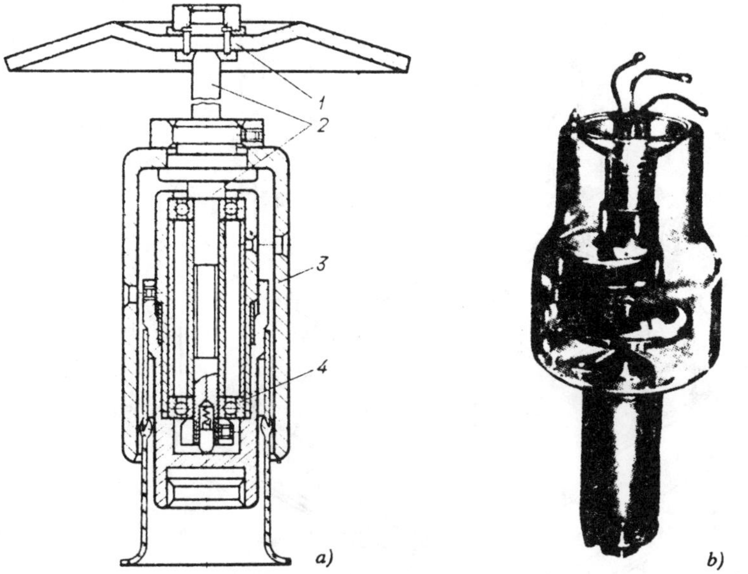


**Figure 8.9** *a)* Construction of rotary target. *b)* External view of x-ray tube. *1*—Rotary tungsten disc (usually 50 to 125 mm in diameter); *2*—shaft; *3*—rotor of induction motor (copper cylinder); *4*—ball bearing.

vaporization of the target, and the plasma arises (anode jet) under the action of the electrons. This results in compensation for the space charge of the electrons, reduction of the cathode-anode distance with a respective increase of the tube current, and, lastly, decrease of pulse length.

The targets penetrable by bremsstrahlung in individual cases are also double-layered (Fig. 8.8*d*). Commonly used as substrate is beryllium, which features fairly high heat conduction and light specific weight. The substrate may serve as the extracting window of an accelerating device. The targets are practically always used in linear and other electron accelerators with energies exceeding 2 to 3 MeV which are used for generating bremsstrahlung. The optimum thickness of a tungsten target is specified in Section 2.2.*

## 8.3 RADIATION ZONES AND RADIATION SHIELDINGS

A standard radiation zone includes a radiation-shielded space and accomodated in it a radiation source, an object to be radiation-treated and a system for moving the latter. Figure 8.10 shows a radiation zone (10) of a processing installation based on the dyna-

* See new material added in proof on page 222.

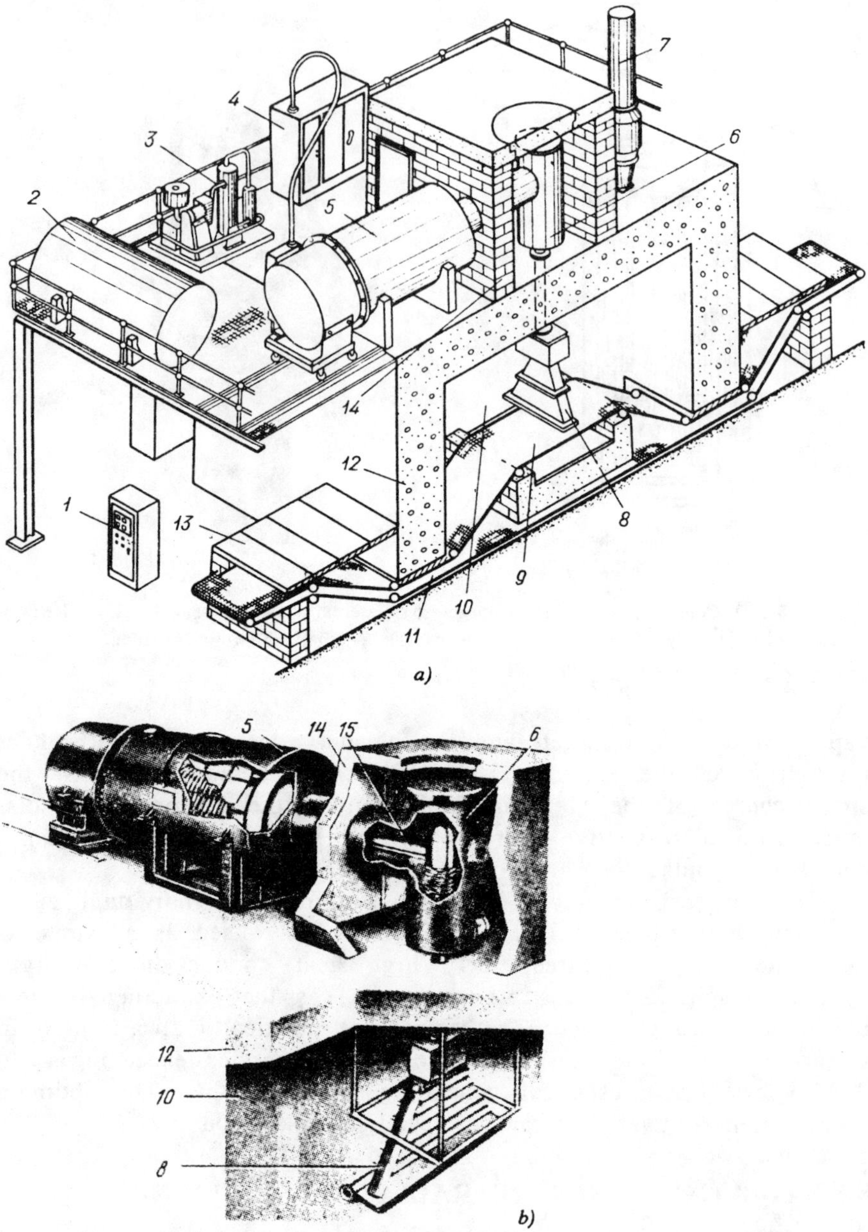


**Figure 8.10** Dynamitron-based technological installation. *1*—Control panel; *2*—gas tank; *3*—components of gas system; *4*—high-frequency generator; *5*—high-voltage generator; *6*—accelerating device; *7*—vent pipe; *8*—horn; *9*—article under irradiation; *10*—radiation zone; *11*—channel; *12*—radiation shielding; *13*—radiation protection of channel; *14*—radiation shielding of accelerating device; *15*—gas-filled coaxial cable.

mitron-type accelerator. Article (9) to be irradiated is delivered to the radiation zone through channels (11) in radiation shielding (12). The thickness of the radiation shield is calculated to ensure a decrease of the radiation effect by a factor of $10^6$ to $10^8$ (see Table 2.3). The channel inlets (11) are screened by additional shielding (13). An accelerating device (6) is mounted beyond the radiation zone and enclosed in a shielding (14) of somewhat lesser thickness. A high-voltage generator (5) is installed in an unshielded room and is connected to the accelerating device through a gas-filled coaxial cable (15).

Figure 8.11 displays a diagram of the installation employing an ICT accelerator for irradiation of bulk materials. The parallel-operated conveyers feed material (2) to be treated to two radiation zones (7). To obtain two-sided irradiation, provision is made for two accelerating devices (4) with extraction windows (5) in each zone.The radiation zone (7) is provided with radiation shielding (9). The top part of the latter can be removed. Accelerating devices (4) and high-voltage generators (3) are placed beyond the main shielding and are surrounded with a relatively thin protective wall (10). Irradiated materials are transported by conveyers (6).

The radiation shielding protects mainly the radiation zone, since the most serious hazard is bremsstrahlung arising in the place of interaction between the electrons and target. It is possible that accelerated electrons may also penetrate the shield. Having completed their complicated travel (see Fig. 2.1), these electrons may exit outward even through curvilinear holes. The secondary electrons generated by the primary beam are also hazardous. Accelerating tubes or other evacuated envelopes containing parts under high potential may also serve as radiation sources. Radiation in them may arise as a result of random contact of fast-moving electrons with the tube parts or in case of breakdowns, in the vacuum envelopes. In the direct-action accelerators, the voltage generators, where compressed gas, oil, or solid dielectrics are used as insulating media, do not constitute any threat and, therefore, as a rule, are installed beyond the shielding enclosure. In the case of a fixed distance to the object under protection, the thickness of the radiation shielding is practically independent of the distance between the shielding and the radiation source, and therefore, in order to reduce the cost of the protective facilities it is recomended the shielding be at a minimum distance from the radiation sources. In this respect, preference should be given to equipment provided with local protection and used at an electron energy within 0.5 to 0.7 MeV (see, for example, Fig. 6.21). Such installations are usually employed in the production rooms and are built directly into the

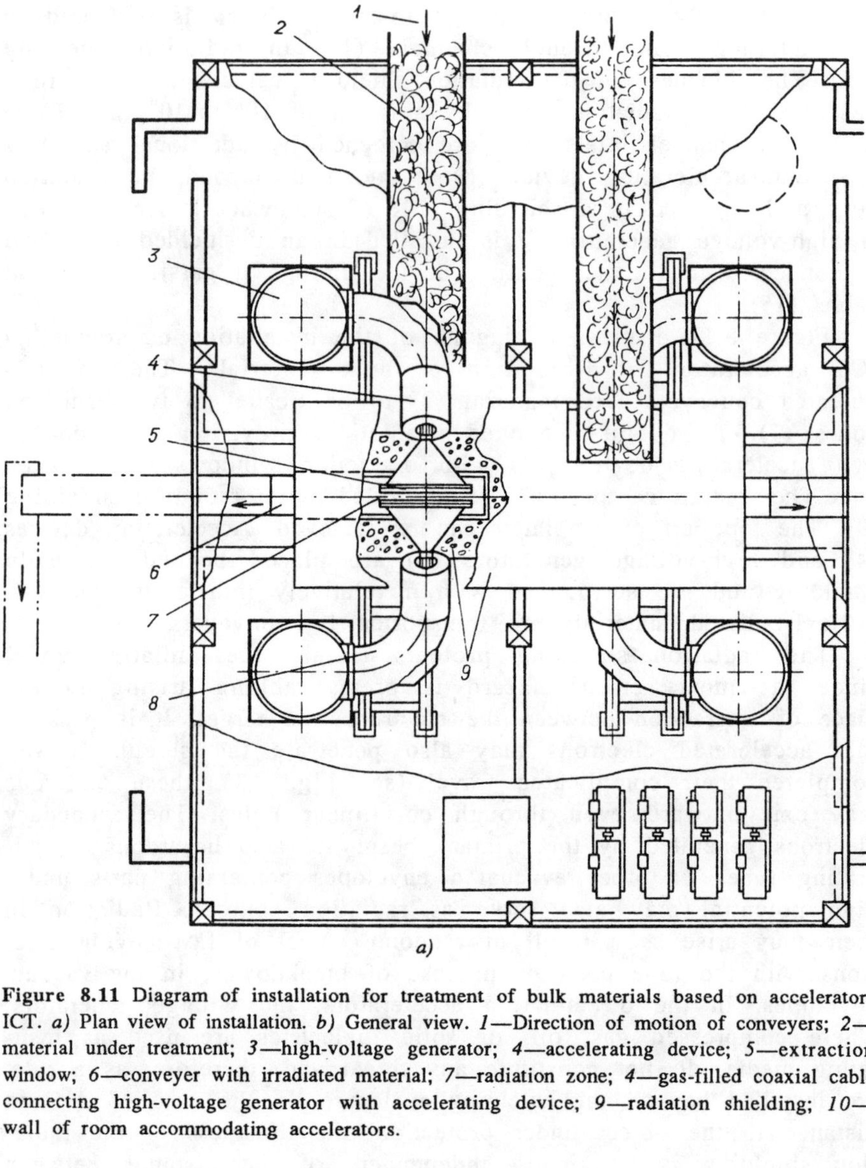


**Figure 8.11** Diagram of installation for treatment of bulk materials based on accelerators ICT. *a)* Plan view of installation. *b)* General view. *1*—Direction of motion of conveyers; *2*—material under treatment; *3*—high-voltage generator; *4*—accelerating device; *5*—extraction window; *6*—conveyer with irradiated material; *7*—radiation zone; *4*—gas-filled coaxial cable connecting high-voltage generator with accelerating device; *9*—radiation shielding; *10*—wall of room accommodating accelerators.

feeding conveyers. A number of varients of radiation zones and constructions of local radiation shielding are presented in Chapter 9 containing description of standard radiation installations.

The tolerable radiation dose for the attending personnel operating with hazardous radiation sources in the USSR is $10^{-3}$ z/yr (0.1 rem/week) or $5 \times 10^{-2}$ z/yr (5 rem/yr). People who do not directly

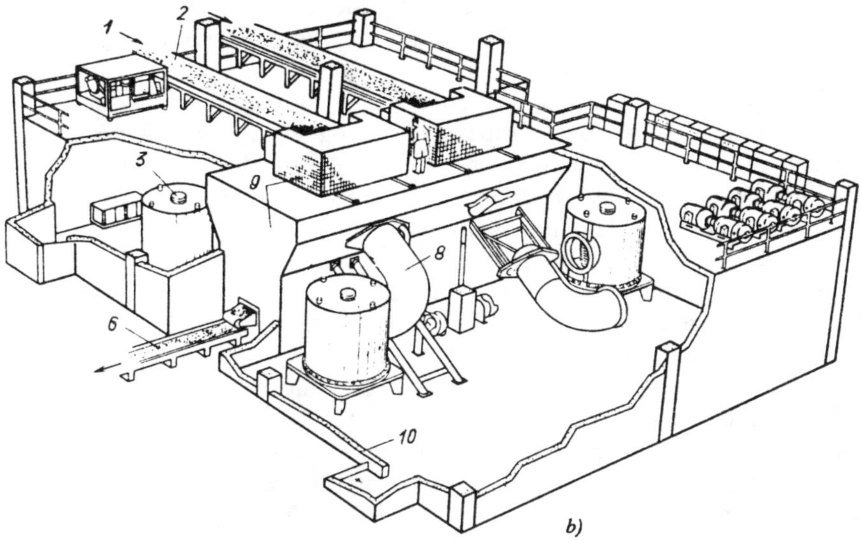


**Figure 8.11** (*Continued*) (*b*)

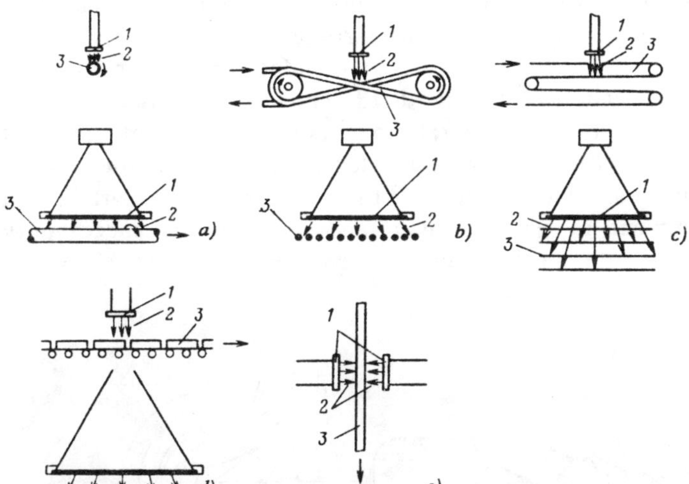


**Figure 8.12** Diagrams of motion of articles in zone of beam radiation. *a)* and *b)* Irradiation of pipes or cable insulation. *c)* Irradiation of thin-film materials. *d)* Irradiation of parts on conveyer. *e)* Double-sided irradiation of sheet material. Direction of movement of articles is shown by arrows *1*—Extraction window; *2*—electron beam; *3*—article under radiation treatment.

handle the radiation facilities should not be exposed to radiation levels exceeding that of natural radiation, which according to international standards is $2 \times 10^{-3}$ z/2000 h per year.

It should be noted that radiation zones are equipped with ventilation systems to remove ozone and nitrogen oxides arising in the atmosphere under the action of electron beam. The maximum permissible concentrations in inhabited rooms shall be for ozone, 0.5 mg/m$^3$ and for nitrogen, 0.1 mg/m$^3$.

Figure 8.12 shows several widely used methods of transfer of articles under treatment within the radiation zones. Thus, for overall uniform irradiation of round articles (pipes, cables, etc., the articles are revolved (Fig. 8.12*a*). Other versions of multisided radiation treatment are discussed in Section 8.1 (Figs. 8.4*b* and 8.5). In the diagram depicted in Fig. 8.12*c*, the thickness of the material under treatment is several times less than the length of the electron path and therefore the material receives the necessary radiation dose only after several passes. This principle enables achievement of a higher coefficient of utilization of the beam and easier cooling of the material under treatment. In the case illustrated in Fig. 8.12*e*, the thickness of the material exceeds the path of the electrons. This system permits more than twice the reduction of accelerator energy in treating the article throughout its thickness (see Fig. 2.4*b*) and also enables two-sided processing of surfaces of massive articles. Figure 8.13 shows the irradiation of large-sized objects by bremsstrahlung (sterilization of articles packed in boxes) [128]. Each object undergoes four radiation treatments provided that each side is turned toward the radiation source. For treatment of gaseous materials, use may be successfully made of a system with an extracted concentrated beam, as shown in

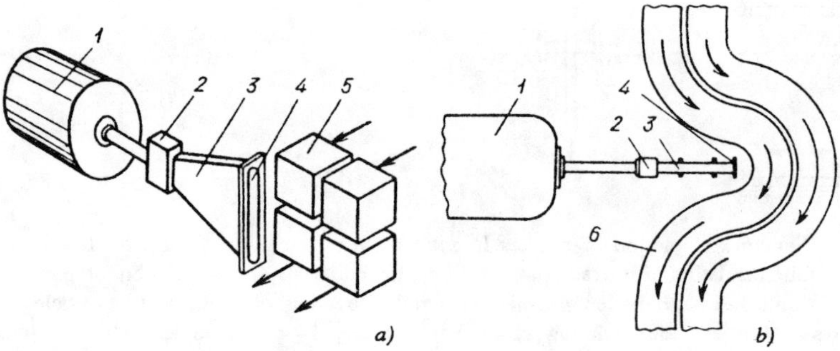


**Figure 8.13** Diagram of treatment of massive articles by bremsstrahlung radiation. *1*—Accelerator; *2*—beam scanning system; *4*—horn; *4*—extraction window; *5*—object under radiation; *6*—conveyer.

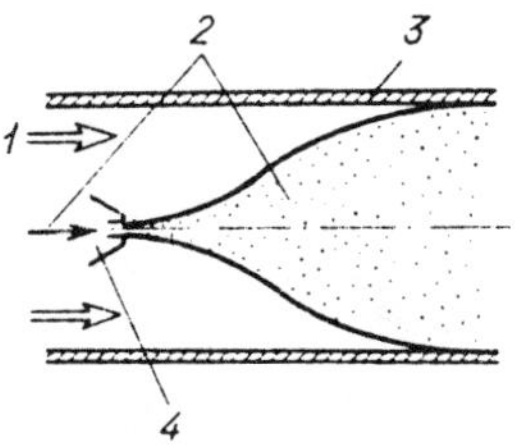


**Figure 8.14** Version of installation for radiation treatment of gaseous products. *1*—Direction of gas flow; 2—electron beam; *3*—reactor tube; *4*—system for extracting concentrated beam.

Fig. 8.14. Decrease of divergence of the electrons in the transverse direction may be attained only by the use of a longitudinal magnetic field.

In some processes, it is preferable to dispense with extraction of the electrons into the atmosphere, but conversely to introduce the material under treatment into the vacuum space. At the very outset of development of the radiation technology, the problem of insertion of round or flat articles (for instance, cables) into the vacuum chamber (which above all eliminates interaction of the material under treatment and air oxygen). However this method proved to be more difficult for radiation technology than extraction of the electron beam. That is why vacuum treatment by electrons is practiced nowadays only in the case of minor energies of the particles, i.e., when extraction of the beam from vacuum is impractical. Mainly this concerns the heat treatment of material, and the articles to be treated are fed into the vacuum chamber periodically. For welding large-sized parts this problem is often solved by constructing vacuum chambers that can be advanced along the weld. Figure 8.15 shows a facility used for continuous feeding of articles to be treated into the vacuum chamber. The steel band is inserted into the vacuum chamber where a coating is deposited by spraying under the action of the electron beam [38].

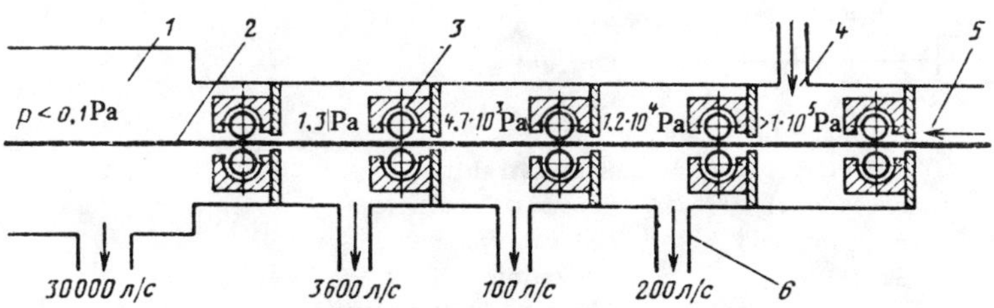


**Figure 8.15** Vacuum lock with pressure rollers for continuous feed of steel band into vacuum. *1*—Treatment zone (vacuum chamber); 2—steel band; *3*—pressure roller; *4*—delivery of gas; *5*—direction of advance of steel band; *6*—branch pipe for connection to vacuum pumps.

## 8.4 MEASUREMENT OF BEAM PARAMETERS AND RADIATION DOSE

Figure 8.16 demonstrates principle characteristics subject to check and some methods used for checking the parameters of radiation generated by an accelerator. Some methods are applicable only to set up operations, others are useful for continuous monitoring of the process. In some industrial accelerators provision is made for automatic control systems that maintain the preset radiation intensity. Used as pickups are some of the systems for measuring the radiation parameters described below. Methods of measurement of electron beam characteristics shall be considered first.

### 8.4.1 Measuring Electron Beam Current

One of the extensively applied methods of measurement of electron beam current consists in reception of emitted electrons by the target and measurement of the current in the target circuit. Due to the fact that a flux of secondary electrons arises on the target, the value of the current to be measured will be distorted to a level of tens of percent; therefore, targets in the shape of the Faraday cylinder are usually used (Fig. 8.17*a*). The inner side of the cylinder is sometimes made of graphite featuring a small coefficient of secondary emmission. With the use of such a target a secondary-electron effect may be minimized and measurement error may be diminished to ~0.5 percent. For high-magnitude currents of short-duration beams, primary consideration is given to decreasing inductance of the measuring instrument. It is essential that the time constant of the measuring resistor should be much less than the current front length: $L/R \ll \tau_{front}$. To obtain this, use is made of

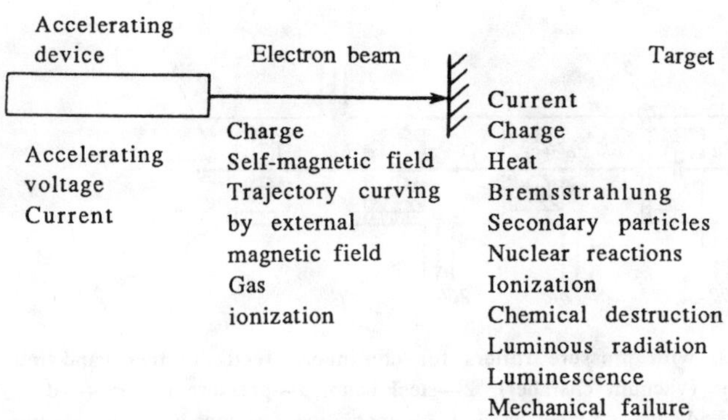


**Figure 8.16** Main characteristics subject to inspection and some inspection methods.

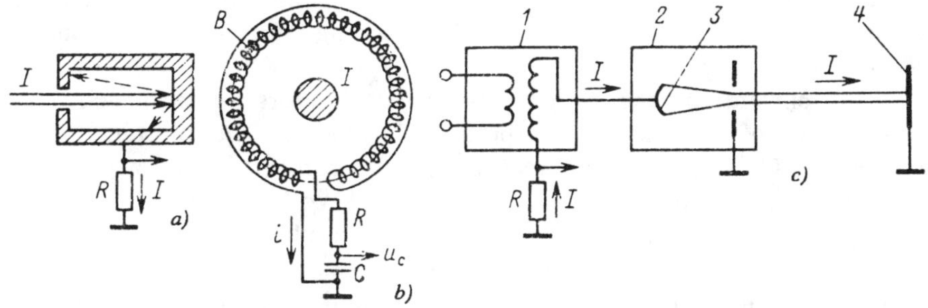


**Figure 8.17** Diagram of unit for measuring beam current. *a)* Faraday cylinder. *b)* Rogovsky belt. *c)* Pickup of current in accelerator. *1*—High-voltage generator; *4*—accelerating device; *4*—cathode; *4*—target; *B*—magnetic field of electron beam. Dotted lines show possible trajectories of secondary particles.

shunts composed of a number of parallel-connected resistors which constitute a coaxial line. The Faraday cylinder is also used for finding the total energy released by the electron beam (calorimetric method). The merits of measurements by use of the Faraday cylinder are simplicity and high accuracy; the disadvantage is the impossibility of determining the beam parameters without distributing the beam itself.

This shortcoming is not characteristic of the circuit for measuring pulsed current with the help of the Rogovsky belt which is essentially a torroidal winding enveloping the electron beam (Fig. 8.17*b*). The electromotive force generated on the winding is induced by the magnetic field of beam *B*:

$$u = -M \frac{dI}{dt}$$

where $M$ is the coefficient of mutual induction, and $I$ is the total current of the beam. In order to obtain a signal proportional to $I$, provision is made for an integrating circuit at the output of the measuring belt. Current $I$ flowing in the circuit made up of Rogovsky belt, resistor $R$, and capacitor $C$ is derived from the equation

$$-M \frac{dI}{dt} = L \frac{di}{dt} + iR + \frac{1}{C} \int i \, dt$$

where $L$ stands for the inductance of the circuit.

Depending on selection of $L$, $R$, and $C$, use may be made of two modifications of the integrating measuring circuits, namely,

1. $C = 0$, $\omega L/R >> 1$, then $u_R = IM \ R/L$
2. $R \gg \omega L$, $1/\omega C$, then $u_C = IM/RC$

In general, it is necessary to take into consideration the natural capacitance and the resistance of the winding and other components of the measuring circuit.

The current in the industrial-use accelerators often is checked by the use of a pick-up incorporated in the cathode circuit. Figure 8.17*c* shows a circuit for measuring the current in an accelerator employing a set-up transformer as a high-voltage generator. Measuring resistor $R$ is connected to the circuit of the transformer secondary winding from the side of the chassis. With such a circuit, it is essential that the capacitive current in the parasitic circuits in the high voltage generator, particularly from the secondary winding to ground, be rather small.

For relative measurements of the electron beam current, it is possible to determine the current of the ions and electrons produced as a result of ionization by bunched atoms of the residual gas. The current of the beam coming to the target is characterized, among other things, by the secondary emission current and intensity of the bremsstrahlung (see Fig. 8.16).

### 8.4.2 Measuring Electron Energy

The energy $W$ in the direct-action accelerators is found by measuring the accelerating voltage in the accelerator. (In pulsed accelerators, particularly in nanosecond ones, respective correction is introduced for inductive drop of voltage between electrodes.) The other method of determining energy of accelerated electrons is based on the principle of measurement of radius of curvature of trajectory $R$ in the external transversal magnetic field $B$. For relativistic electrons, $W$ is related to $R$ by

$$W = \sqrt{9 \times 10^4 B^2 R^2 + (0.511)^2} - 0.511$$

where $W$ is expressed in MeV, $B$ in T, and $R$ in meters. The configuration of transverse magnetic fields in spectrometers used for determining energy ensures separation of trajectories of electrons differing in energy and provides focusing of electrons possessing similar energy. Figure 8.18 illustrates diagrammatically a spectrometer with semicircular focusing of electrons in plan. The spectrometer is designed for measuring the spectrum of a pulsed electron beam. Electrons (1) after turning in homogeneous magnetic field $B$ through 180° are fed to insulated current collectors (2). Each of the collectors is coupled to a capacitor ($C$). On the basis of the charging voltage of the capacitors after applying an electron pulse, it is made possible to find distribution of the particles with different energies. In the case of a time-constant beam, capacitors are sub-

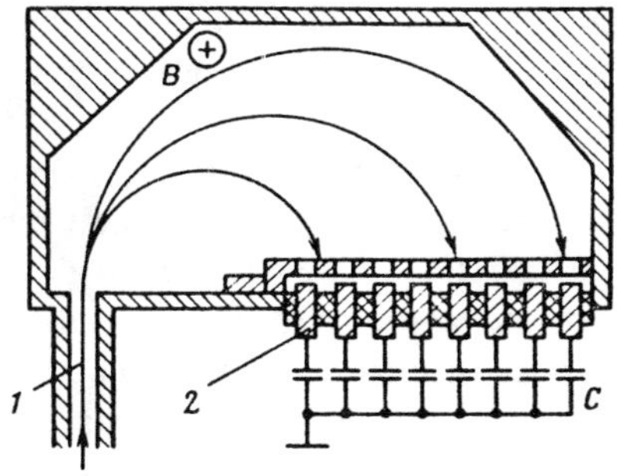


**Figure 8.18** Measurement of electron energy by spectrometer with semicircular focusing. *1*—Electron beam; *2*—current collectors; *B*—magnetic field of spectrometer.

stituted by resistors. To increase the illuminating power of the spectrometers, use is made of a magnetic field which tends to attenuate with increase of the radius and ensures proper focusing of the particles in two directions (in the plane and at a right angle). The electrons in such spectrometers are turned approximately through 250° and the illumination power of the instrument is much higher than in the case of a uniform magnetic field.

Figure 8.19 shows the methods of measuring electron energy both in path and in absorption in the matter. Absorption of electrons in substances has been adequately studied; therefore, electron energy can be determined by finding the relationship between the current on the Faraday cylinder and the thickness of the foil interposed in the beam. The value of $R_{ex}$ is derived by the intersection of the tangent to the sharply dropping section of the absorption curve with the abcissa axis. For the path of electrons in aluminum in the energy range of 4 to 12 MeV, $R_{ex} = 0.565\ W - 0.423$, where $R_{ex}$ is measured in g/cm², $W$ in MeV.

In the charge-finding spectrometer, the energy spectrum is determined by the charge of the electrons absorbed in each of many foils (Fig. 8.19*b*). The energy of electrons of a monoenergetic beam

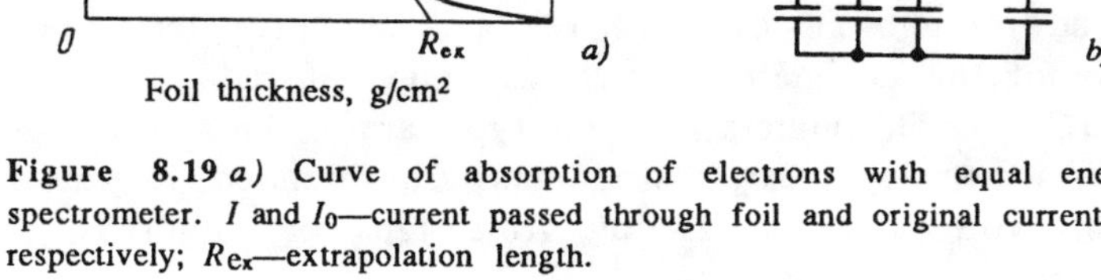


**Figure 8.19** *a)* Curve of absorption of electrons with equal energies. *b)* Charge spectrometer. *I* and $I_0$—current passed through foil and original current of electron beam, respectively; $R_{ex}$—extrapolation length.

(or maximum energy of a nonmonoenergetic beam) is derived on the basis of the *e-n* and $\gamma$-*n* threshold reactions. The maximum energy of bremsstrahlung quanta arising in matter under deceleration of a monoenergetic electron beam is equal to the energy of the electrons. Activation of the target occurs only if the energy of the electrons exceeds the threshold energy (see Section 2.4). The specimens, in this case, are irradiated by electrons of different energy exceeding the threshold energy, which makes it possible to find the reaction yield–energy dependence and then to derive the value of threshold energy.

Energy may also be measured by the calorimetric method (if the current has been determined by another method) and on the basis of bremsstrahlung radiation. In the latter case, the energy can be found from the ratio between the intensity of radiation applied to two similar pickups installed behind different absorbers. This system should first be calculated or calibrated against the source with well-known energy.

### 8.4.3 Measuring Bremsstrahlung Intensity

A most commonly applied method of measuring bremsstrahlung intensity is the ionization method based on the proportionality of the number of ionizing events in a gas space to the value of the absorbed energy. The ionization chamber is a gas space confined in an enclosure with hard walls and built-in electrons. As the latter are connected to an external power source, a current arises that is proportional to the radiation energy absorbed in that chamber. Another version of the ionization chamber is represented by a charged capacitor whose voltage tends to drop as the radiation dose rises. In addition to measurement of bremsstrahlung intensity the ionization method is applied for determining the characteristics of other kinds of radiation featuring high penetrability, including this property of electrons, beginning with the energy of a few megaelectronvolts.

An ionization type dosimeter may be exemplified by the personal monitoring dosimeter type DK-2 (Fig. 8.20). A filament (10) is charged by means of a movable contact (9) relative to a cylindrical electrode (4). The position of the filament depends on its potential. Movement of the filament when its potential is changed can be visually observed through an eyepiece (2) and objective lens (11). The entire scale of the dosimeter covers the range of radiation dose of 0.2 *R* (~50 μC/kg). Dosimeters of this type are manufactured by several firms. As a rule, these instruments are intended for measuring radiation with energies ranging from tens or hundreds of keV to 1 to 2 MeV.

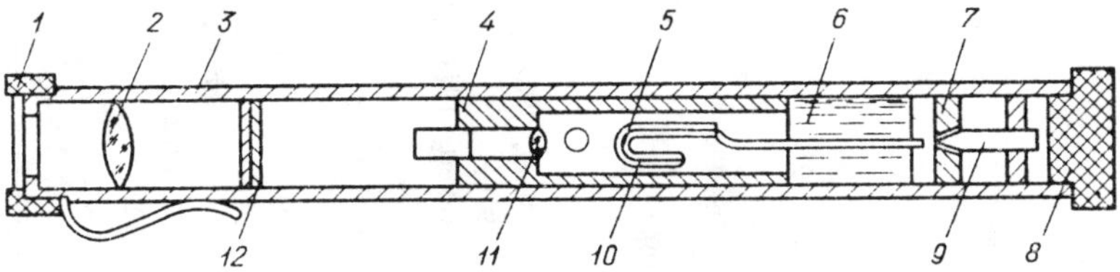


**Figure 8.20** Personal dosimeter K-2. *1*—shielding window; *2*—eyepiece; *4*—dosimeter casing; *4*—external electrode of ionization chamber; *5*—fiber holder of electroscope; *6*—insulator; *7*—membrane; *8*—detachable cap; *9*—movable contact; *10*—quartz fiber; *11*—objective lens; *12*—scale.

## 8.4.4 Measuring Radiation Dose

Now we shall consider methods of determining the absorbed radiation dose which is a principle parameter required in radiation treatment of materials and in radiation therapy [129, 130]. The existing methods cover a wide range of the doses applicable to technological processes: $10^2$ to $10^7$ Gy at dose rate of $10^{-1}$ to $10^5$ Gy/s.

Dosimeters must meet the following main requirements:

1. Independence of dose intensity and temperatures
2. High resolution in determination of topography of beam field
3. High accuracy of measurements, viz., with error not exceeding 10 percent for monitoring technological processes, and not more than 1.5 to 2 percent for beam therapy.

The absorbed radiation dose may also be determined by means of various physical and chemical effects arising in the target under irradiation (see Fig. 8.16).

At present extensive use in radiation technology has been made of dose-metering systems based on the principle of change of chemical structure of substances—solids, liquids and gases. Information on the most commonly applied methods of measurement of the absorbed dose is given in Table 8.2 [131]. The thin-film dosimeters that change color under irradiation and are then analyzed by spectrometry are described in Table 8.3 [131].

**Table 8.2 High-dose dosimetry systems (for doses up to $10^6$ or higher)**

| Dosimeter | Radiation effect and analysis | Approx. useful dose range (Gy) | Authors |
|---|---|---|---|
| $H_2O$, C, metal | Temperature rise measured by thermistor, thermocouple, etc. | $10–10^6$ | Radak and Markovic, 1970 |
| Hydrogen pressure system | Increase of hydrogen pressure inside ampoules containing polyethylene | $10^5–10^7$ | Sheldon, 1970 |
| Sodium formate in aqueous solution | Formate decomposition measured by titration with $KMnO_4$ and colorimetry | $10^3–10^6$ | Janovsky et al., 1963 |
| Oxalic acid in aqueous solution | $(COOH)_2$ decomposition measured by titration with NaOH and colorimetry | $10^4–10^6$ | Holm, 1970 |
| Sugar solutions | Decrease in optical rotation measured by polarimetry | $10^4–10^6$ | Suschny et al. 1967 |
| Polymer solutions | Degradation or cross-linking measured by viscometry, spectrophotometry, gel permeation, chromatography, melt index, tensile strength, elongation, solvent extraction, or conductivity | $10–10^6$ | Feng, 1958<br>Kirchner et al., 1958<br>Taimuty, 1962<br>DwGraff and McLaughlin, 1931 |
| Organic dyes, amino acids, or polymeric solids | Degradation or cross-linking measured by ESR spectrometry or gel determination | $10–10^6$ | Miller, 1959<br>Henglein, 1959<br>Judeikis et al., 1968<br>Ettinger et al., 1981<br>Regulla and Deffner, 1982 |
| Dye solutions and plastics | Degradation and bleaching of dyes measured by spectrophotometry | $10^4–10^6$ | McLaughlin, 1970<br>Hubner and Prokert, 1971 |
| Dye solutions and plastics | Bond unsaturation and dye formation measured by spectrophotometry | $10^3–10^7$ | McLaughlin, 1970<br>Ettinger et al., 1981 |
| Organic phosphors | Degradation of luminescence or trans-cis isomerization, measured by spectrophotometry | $10^4–10^7$ | McLaughlin, 1970<br>Fowler and Attix, 1966<br>Harrah, 1969<br>Ilic-Popovic and Hjortenberg, 1969 |
| Organo-metallic crystals | Chemical shifts measured by Mossbauer spectroscopy or bleaching of optical absorption bands | $10^5–10^8$ | Takashima et al., 1972<br>Chuang et al., 1975 |
| Alkali halide crystals | Color-center formation measured by spectrophotometry | | McLaughlin et al., 1979, 1980 |
| Mn–As–B glass | Color-center formation measured by spectrophotometry | $10^3–10^6$ | Bishay and Arafa, 1967 |

Table 8.3 Thin-film plastic and dyed plastic dosimeters analyzed by UV and VIS spectrophotometry

| Film type | Wavelength analysis (nm) | Typical thickness (mm) | Commercial source |
|---|---|---|---|
| Undyed cellulose triacetate (CTA) | 290, 310 | 125 | Numelek (France) |
| Undyed polychlorostyrene (PCS) | 310, 348 | 50 | Far West Technology (U. S.) |
| Undyed polybromostyrene (PBS) | 320, 350 | 50 | Far West Technology (U. S.) |
| Undyed polyethylene terephthalate (Mylar) | 325 | 12 | DuPont de Nemours (U. S.) |
| Undyed polystyrene (PS) | 332 | 30 | Dow Chemical Co. (U. S.) |
| Blue Celluloid | 385, 590 | 250 | Celanese Chemical Co. (U. S.) |
| Red Celluloid | 385, 510 | 250 | Celanese Chemical Co. (U. S.) |
| Green Celluloid | 305, 410, 635 | 130 | Celanese Chemical Co. (U. S.) |
| Ultraphan blue | 450, 600 | 130 | Lonza AG (Switzerland) |
| Ultraphan red | 405 | 138 | Lonza AG (Switzerland) |
| Ultraphan green | 630 | 146 | Lonza AG (Switzerland) |
| Tohcello blue Cellophane | 605 | 21 | New Tohcello Shoji Co. (Japan) |
| Futamura blue Cellophane | 655 | 19 | Yamashita Insatsu Kamiki (Japan) |
| DuPont blue Cellophane | 655 | 26 | DuPont de Nemours (U. S.) |
| Red cellulose nitrate on Mylar (LR–115) | 410, 515 | 13 | Kodak Pathe (France) |
| Tetrazolium-CL in PVA on Mylar (LR–132) | 490, 550 | 8 | Kodak Pathe (France) |
| Malachite green methoxide in polychlorostyrene (PCS) | 420, 620 | 60 | Far West Technology (U. S.) |
| Pararosanilin cyanide in polyvinyl butyral (PVB) | 545 | 16 | Universidad Nacional Autonoma de Mexico (Mexico) |
| Hexa (hydroxyethyl) pararosaniline cyanide (HPR-CN) in polyvinyl butyral (PVB) | 510, 600 | 70 | Risø Natl. Lab. (Denmark) |
| HPR-CN in polyvinyl butyral (PVB) | 510, 600 | 50 | Far West Technology (U. S.) |
| HPR-CN in cellulose triacetate (CTA) | 590 | 30 | Risø Natl. Lab. (Denmark) |
| HPR-CN in polyvinyl pyrrolidone (PVR) | 595 | 115 | Risø Natl. Lab. (Denmark) |
| Malachite green oxide in polyvinyl chloride (PVC) | 430, 625 | 60 | Risø Natl. Lab. (Denmark) |
| HPR-CN in "Elvamide" Nylon (EVN) | 510, 605 | 45 | Far West Technology (U. S.) |
| HPR-CN in "Hexalon" Nylon (HLN) | 605 | 50 | Far West Technology (U. S.) |
| New fuchsin cyanide in "Trogamid" polyamide (TCR) | 563 | 55 | Far West Technology (U. S.) |
| LiF crystalline chips | 247, 444, 547 | 200–2000 | Harshaw Chem. Co. (U. S.) |

## MATERIAL ADDED IN PROOF

The extended target arranged in the atmosphere and designed to generate the bremsstrahlung radiation using an electron beam of 2 MeV energy and 60 kW power (Fig. 8.9*c*) is described in [182]. On the basis of the described experiments, the target is developed for converting the electron energy 5 MeV, 300 kW into the bremsstrahlung radiation (Fig. 6.24*b*). The target length is near 1.5 m. The conversion efficiency for a 3.7 g/cm$^2$ thick target will be 10.8%. The target is supposed to be used at the electron accelerator with the Cockroft-Walton generator (Sec. 6.6) for the sterilization of medical appliances.

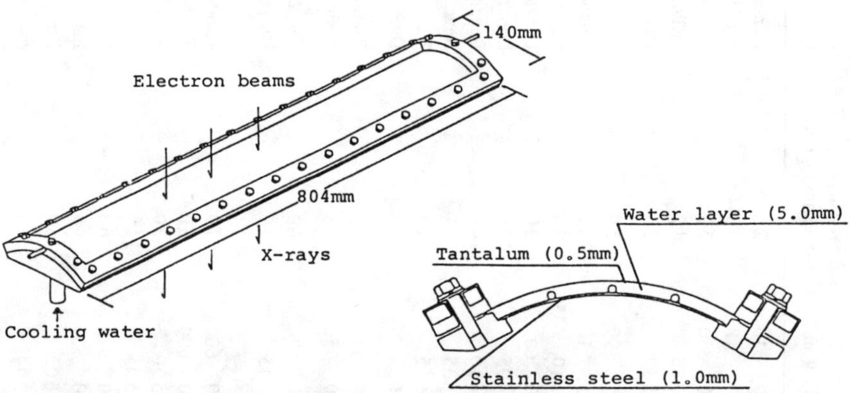


Fig. 8.9*c* X-ray target for the 2 MeV 60 kW accelerator.

CHAPTER

# NINE

# TYPICAL RADIATION FACILITIES BASED ON ELECTRON ACCELERATORS

## 9.1 INDUSTRIAL INSTALLATIONS FOR ELECTRON-BEAM CROSS-LINKING OF POLYETHYLENE

This chapter describes typical radiation installations based on electron accelerators, that are used in radiation chemical processing, flaw detection, agriculture, and medicine and partially being under tests and experimental application in industrial fields of production. The information on the processes with the use of accelerated electrons and on the components of the radiation processing devices are set forth above. This chapter deals mainly with composition of production lines, their performance characteristics, as well as economical efficiency of production processes. Unfortunately, many details of the technology and construction features of many installations used in industry have not been published, which presents difficulties for a more comprehensive description of the present-day level of development in radiation technology.

### 9.1.1 Radiation Modification of Cable Insulation

One of the early technologies applied by the use of a beam of accelerated electrons was the process of radiation modification of cable insulations (see Section 3.2). The production of cross-linked polyolefin insulation under irradiation with the aid of electron accelerators began in the 1960s by a number of firms. The first industrial installation under consideration in the USSR was set into operation in 1968 in the Research Institute of the Cable Industry [132, 133]. A diagram of this equipment is presented in Fig. 9.1. The room accommodating the accelerator beam extractor (3) and multiroller feeding conveyer (5) is surrounded with radiation shielding (4). The room in which the accelerator (2) is mounted is

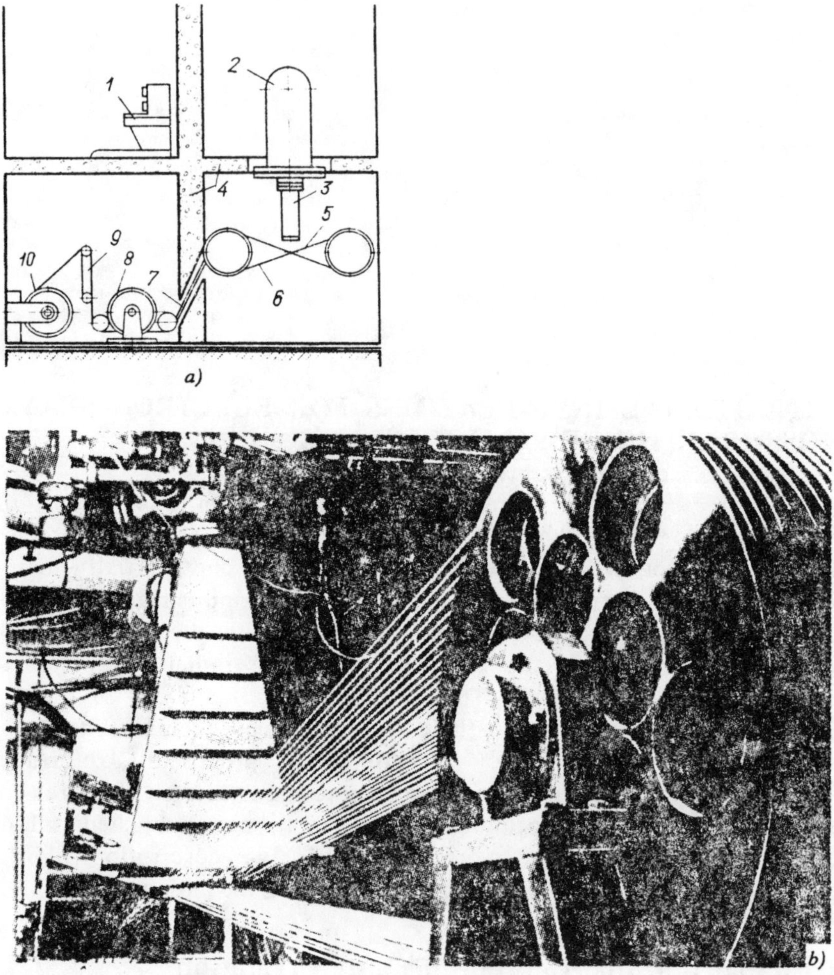


**Figure 9.1** The first Soviet installation of electronic accelerator for industrial radiation processing. *a)* Diagram of installation. *b)* Radiation zone. *c)* Multiroller conveyer. *d)* Appliance with drums. *1*—Control panel for accelerator and technological equipment; *2*—electron accelerator ELT 1.5; *3*—extraction device of accelerator; *4*—radiation shielding; *5*—multiroller conveyer; *6*—cable; *7*—slanted channels in radiation shielding; *8*—feeding device; *9*—compensator; *10*—receiving device.

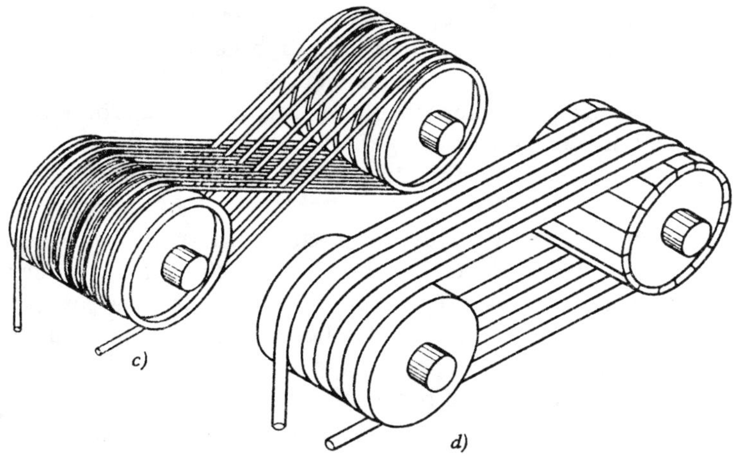


**Figure 9.1** (*Continued*) (*c, d*)

also partially shielded. The cable (6) to be processed is fed to the radiation zone and removed after treatment through sloping ducts (7) provided in the radiation shielding. In front of the intake (10), provision is made for a compensator (9) responsible for control of cable tension. An ELT-1.5 accelerator [165], featuring a maximum energy of 1.5 MeV and a power of 25 kW is used as the electron generating source. The extraction window measures 400 × 40 mm.

The cable under treatment is moved forward along the shorter side of the beam extraction window at a distance of a few centimeters from the electron extracting foil. Substantial problems are presented by the process of uniform irradiation of round cables due to shielding of the cable insulation coating by the conductive core and because of different thickness of the insulation in the direction of the electron beam line of travel. The multiroller conveyer enables the cable under treatment to be turned 180° after each processing pass. Due to this fact and owing to change of initial direction of a part of the electrons caused by scattering in the air, the nonuniformity of the absorbed radiation dose in 80 percent of the cable insulation volume does not exceed ±20 percent. The maximum radial thickness of the cable insulation was ~2 mm; therefore the energy of the electron beam generally was within 0.6 to 0.7 MeV. The speed of delivery of the cable was 25 to 100 m/min, which approximately corresponds to the rate of manufacture of cables on production lines at the plant. A radiation dose of some 100 kGy is injected to the cable insulation in many radiation cycles (passes). The heating after each pass of the cable amounts to a few degrees centigrade. The cable is cooled during the intervals between successive radiation treatments. This process makes it possible to avoid oxidation of polyethylene at a relatively high dosage rate of 100 kGy/s and to perform electron treatment of the insulation in the ordinary atmosphere. Some problems also arise from the possibility of breakdowns under the effect of the changes accumulated in the insulation during the radiation process (see Section 2.5).

The installation under consideration is furnished with an interlocking system which reacts to breaks of the cable under treatment and immediately shuts off the electron beam by the use of a protective shutter. Closing a part of the beam with such a shutter allows additional control of the radiation dose in case of a change in speed of feed of the cable on the conveyer. In this case, the shutter is fabricated of aluminum and is cooled with running water. The coefficient of utilization of the radiation is 35 to 65 percent. The conveyer is stopped and kept shut down primarily because of the limited capacity of the drums of the feeding and take-up appliances, and also during replacement or repair of accelerator parts and associated equipment. After switching on the installation, the nominal operating conditions can be attained in a few minutes. Thus the leading part of the cable is not exposed sufficiently for obtaining the specified radiation dose. The cost of radiation is 0.5 ruble per kilogram, which does not exceed 10 percent of the full price of the final product. The installation discussed was used for about 10 years, with an average annual operation time of 3500 h.

Experience in operation of the cable radiation installation under consideration gained by the other firms provided a means for a wide application of electron radiation in cable manufacture in the USSR. In the mid-1970s, radiation sources were provided by second generation industrial electronic accelerators i.e., electron transformer-rectifier ELV, and ILU-6 accelerators [134]. A newly designed conveyer for feeding cables into the radiation zone has been developed. The first version of the conveyer, where the cable is arranged in the form of figure 8 (Fig 9.1*c*) has a number of disadvantages: comparatively low coefficient of utilization of electron flux (due to considerable gaps between adjacent turns); necessity of replacement of rollers in case of pronounced differences between diameters of articles under treatment; high tension force and consequently systematic breaks of cables under treatment, particularly with cables of small diameter. These shortcomings are practically absent in the conveyer illustrated in Fig. 9.1*d*. This facility incorporates two drums, and the spacings with which cables or wires are laid for processing are controlled remotely. Each turn is moved independently. The neighboring turns nearly adjoin one another, as a result of which the efficiency of the beam approximates 100 percent. The linear speed of articles under treatment can be regulated in the range of 50 to 500 m/min, without replacing the drums.

At the time of the early installations for irradiation of cable insulation every effort was made to develop versatile systems applicable to radiation treatment of articles with a vast range of parameters. However, as the number of modifications of the installations grew, it became common practice to have these facilities specialized for treatment of limited types of articles. Thus, for irradiation of high-voltage cables with thick insulating coatings, a promising processing method is application of several electron beams to be directed to the cable from different sides (see, for example, Fig. 8.4*b*).

The problem of uniform radiation of large articles may be solved with the use of effects of bremsstrahlung obtained by conversion of energies of intense electron beams of 4 to 5 MeV and higher.

At present cables with irradiated insulation are produced in the United States by Western Electric Co., International Telephone and Telegraph Co., Ranchem Corp., General Cable Corp.; in Japan by Sumitomo Electric Industry; in the Federal Republic of Germany by AEC-Telefunken; in France by ACOME; in East Germany by FEB Kabelwerke Oberspree; and by other companies.

Figure 9.2 represents a chart of approximate expenses for manufacturing a radiochemical processing installation (RCI) which

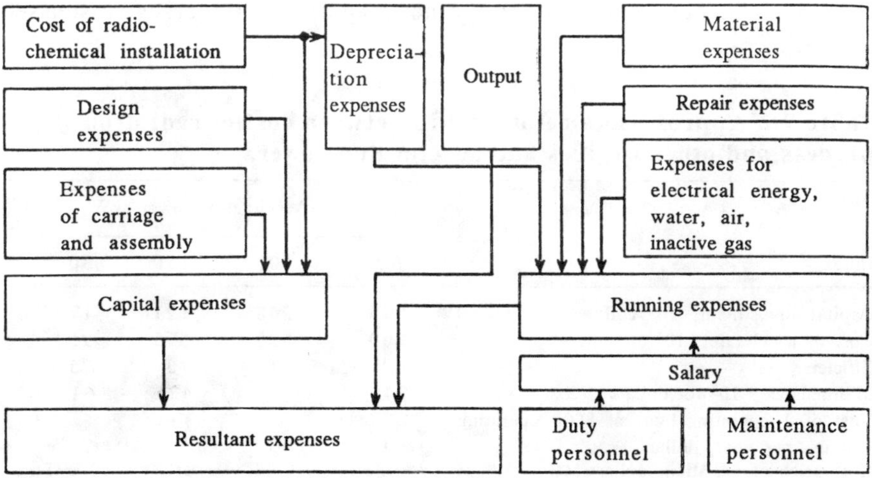


**Figure 9.2** Chart of approximate expenses for radiation processing installation.

**Table 9.1 Comparison of cost and other indices of radiation and chemical cross-linking of cable insulation (in dollars)**

| Index | Radiation cross-linking | Chemical cross-linking |
|---|---|---|
| Capital investment | 260,000 | 200,500 |
| Amortization (5 years) | 52,000 | 40,000 |
| Wages | 24,000 | 36,000 |
| Overhead | 25,000 | 36,000 |
| Maintenance | 12,000 | 16,000 |
| Loss due to rejects | 3,600 (3%) | 18,400 (10%) |
| Cost of 1 h of operation at 6000 operating hours per year | 20.6 | 33.6 |
| Cost of cross-linking, 1-km insulation length | 1.35 | 3.67 |

corresponds to the cost of cable treatment installations and of other versions of radiochemical processing plants [135]. The economical efficiency of the radiation cross-linking operations by use of electron beam radiation are apparent from Tables 9.1 and 9.2 [136].

As a rule, the cost of radiation cross-linking of plastics is one-third to one-half the cost of the chemical method of this process. In addition, heating resistance and some other characteristics of polyethylene stitched by radiation are superior to the characteristics of welded chemically material (see Table 3.3). The lower cost of the radiation technological process as compared with the traditional technology is mainly due to reduced consumption of energy. As seen from Table 9.2, expenses for radiation treatment tends to diminish with increase of power of the installation. This relationship is also characteristic of other radiation processes discussed below.

**Table 9.2 Approximate relationship between cost of radiation process and other indices and accelerator power.**

| | Accelerator power, kW | | | |
|---|---|---|---|---|
| Index | 50 | 100 | 150 | 300 |
| Capital investment, $10^3$ dollars | 595 | 948 | 1218 | 1725 |
| Cost of accelerator, $10^3$ of dollars | 300 | 525 | 675 | 950 |
| Efficiency, % | 62 | 71 | 73 | 75 |
| Maintenance, $10^3$ of dollars | 248 | 344 | 425 | 611 |
| Cost of 1 operating hour at 8000 operating hours per year, dollars | 31 | 43 | 53 | 76.5 |
| Cost price of radiation, dollars/MGy · kg | 0.28 | 0.17 | 0.13 | 0.095 |

### 9.1.2 Production of Polyethylene Film

Production of heat-shrunk tubular (sleeve) polyethylene film thermoplen in the USSR has been in full swing since 1975 in the scientific-industrial enterprise "Plastic" (for the characteristics of the thermoplen, see Table 3.5). The installation for thermoplen processing is diagrammed in Fig. 9.3 [137]. The electron source in this installation is the electronic accelerator Electron-3M rated at a nominal beam energy of 0.7 MeV and a power of ~0.7 kW. The scanning length is 0.5 to 1 m, the nonuniformity of the beam density along the scanning line is not more than ±5 percent. The service life of the foil is 500 to 1000h. The treatment process can be described in brief as follows. Air under low pressure is delivered into a polyethylene sleeve (2) vertically forced out of an extruder (1). Because the speed of pull rollers (4) exceeds that of extrusion of sleeve, the film attains its original stretch ratio. Rollers (3) provide uniform folding of the sleeve. The radiation chamber (5) has labyrinth ducts with an inlet and an outlet for the film to be radiated. In the radiation zone, the film is arranged in meandering manner (7). The radiation dose is 100 to 200 kGy. The film thickness is "less than or equal to symbol" ≤1 mm, and the width of the folded sleeve is ≤1 m. Upon completing the radiation treatment in chamber (5), the sleeve is delivered to an orientation chamber (9) filled with hot fluid that is inert to polymers. The fluid temperature is ~100°C. The feed rollers (8) together with pull rollers enable the sleeve to advance and at the same time serve as a lock for creating the next air bubble. A take-up winder (10) ensures longitudinal stretching of the film by pulling it at a higher rate—6 to 24 m/min—while the speed of the feed rollers is only 2.4 to 12 m/min. After the sleeve is brought out of the radiation chamber (9), it

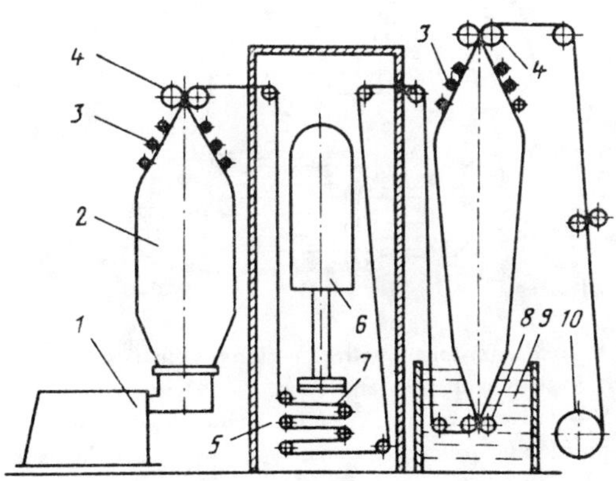


**Fuigure 9.3** Diagram of installation for production of heat-shrunk polyethylene film. *1*—extruder; *2*—sleeve; *3*—folding rollers; *4*—pull rollers; *5*—radiation zone; *6*—electron accelerator; *7*—film in radiation zone; *8*—feed rollers; *9*—orientation chamber; *10*—take-up winder.

undergoes orientation in the transverse direction under the action of the inflated air. Then the film sleeve is cooled down by compressed air. By varying the air pressure inside the sleeve, it is possible to establish different stretch ratios longitudinally and transversely. Usually this ratio ranges from 3:1 to 5:1.

One of the basic applications of this polyethylene film is for manufacturing bags for packing cooled foodstuffs in vacuum. The food is placed into sections of film sleeve, and then these sections are evacuated and welded. The closed film bags are blown out with warm air or placed in hot water for a short time. As a result, the film will tightly envelop the enclosed matter.

The installation shown in Fig. 9.3 can be modified to be applied for production of heat-shrunk tubes used for termination of wires and cables. The installation capacity in treating film is 7000 kg/h kGy, or 800 to 5000 kg/h kGy when processing tubes (depending on the tube diameters). One of the main problems to be solved in production of the tubes is the need for minimum longitudinal shrinkage to prevent the tubes from slipping from the protected places in the process of shrinkage. This is achieved by delivering heated gas or liquid under pressure into the tube with simultaneous evacuation from the outside. Some other methods of production of the tubes are demonstrated in Fig. 9.4 [138]. In the case depicted in Fig. 9.4*a*, an irradiated tube (2) is inserted into the heated part of a die (3) in which it is expanded by compressed air. Then this tube is cooled in the subsequent part of die 4, and thus remains unchanged in size.

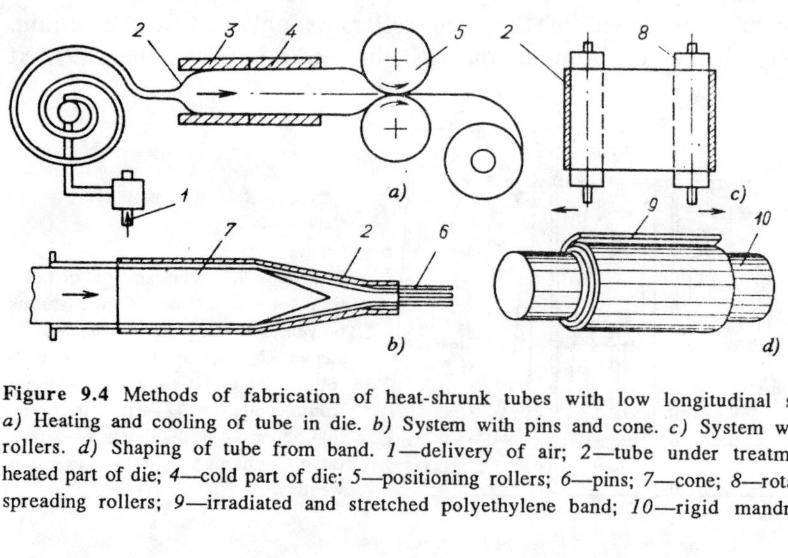


**Figure 9.4** Methods of fabrication of heat-shrunk tubes with low longitudinal shrinkage. *a)* Heating and cooling of tube in die. *b)* System with pins and cone. *c)* System with rotary rollers. *d)* Shaping of tube from band. *1*—delivery of air; *2*—tube under treatment; *3*—heated part of die; *4*—cold part of die; *5*—positioning rollers; *6*—pins; *7*—cone; *8*—rotating and spreading rollers; *9*—irradiated and stretched polyethylene band; *10*—rigid mandrel.

**Table 9.3. Economic efficiency of radiated polyethylene bags [20].**

| Thickness, µm | Price of bag, dollars | Cost of radiation, dollars | Price of radiated bag, dollars | Economical efficiency(% of cost of non-radiated bags) |
|---|---|---|---|---|
| 100 | 0.3528 | 0.0117 | 0.3645 | 17.4 |
| 75 | 0.2645 | 0.0088 | 0.2733 | 38.00 |
| 50 | 0.1764 | 0.0059 | 0.1823 | 58.00 |

This done, the tube is laid by the use of rollers (5). In Fig. 9.4*b*, the tube (2) is fitted on a bundle of pins (6) with a cone (7) inserted into the center of the pins. The cone moves the pins apart and thereby expands the tube. In the version depicted in Fig. 9.4, inside tube 2, there are two revolving rollers (8) which are brought away from each other, thereby increasing the diameter of tube 2. Figure 9.4*d* shows fabrication of a tube from a radiated stretched band (9) wound on a rigid mandrel (10). The wound band is warmed up until the edge layers are glued together and a tube is formed. Then the tube is cooled down and removed from the mandrel. Heat-shrunk polyethylene is produced by many companies including Alpha Wire Corp. (USA), Nitto Electric Industrial Corp. (Japan), and Compagne Francaise des Isolant (France). Table 9.3 presents data on the economic efficiency of radiated polyethylene bags [20].

## 9.2 INSTALLATIONS OF ELECTRON BEAM CURING

The first experimental, industrial installations for curing coatings applied to paper, wood, metal, textile, and other materials were developed in the mid-1960s by Radiation Dynamics Inc. (on the basis of an electronic accelerator of 0.3 MeV, 7.5 kW), Ford Motor Co. (on the basis of the ICT accelerator with a beam of 0.5 MeV and a power of 10 KW), High Voltage Engineering Corp., Buls Cascade Corp. (U.S.), Tube Instrument in cooperation with Forter Paints (U.K.), Brown Boveri (Switzerland), and other companies. One of the leading firms that is known as a designer and supplier of the accelerators and technological equipment for the process under consideration at present is Energy Science, Inc. (U.S.).

By virtue of a minor thickness of coating (hundreds of micrometers) and correspondingly small energies of electrons, the materials are usually processed under local radiation shielding. In some cases, a need arises to provide an inert atmosphere. Figure 9.5*a* shows a typical Electrocurtain type accelerating device and radiation zone of the installations, manufactured by Energy Science, Inc. The sheet beam (6) of accelerated electrons hits the material under treatment after passing through the foil of the extraction window (1). Figure 9.5*c* is a diagram of a radiation plant used for hardening adhesive-type coatings. This installation was assembled and mounted in 1975 for Bixby International, Inc. [139]. The energy of

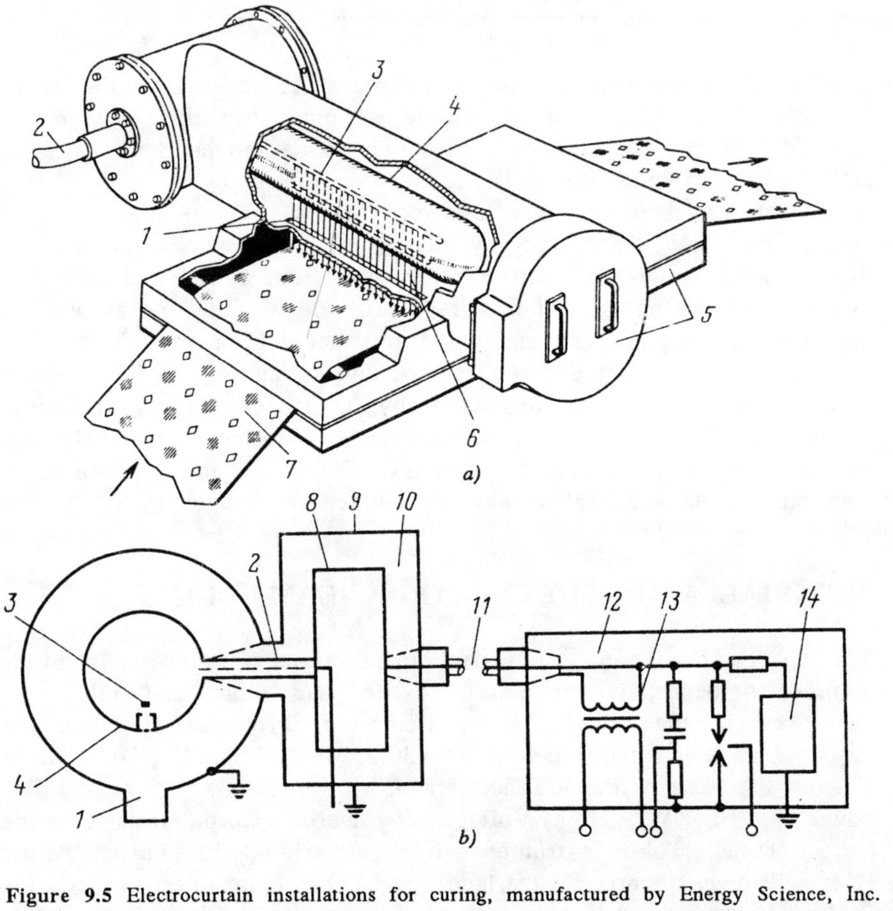


**Figure 9.5** Electrocurtain installations for curing, manufactured by Energy Science, Inc. (U.S.). *a)* Typical construction of accelerating device and radiation zone (see also Figure 6.43*e*). *b)* Schematic diagram of electrocurtain installations. *c)* Diagram of installation for curing adhesive coatings. *d)* Diagram of installations with four electron beams. *e)* General view of installation for treatment of coatings of materials of up to 1.7 m. wide. *1*—Extraction window; *2*—power supply cable of accelerating device; *3*—cathode; *4*—cathode electrode; *5*—elements of radiation shielding; *6*—electron beam; *7*—material under treatment; *8*—modulator and heater transformer; *9*—optical monitoring; *10*—gas insulation; *11*—cable; *12*—oil insulation; *13*—dividing transformer for supplying cathode; *14*—high-voltage generator; *15*—rewinding drums; *16*—rollers; *17*—apparatus for brushing materials under treatment; *18*—accelerator; *19*—radiation zone; *20*—evacuation zone; *21*—flakes; *22*—slitting roll; *23*—stretcher; *24*—cooled drum.

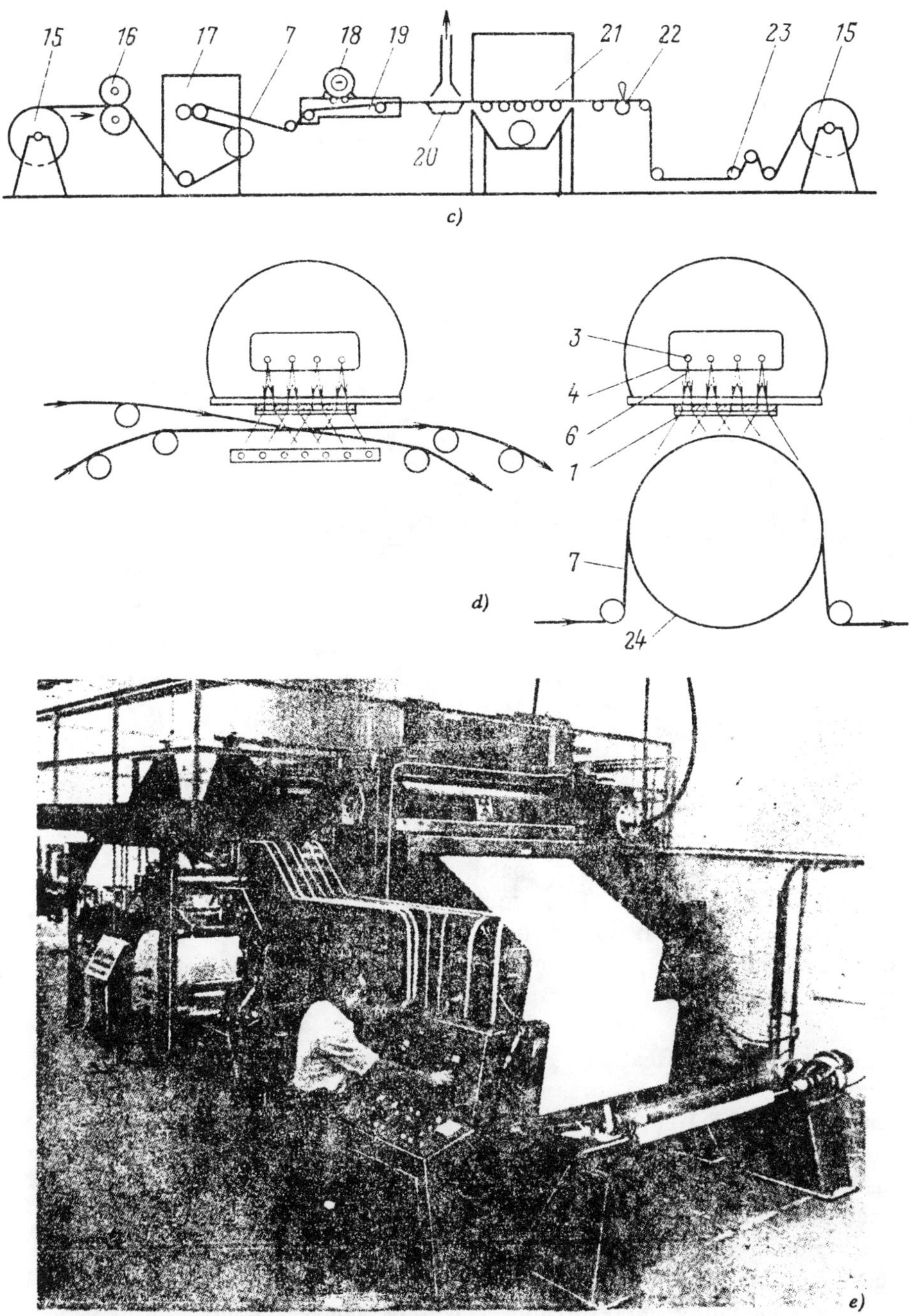


Figure 9.5 (*Continued*) (*c, d, e*).

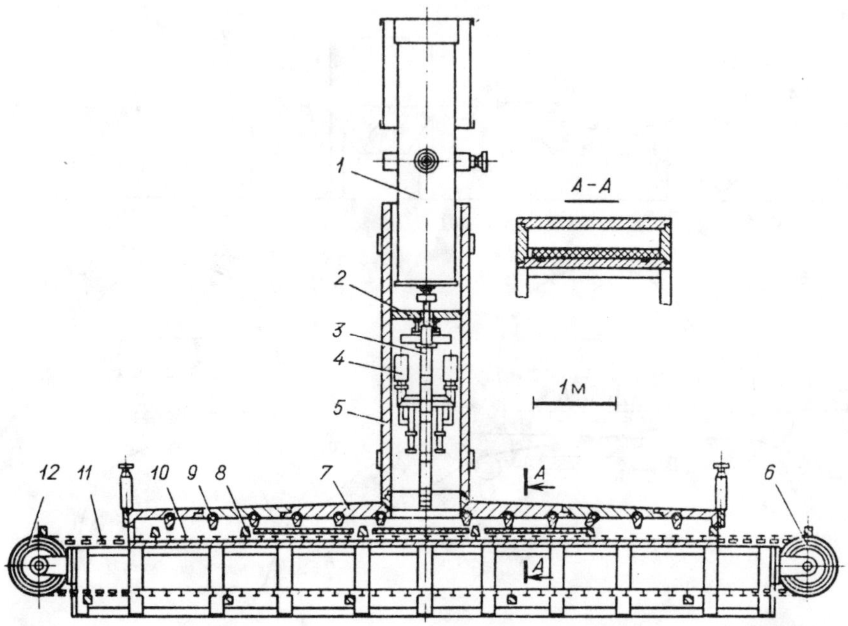


**Figure 9.6** Sketch of installation for curing varnish and paint coatings (based on Electron-3 accelerator). *1*—Electron accelerator. *2*, *5*, and *7*—elements of radiation shielding of accelerator and conveyer; *3*—radiator; *4*—vacuum system; *6* and *12*—driving and driven drums of conveyer; *11*—conveyer drive chain.

electrons is up to 200 keV, and the width of the beam is up to 1.22 m. The conveyer speed amounts to a few tens of meters per second. Irrespective of the conveyer speed, a constant radiation dose is provided during operation. This equipment is used by the shoe industry. The total area occupied by the installation is 9 $m^2$; the length along the conveyer is 1.5 m. Figure 9.5*e* shows an Electrocurtain model used for treatment of coatings applied on materials of up to 1.7 m wide.

In some modifications of the Electrocurtain model Energy Science, Inc., makes use of two or three accelerating devices with a beam power that reaches 1.2 kW/cm of window length. The energy is controlled within 150 and 300 keV [140]. These installations ensure that, within one radiation pass, the article under exposure receives a dose of the order of 10 kGy at the conveyer speed of several hundred meters per minute. Frutiger and Nablo [140], stress that it is practical to use these installations to perform such operations as annealing metal bands and other heat-treatment processes.

New modifications of the Electrocurtain units employ several parallel electron beams [141]. Figure 9.5*d* shows a unit with four cathodes (and correspondingly with four extraction windows). The electron energy equals 300 keV, and the current of each beam is about 150 mA. For removal of heat from the material under treatment (7) use is made of a cooled drum (24).

Figure 9.6 shows an early Soviet installation designed for curing. The Electron-3 accelerator serves here as the electron source. The extraction window of the installation measures 1000 × 40 mm, the energy stability being within ±5 percent. Provision is made for appliances creating an inert atmosphere in the radiation zone. The distance from the extraction window to the sheets under treatment can be varied from 50 to 400 mm, with the conveyer speed of 0.5 to 20 m/min. The overall dimensions of the coated sheets under radiation are 1500 × 1000 × 50 mm. Radiation of coatings applied on rolled materials of 1000 mm wide is also possible. Proximal radiation protection (2, 5, 7) is made of shields from 120-mm thick sheet steel. The lower part of the shielding close to the main source of the bremsstrahlung radiation carries an additional built-in layer of lead. The feed conveyer comprises two closed roller chains fitted on the gear rims of drums (6 and 12) of which one is driving. The production capacity of the installation amounts to $10^5$ m$^2$/year. The overall dimensions of the facility are 4.5 × 7.5 × 7 m.

Figure 9.7*a* depicts a general view of the installation developed on the basis of the Aurora accelerator and intended for hardening varnish and paint coatings on television set cases [142]. The maxi-

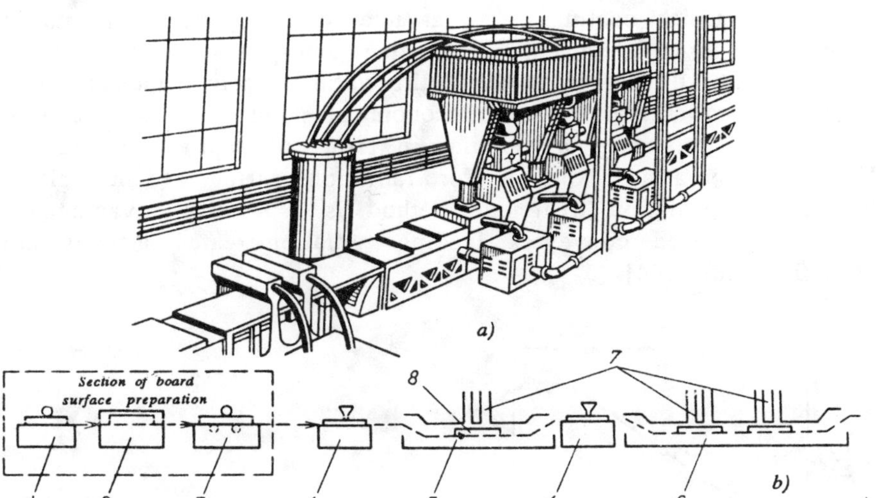


**Figure 9.7** Installations for curing varnish and paint coatings. *a)* General view of installation for treatment of television set casings. *b)* Diagram of installation for treatment of furniture wooden boards. *1*—Dust-removing machine; *2*—chamber for heating furniture boards; *3*—rolls for prime coating; *4*—varnishing machine; *5*—first radiation zone; *6*—second radiation zone; *7*–radiators; *8*—article under treatment.

mum electron energy is 0.5 MeV, and the total current of the electrons in three tubes goes up to 50 mA. The extraction windows measure 700 × 40 mm. The maximum dimensions of the articles suitable for radiation treatment are 800 × 550 × 25 mm. The speed at which the articles are fed across the radiation zone reaches 10 m/min. The overall dimensions of the installation are 16 × 1 × 4.5 m. The local biological shielding is composed of individual sections. The unit has a production capacity of $3 \times 10^5$ $m^2$/year.

Figure 9.7*b* shows the equipment used for hardening varnish and paint coatings on furniture wooden boards. The installation is based on two USSR electronic accelerators EOL [143]. It can treat articles up to 900 × 2000 × 25 mm in size. This installation incorporates three radiators (7) of which the first is mounted separately. After the boards pass the preparation section, they are brought to the coating section where by the use of a varnishing machine (4) they are coated with a layer of varnish. The varnish coating is hardened in the first radiation zone (5). Then the article is conveyed to another varnishing machine where the second layer of varnish is applied to the surface. The second layer of the varnish coating is hardened in radiation zone (6). The overall dimensions of radiation zone 5 are 10 × 2 × 4 m. The dimensions of radiation zone 6 are 15 × 2 × 4 m. The productivity of this production line reaches $9.5 \times 10^5$ $m^2$/year.

At present in the USSR, there are several operating lines for hardening varnish coatings, e.g., at television engineering plants in Lvov, Simpheropol, and at some other enterprises. In the United States, about 10 percent of the entire varnish coating industry uses electron beam curing. From an economical point of view, radiation and thermal processes are similar in terms of capital investments. However, operating expenses for radiation curing are materially lower. The radiation treatment method is especially advantageous where high-speed conveyers feeding radiation-treated articles are used (Fig 9.8) [144].

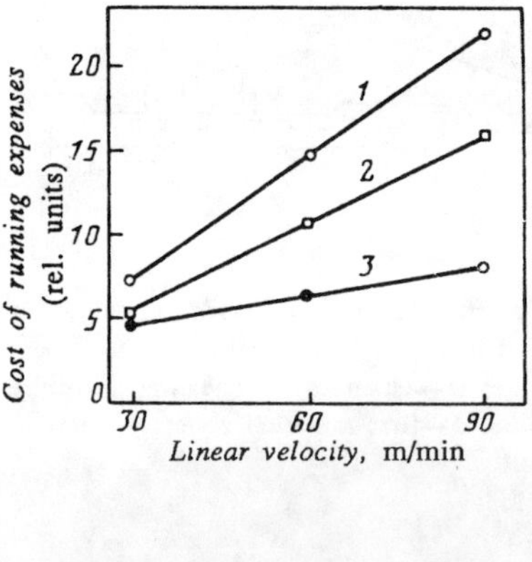


**Figure 9.8** Comparative current expenses depending on linear speed of conveyer for processes of heat (*1*), ultraviolet (*2*), and radiation (*3*) curing of coatings.

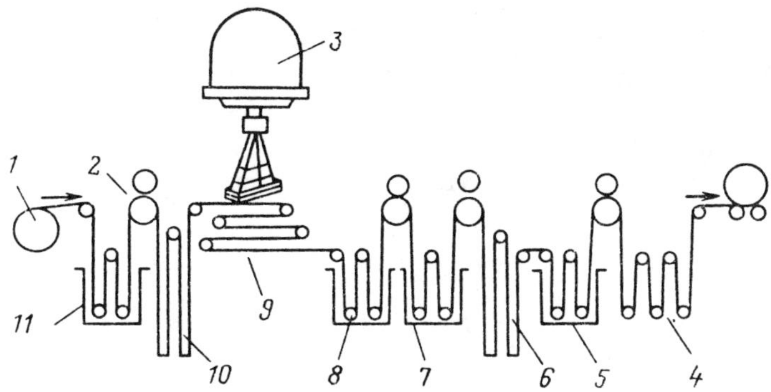


**Figure 9.9** Experimental-industrial installation LRKh-140 for radiochemical treatment of fabric. *1*—Fabric; *2*—squeezing devices; *3*—accelerator; *4*—dryer; *5*—washer; *6* and *10*—ager; *7*—impregnation with cupric oxide solution; *8*—washer; *9*—radiation zone; *11*—impregnation bath.

## 9.3 INSTALLATIONS FOR RADIATION TREATMENT OF FABRICS

One of the first industrial installations for radiation processing of fabrics to improve fabric properties was put into operation in 1966 by the Deering-Milliken Corp. (USA). The form of the material treated by this method remained unchanged even after many washings and displayed adequate resistance to contamination. Use was made of five ICT-500 accelerators. The total productive capacity of this installation amounted to $\sim 4 \times 10^6$ m of fabric per year [145].

The diagram of the Soviet experimental industrial installation LRKh-140 for radiochemical treatment of fabrics is shown in Fig. 9.9 [146]. This installation was put into use at the Lenin cotton fabric factory in Glukhov town in 1976. The Electron-3M accelerator and radiation zone are guarded with a local radiation shielding made of steel and iron, 150 to 250 mm thick.

The width of the material under irradiation is 1 m; the speed of advance of the fabric under treatment is 10 to 100 m/min. In radiation of nine layers of fabric at once, the losses of electron beam (including losses in the foil) amount approximately to 35 percent.

The experimental industrial installation for treatment designed on the basis of the ELT-1.5 accelerator, with the extraction window 2.5-m long, began operation in the early 1970s in Dresden (GDR). The experimental industrial, pilot installation based on the Aurora-3 accelerator has been operating at the "Lautex" textile enterprise (Grossenheim, GDR) since 1981. The width of the fabric under treatment is 0.8 to 1.8 m, the density is 50 to 200 g/m$^2$, the nonuniformity of the beam current intensity at a distance of 150 mm from the extraction foil at the window length of 1.8 m amounts to $\pm 10$

percent. The instability of the current intensity per operating hour is ±5 percent. In the production line that is being designed on the basis of the Aurora-3 accelerator average beam power will amount to 50 to 70 kW with the speed of advance of fabric to be treated being 50 to 60 m/min.

Table 9.4 describes characteristics of radiation treatment processing of fabrics.

## 9.4 INSTALLATIONS FOR STERILIZATION AND DEINSECTIZATION

Figure 9.10 shows an installation designed for sterilizing medical articles. The equipment that was mounted at the plant in Kurgan had been developed at the Institute of Radiation Technology in cooperation with the Efremov Institute of Electrophysical Apparatuses [148]. The installation incorporates two independent conveyers. The electron source of each conveyer is the linear accelerator LUE-8-5B, which has the following parameters: electron energy of 8 to 10 MeV, mean power of the electron beam of 5 kW. The beam pulse duration is 2.8 μs, the maximum repetition frequency is 500 Hz, and the non-monoenergeticity is ±7 percent. The accelerator is supplied with power from three-phase AC mains of 300 V. The efficiency of the accelerator is ~10 percent.

**Table 9.4 Main characteristics of some processes of radiation treatment of fabrics**

| Description of process | Characteristic: Average dose of radiation, kGy | Maximum speed of conveyer, m/min | Surface density of material, g/m² | Width of material, m | Radiation power absorbed by material, kW |
|---|---|---|---|---|---|
| Germicidal treatment | 25 | 60 | 150–300 | 0.75–1.45 | 3–11 |
| Crease resistance and low-shrinkage treatment | | | | | |
| With predrying | 10 | 90 | 70–200 | 0.75–1.2 | 0.8–36 |
| Wet fabric on stretching frame | 10 | 90 | 140–400 | 0.75–1.2 | 1.6–7 |
| Hydrophilic and antistatic treatment | 20 | 60 | 120–400 | 1–2 | 2.5–16 |
| Local modification | | | | | |
| With templet | 150 | 30 | 80–200 | 1–2 | 6–30 |
| With thermal rollers | 100 | 30 | 80–200 | 1–2 | 4–20 |

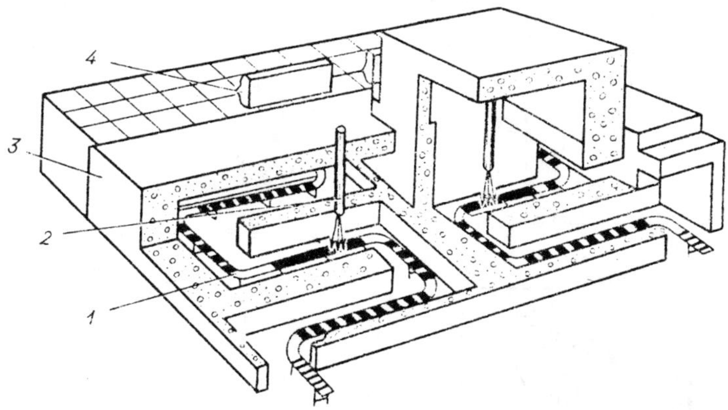


**Figure 9.10** Installation for sterilization of medical articles (in Kurgantown, USSR). *1*—Conveyer; *2*—accelerator; *3*—radiation shielding; *4*—control panel.

The length of the extraction window and the conveyer width are ~0.5 m. The accelerators are mounted vertically. The height of the conveyer from the floor is 0.8 m. The conveyer speed lies within 0.03 and 3 cm/s, the stability of the speed being no worse than ±2 percent. The characteristic radiation conditions are conveyer speed, 0.8 cm/s, and radiation dose, 25 kGy. The dimension specifications for the product under irradiation are 60 × 50 × 20 cm, the instability of the dose reaches 25 percent, and the coefficient of utilization of radiation is 60 percent. The component parts of the conveyers are made of stainless steel. The radiation shielding is made of concrete 2.1 to 2.3 $t/m^3$ in specific weight. The thickness of the concrete walls is 2.8 m. The area of each radiation chamber is 120 $m^2$, the height is 9.5 m. The total power required for operation of each radiation device amounts to 100 kW. The flow rate of cooling water is 10 $m^3/h$. Television sets are used for visual monitoring of the radiation treatment procedure.

Figure 9.11 shows a portable sterilization unit Electropulse-208 manufactured by Energy Science, Inc. [149]. A high-voltage generator (1) is a pulsed voltage generator, rated at 200 kV with an average power of 0.5 kW. Two accelerating devices (2) contain extended cold cathodes. An object (3) under sterilization (moving at right angle to the picture ) undergoes sterilization from both sides. A vacuum pump (5) ensures the needed rarefaction in the accelerating devices and in the power supply feeder (4). The dimensions of the installation are given in Fig. 9.11. The totally consumed power is ~1 kW. The installation is provided with local radiation shielding and enables sterilization with a dose of 25 kGy. Subject to sterilization with this dose are objects 15 cm in width moving in the radiation zone at a speed of 4 m/min. According to the estima-

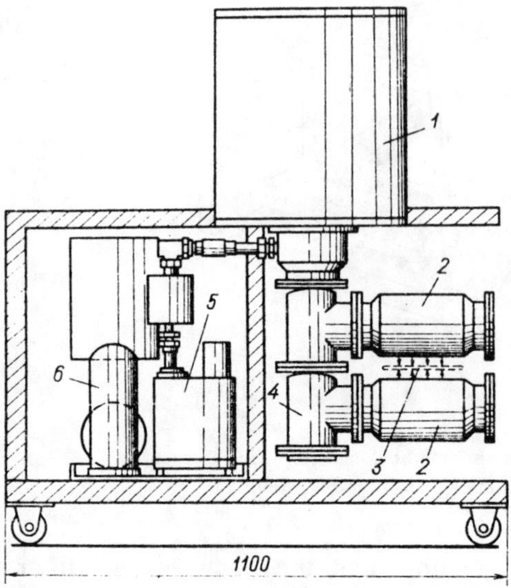


**Figure 9.11** Portable sterilization unit Electropulse-208. *1*—High-voltage generator; *2*— accelerating device; *3*—object under sterilization; *4*—feeder; *5*—vacuum pump; *6*—forced cooling system.

tions of some authors the cost of sterilization of medical articles amounts to 5 to 20 dollars per cubic meter.

Figure 9.12 demonstrates an experimental industrial pilot installation based on the ELV-2 accelerator intended for disinsectization of grain. This installation was put into action at the commercial elevator in Odessa in 1980 [150]. Electron energy is 1.4 MeV; the beam power is 20 kW. The productive capacity of the installation with a radiation dose of 0.2 to 0.4 kGy amounts to ~200 t/h. The grain to be irradiated from a receiving hopper (3) (10 $m^3$ capacity) falls by gravity, flows over a chute (7), and enters the radiation chamber (8). In the spreading hatch, the grain flows at a width of 1.5 m. The speed of the grain flow across the radiation chamber is ~6 m/s, with the thickness of the grain layer being 7 to 9 mm. The width of the radiation chamber, and respectively, the beam scanning width, is ~1.5 m. The radiation chamber is provided from the outside with a water-cooling system (9 and 13). An aspiration chamber (17) is provided for suction of dust and ozone. Pickups (2 and 15) monitor the grain level and produce a signal for adjusting the grain flow rate by means of a gate (4). If there is a shutdown of the accelerator, the grain flow is immediately shut off with the aid of fast-operating gate (5). Experience gained as a result of operation of this equipment has confirmed the high efficiency and economic profits of this installation.

The estimations of radioactive deinsectization of the grain with the use of different radiation sources are detailed in a number of research papers. The costs of deinsectization of grain by use of various dynamitrons with different parameters are given in Table 9.5 [32].

As Table 9.5 shows, the cost of radiation processing of grain tends to diminish with increase of both the rated productive capacity and the intensity of use of the installation. Comparative evaluations of grain deinsectization by use of electronic accelerators and isotope sources point to the fact that the accelerators with a high productive capacity, i.e., 100 t/h and above, are preferable [32]. The

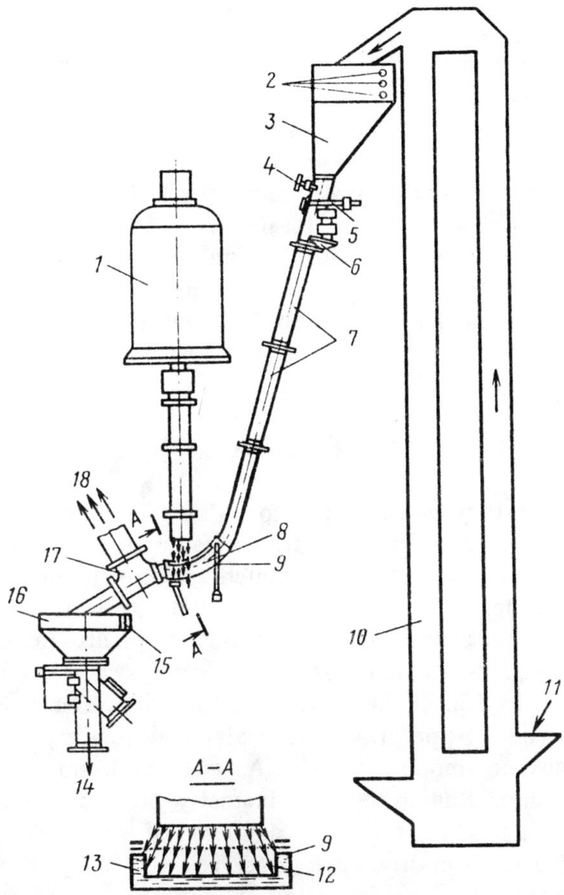


**Figure 9.12** Pilot installation for deinsectization of grain. *1*—Electron accelerator; *2*—grain level pickups; *3*—receiving hopper; *4*—grain flow control gate; *5*—fast-operating gage; *6*—grain spreading unit; *7*—speed-up channel; *8*—radiation chamber; *9*—water cooler; *10*—bucket conveyer; *11*—feed of grain for treatment; *12*—grain in radiation chamber; *13*—water; *14*—treated grain; *15*—grain level pickups; *16*—damping hopper; *17*—aspiration chamber; *18*—aspiration dust.

**Table 9.5 Cost of deinsectization of grain by radiation dose of 160 Gy applied by means of Dynamitrons**

| Parameters of electron beam | | | Cost of irradiation of 1 t of grain at different intensities of use, dollars | | |
|---|---|---|---|---|---|
| Energy, MeV | Power, kW | Productive capacity, t/h | 2000 h/year | 4000 h/year | 6000 h/year |
| 1.5 | 5 | 67 | 0.45 | 0.26 | 0.19 |
| 1.5 | 10 | 133 | 0.23 | 0.14 | 0.11 |
| 1.5 | 15 | 200 | 0.17 | 0.10 | 0.08 |
| 3 | 20 | 266 | 0.17 | 0.10 | 0.68 |
| 3 | 30 | 400 | 0.12 | 0.07 | 0.03 |

cost of deinsectization by the use of accelerators is lower than with chemical treatment of grain.

The practicability of using high-power bremsstrahlung generators for treatment of fish to prolong their storage life has also been studied [27]. The estimated cost of operation during 400 h with the use of two generators operating in parallel with electron beams of 300 keV and power of up to 500 kW is 100 to 150 marks per ton of fish.

## 9.5 INSTALLATIONS FOR FLAW INSPECTION

Figure 9.13 shows an installation intended for flaw detection of 600-mm thick steel parts by the use of bremsstrahlung radiation [151]. The installation is assembled and mounted at the Izhorsky works in Leningrad and is designed for flaw inspection of welds of reactor casings and steam generators of atomic power plants rated at powers up to 3 MW. Used as the main source of bremsstrahlung is a linear accelerator LUE-15-15000 D developed at the Efremov Institute for Electrophysical Apparatus. The electron energy is 18.5 MeV, the average current amounts to 80 μA, and the beam diameter on the target is ~3 mm. Radiation dose intensity at a distance of 1 m from the target is 3.7 C/kg min; the diameter of the bremsstrahlung field at a distance of 1 m from the target is ~400 mm. The overall dimensions of the radiator are indicated in Fig. 9.13*c* and *d*. The mass of the radiator is ~2.5 t (or, including suspension 11, 6.5 t). The exposure time for radiographic examination of articles of 600 mm thickness is tens of minutes, while inspection of an article of 100 mm thickness requires about 1 s. Installed adjacent to the accelerator are the modulator unit of the electron source, power

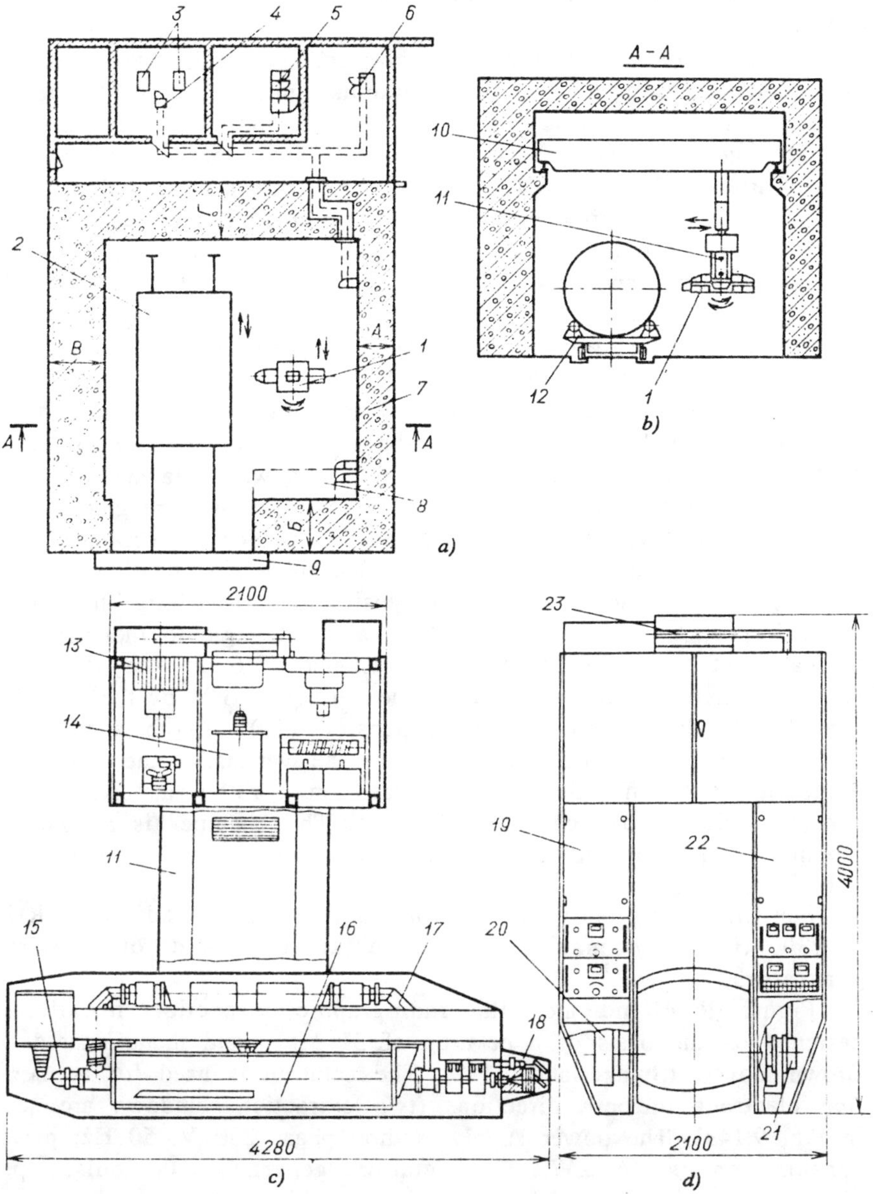


**Figure 9.13** Installation at Izhora plant for radiation flaw inspection of articles for nuclear power plants. *a)* Plan view. *b)* Vertical section. *c)* and *d)* Accelerator LUE-15-15000 D. *1*—Accelerator LUE-15-15000 D; *2*—article under inspection; *3*—frequency converter; *4* and *5*—control and power supply cabinets; *6*—control panel; *7*—radiation shielding; *8*—ground for repair of accelerator; *9*—sliding gate; *10*—crane positioner; *11*—accelerator suspension; *12*—tilting car; *13*—SHF generator; *14*—modulator of SHF generator; *15*—electron source; *16*—accelerating device; *17*—vacuum system; *18*—braking head; *19*—power distribution cabinet; *20*—suspension drive; *21*—rotary wave-guide joint; *22*—control cabinet; *23*—wave-guide channel.

supply rectifiers of the focusing coil, and other devices. Such an arrangement allows the use of a minimum number of cables to be connected to the movable part of the installation. The magnetron of the SHF generator and accelerating device are cooled with water at temperature of 36°C ± 0.5°C. The total flow rate of the cooling water in the primary cooling circuit is 170 L/min. The power for the accelerator is supplied from two frequency converters with a total power of 100 kW and an output voltage of 3 × 220 V ± 1 percent, 400 Hz.

Radiation shielding (7) is made concrete having a specific weight of 2.3 t/m$^3$. The thickness of the concrete wall *A* is 2.3 m, that of walls *B*, *C*, and *D* are each 3.6 m. Sliding gate (9) is made of steel, 0.9 m thick. The exit from the chamber to the control panel room is made in the form of a labyrinth with walls plated with sheet steel (0.2 m thick) and a door of 0.1 m in thickness. The gate is also coated with a layer of paraffin (0.3 m thick) for protection against the effects of neutrons. Provision is also made for hauling the accelerator (1) by means of a crane-positioner (10) along the article under inspection (2) as long as 10 m, at right angle to the article in the range of 4 m, and upward to a distance of 6 m. The accelerator can be turned in the horizontal plane through ±35° by the use of a motorized reducer mounted in suspension (11) of the accelerator. The accelerator can also be rotated in the vertical plane within an angle of 135°. An article (2) is placed on a roller-supported tilting car (12) with a load capacity of 400 t. The chamber is adapted to mount a robot for replacing 40 cameras with x-ray film measuring 0.3 × 0.4.

For inspecting thin welds, the unit is equipped with RUP 150/300-10 x-ray machines, isotope sources, and other x-ray generators.

Figure 9.14 indicates the radiographic inspection installation designed on the basis of a betatron of 31 MeV, and manufactured by Brown Boveri (Switzerland) [152]. The betatron is used for accelerating electrons in both directions (two extraction windows are seen in Fig. 9.14*a*). The power supply is three-phase 330 V, 50 Hz; power consumption is 16 kW. The betatron generates 100 pulses per second. Irradiation at a distance of 1 m from the focus covers an area of 200 × 200 m. The turn angle of the accelerator lies within 240° in the horizontal plane and up to 360° in the vertical plane.

A portable flaw-detection unit with a high-voltage transformer and an evacuated accelerating tube manufactured by Andrex (Denmark) is shown in Fig. 9.15. The accelerating voltage control range is 90 to 300 kV and current is 1 to 5 mA. The focal spot measures 2 × 5.5 mm. Single-phase mains of 110, 220, 380, 440, or 480 V, 50 or 60 Hz, can be used as a primary power supply source. The overall

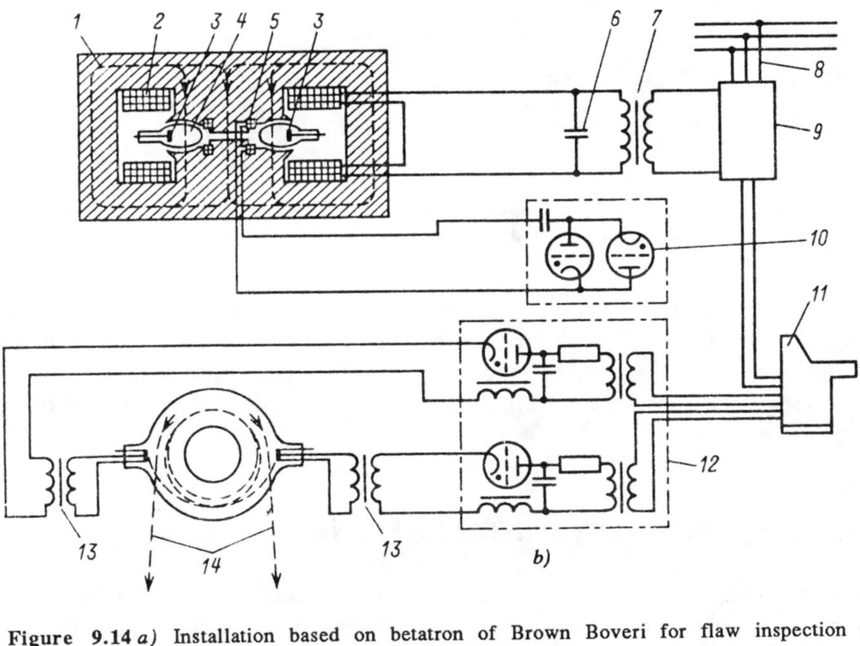


**Figure 9.14** *a)* Installation based on betatron of Brown Boveri for flaw inspection of articles. *b)* Circuit diagram of betatron. *1*—Magnetic circuit; *2*—windings; *3*—injector; *4*—vacuum chamber; *5*—winding for deflection of electrons; *6*—capacitor battery; *7*— transformer; *8*—three-phase supply voltage, 50 Hz; *9*—converter; *10*—supply circuit of extraction windings; *11*—control panel; *12*—supply circuit of injectors; *13*—step up pulsed transformer; *14*—bremsstrahlung.

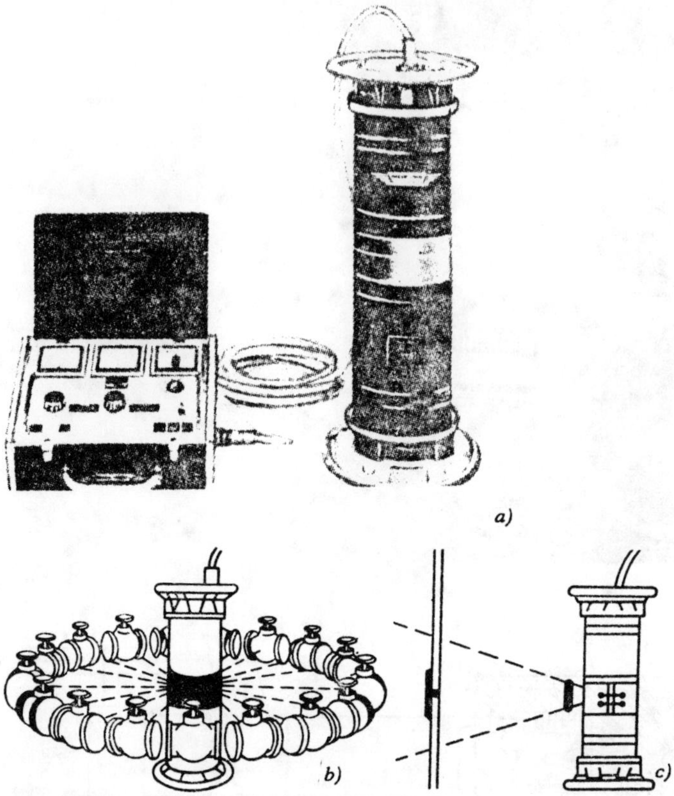

Figure 9.15 Portable unit of Andrex for flaw inspection *a)* General view. *b)* and *c)* Diagrams of panoramic and directed x-raying.

dimensions of the emitters are height, 1 m, and diameter, 0.4 m. The weight is 97 kg. The control panel weighs 29 kg. Figure 9.15*b* and *c* illustrates application of the facility for panoramic and directed x-raying. The optimum thickness of the metal under inspection is ~5 cm of steel.

## 9.6 RADIOTHERAPY INSTALLATIONS

Units designed for radiotherapy by the effect of accelerated electrons and bremsstrahlung employ linear wave-guide accelerators, betatrons and sometimes microtrons. In addition to the accelerator, this type of equipment incorporates the necessary appliances for moving the patient and the radiation source, a system for directing the electron beam, the diagnostic apparatus, the automatic control system, and radiation shielding. In many installations of this type provision is made both for generating bremsstrahlung and for extraction of electron beams from the vacuum chamber. Selection of the most suitable kind of radiation is dictated by the pathology and individual characteristics of the patient. One of the principle conditions of beam therapy is the high-dose of radiation exerted to the pathological focus and minimum irradiation of surrounding healthy tissues and the patient's organism as a whole (see Section 4.8). In some cases, the beam energy and current can be continuously controlled and varied in accordance with the present program, with the radiation source being rotated at the same time about the object under irradiation. As a rule, radiation coverage can be controlled within a range of 0.5 × 0.5 cm to 40 × 40 cm.

Figure 9.16 presents a typical beam therapy device based on a model that has been manufactured since the early 1960s by Brown

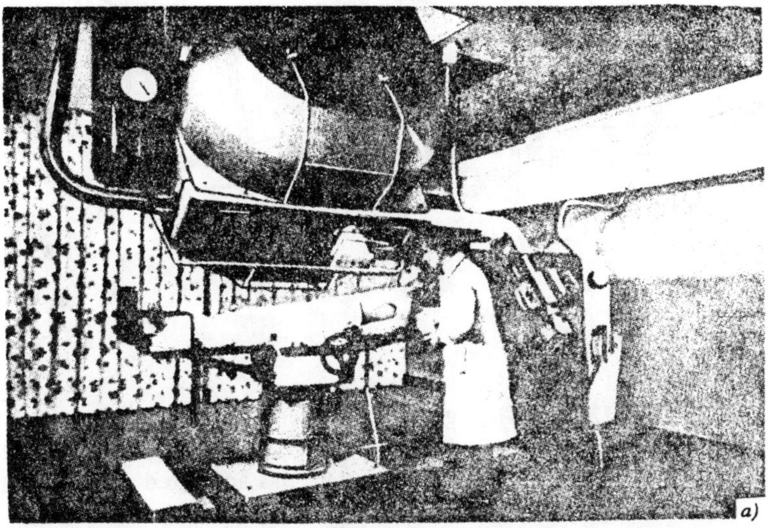

*a)*

**Figure 9.16** Brown Boveri betatron for radiotherapy. *a)* General view. *b)* Possible positions of betatron relative to irradiation table; directions of motion of betatron and table are shown with arrows. *c)* Channel for transport of electron beam extracted into atmosphere. *d)* and *e)* Absorbed dose in relative units versus depth of matter of 1 g/cm$^3$ density for electrons (*e*) and bremsstrahlung ($\gamma$). In (*e*), for comparison, there are curves for $^{60}$Co (dotted line). *f)* Curves of isodoses for absorption of nonfocused electron beam of 8 cm in diameter with different energies in matter of unit density. *g)* and *h)* Isodoses and radiography of focused electron beam with energy of 30 MeV (in similar scale) in matter of unit density. Focus is about 6 cm deep from surface. *1*—Betatron; *2*—irradiation table; *3*—center of rotation of electron beam; *4*—electron beam; *5*—fluorescent screen for diagnosis; *6*—counterweight cord; 7—vacuum chamber of betatron; *8*—extraction window; *9*—axis of vacuum chamber; *10*—betatron yoke; *11*—collimator; *12*—lead shield; *13*—ionization chamber; *14*—permanent magnetic lens; *15*—shields for restriction of electron beam; *16*—object under irradiation.

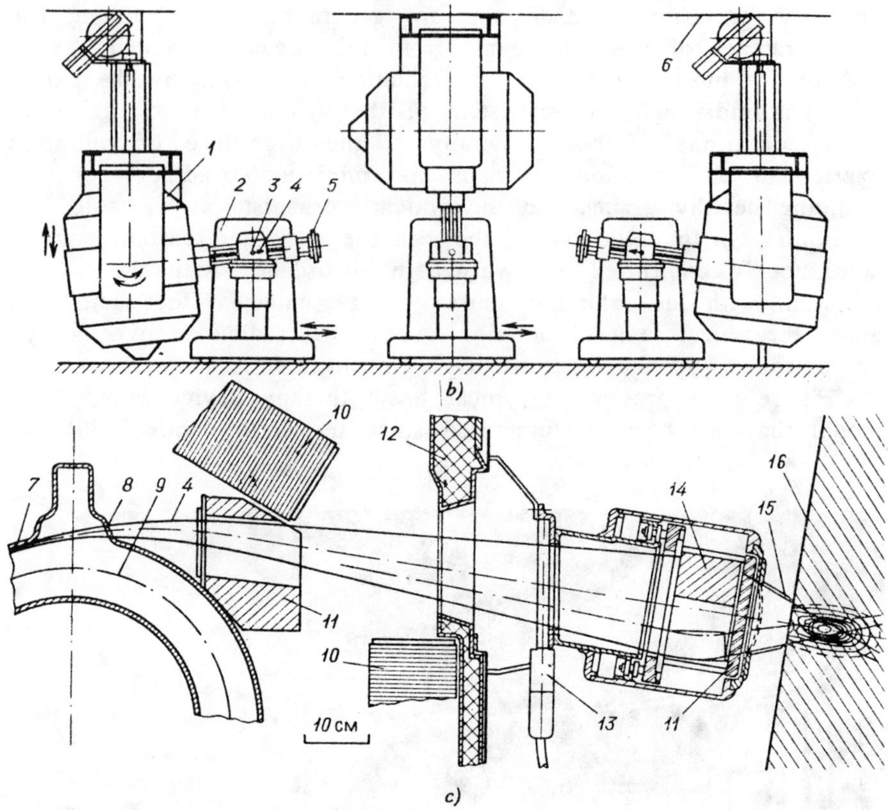


Figure 9.16 (*Continued*) (*b*, *c*)

Boveri (Switzerland) [153]. The guaranteed service life of betatron without replacing the vacuum chamber in 2000 h. The vacuum chamber contains an additional injector that is activated after the expiration of the service lifetime of the first injector. In the existing installations this time sometimes lasted 8000 to 10,000 hours. The maximum electron energy is 35 MeV. A movable betatron (1) and radiation table (2) (Fig. 9.16*b*) allow for different radiation conditions. Provision has also been made for focusing the beam exactly in the location of a neoplasm (Fig. 9.16*c*, *f*, *h*). The curves in Fig. 9.16*d* and *e* illustrate the difference in distribution of the radiation doses in curing by use of electrons and bremsstrahlung.

The absolute intensity of the electronic radiation in the object under irradiation may reach a few grays (hundreds of roentgens) per minute, which ensures the exposure duration of the order of 1 min. The intensity of the bremsstrahlung is approximately one order lower. Figure 9.16*f* shows the isodoses when a poorly diver-

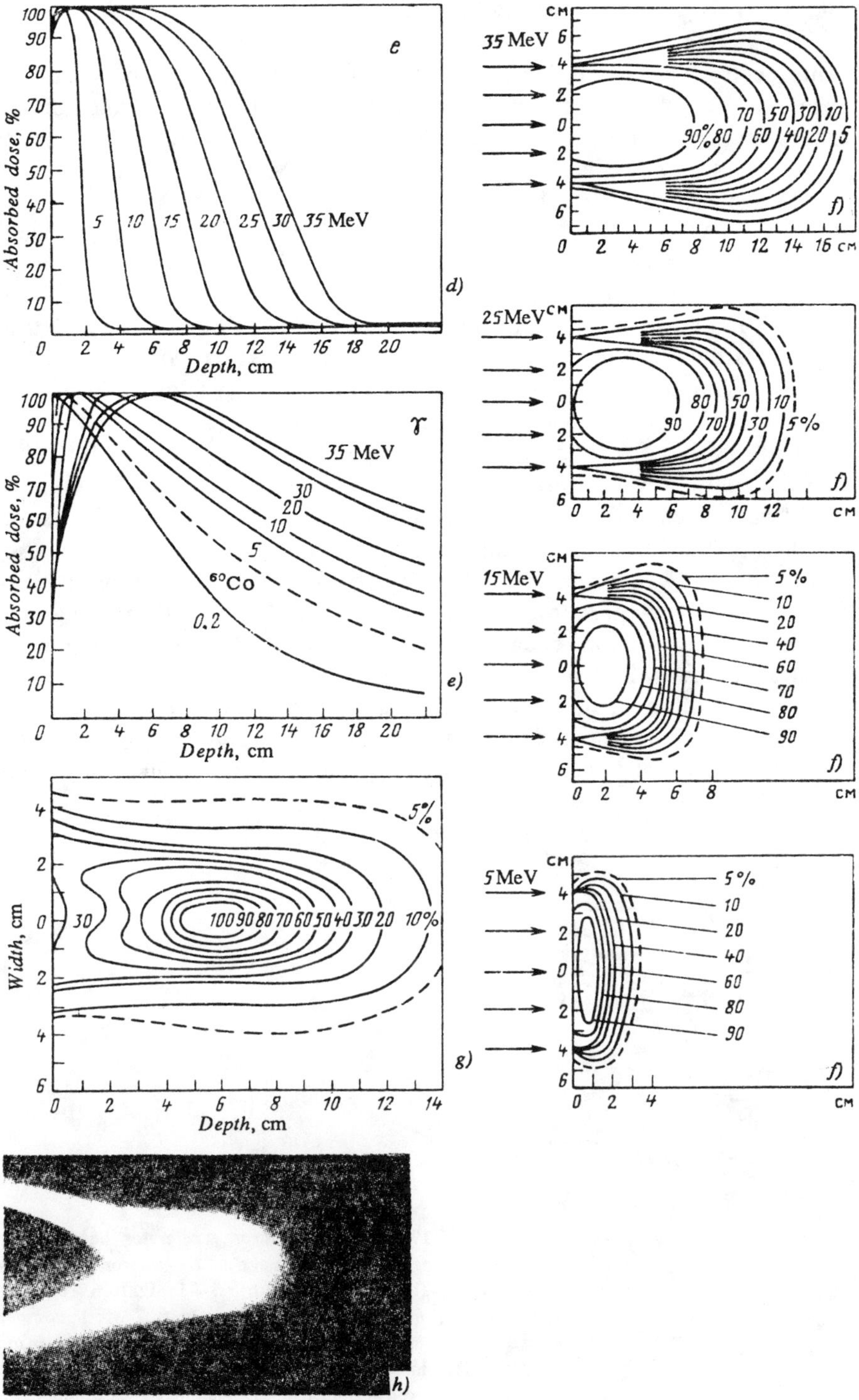


Figure 9.16 (*Continued*) (*d, e, f, g, h*)

gent electron beam of ~8 cm in diameter and different in energy is absorbed by unit-density matter. Focusing the beam in front of the object under treatment enables the distribution of the isodoses to be substantially varied (Fig. 9.16*f*, *h*).

The parameters of the Soviet betatrons used for therapy—types B5-25M, PMB-6, 5-32/6, etc.—are given in the appendix in Table AS.14. At present the linear accelerators find prevailing application for therapeutic purposes, since they provide highest intensities of radiation doses and incorporate rather simple beam extraction systems. The performance characteristics of medicinal-purpose linear accelerators manufactured by various firms are given in Table A.16.

One problem occurring with these units is establishment of a uniform radiation field (not worse than ±2.5 percent) at a field size of 40 × 40 cm. Figure 9.17 demonstrates a version of the linear accelerator head in which at first a dissipating foil (3) was used and then use was made of a foil-compensator (7) capable of attenuating the central part of the electron flux [154]. Dissipating foil decreases the electron energy approximately by 1 MeV, changes the spectral composition of the beam and builds up a flux of bremsstrahlung which also effects the distribution of the doses. For shielding some irradiated regions, there is a diaphragm (11) consisting of 15 pairs of plates each of 1.8 cm wide. The plates can

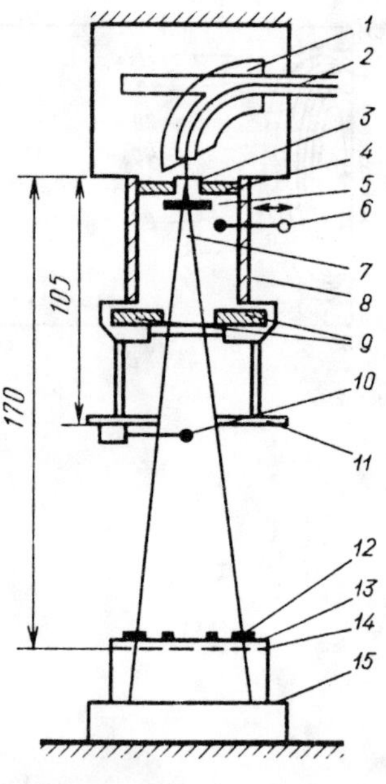


**Figure 9.17** Version of linear accelerator head (dimensions in mm). *1*—turning magnet; *2*—electron beam; *3*—scattering foil; *4*—collimator; *5*—ionization chamber (monitor); *6*—field light centering unit; *7*—foil compensator; *8*—focusing barrel; *9* and *11*—diaphragms; *10*— uniform field indicator; *12*—protection sections; *13*—support table; *14*—exposure surface; *15*—table for accommodation of patients to be treated.

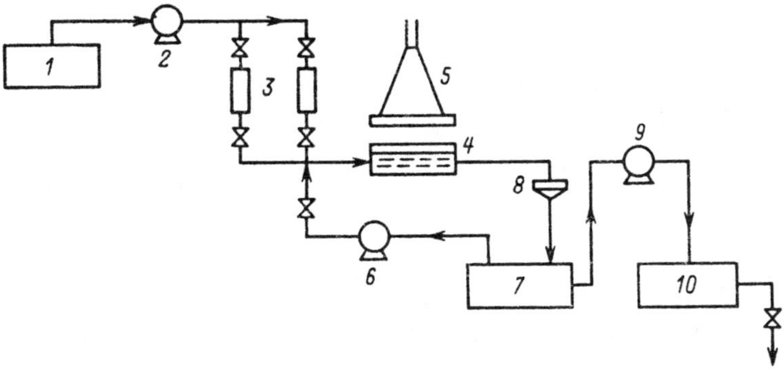


**Figure 9.18** Diagram of device used for radiation treatment of latex. *1*—Hopper with latex; *2*—feed pump; *3*—filters; *4*—chamber for irradiation of latex in flow; *5*—electron accelerator; *6*—return pump; *7*—return hopper; *8*—filter; *9*—pump; *10*—sump.

move in the horizontal plane. The final formation of the field in each particular case is effected by shape shielding sections (12). For directing the electron beam use is made of a field centering device (6) which can be shifted and arranged along the head axis in the absence of the electron beam. Determination of the radiation zone by the use of a light simulator is widely used in beam therapy units.

## 9.7 MISCELLANEOUS RADIATION INSTALLATIONS

### 9.7.1 Irradiation of Latex

Irradiation of rubber in the form of a latex results in cross-linking of polymer inside the globules. The film obtained in this way displays much better properties than that treated by hot vulcanization or other traditional methods of production. The strength of the films may reach 20 MPa at a relative elongation of 900 to 950 percent. Besides the films feature high heat resistance and transparency. The diagram of an installation for irradiation of latex is shown in Fig. 9.18 [155].

### 9.7.2 Radiation Regeneration of Butyl-Based Rubbers

The technological process of radiation regeneration of butyl-based rubbers is illustrated in Fig.9.19 [155]. The bladders of waste shapers-vulcanizers are cut into pieces 20 to 40 mm wide. The latter are crushed and dried. The dry rubber crumbs are then moved

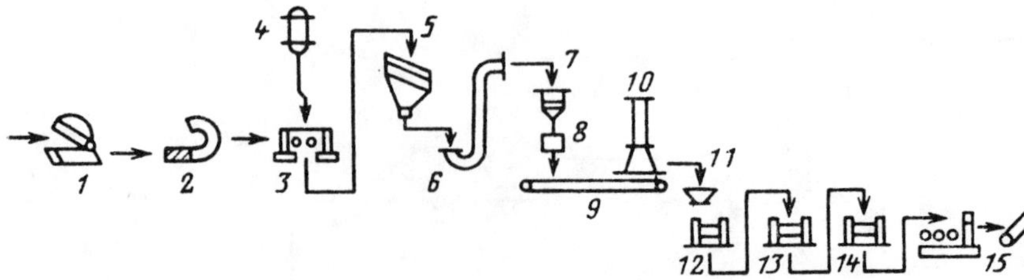


**Figure 9.19** Process chart of radiation process in fabrication of butyl rubber reclaim. *1*—Power shears; *2*—tire chopper; *3*—cracker; *4*—reactor for preparation of 0.2% aqueous solution of alkyl sulfates of primary alcohols; *5*—single-row shaking screen; *6*—pneumatic dryer; *7*—strirrer feeder; *8*—measuring device; *9*—conveyer; *10*—electron accelerator; *11*—automatic balance; *12*—mixing rolls; *13*—feed rolls; *14*—refiners; *15*—packing unit.

on a conveyer under exposure to the electron beam, therapy irradiating the material with a dose of 50 to 100 kGy. Then the crushed rubber is delivered to an automatic balance and to the mixing, feeding and refining mills. The obtained reclaim is then wound into rolls.

### 9.7.3 Irradiation of Blanks of Automobile Tire Components

Treads, tire casings, and inner layers of tubeless tires are irradiated with a dose of 10 to 100 kGy resulting in formation of sparse space lattice in skin layers of the blanks, and thereby improving their strength in nonvulcanized state. During tire manufacture, the blanks treated in this way are less liable to rejection, but can be readily shaped and subsequently heat-cured. This method excludes defects in building up the tires, and the total time required for the building-up procedure diminishes by 20 percent. The radiation-treated blanks can be stored for a long time. Tires manufactured by this method feature higher resistance to fatigue and cracking and, in the final analysis, wear resistance is improved. Figure 9.20 shows an assembly line designed for radiation treatment of blanks of tire parts.

### 9.7.4 Sewage Treatment

A diagram of the experimental installation for sewage treatment which was set into operation in 1976 near Boston, Massachusetts, is shown in Fig. 9.21*a* [156]. Used as an electron source of this installation is the ICT accelerator, produced by the firm HVEC, with an energy of 0.85 MeV and a power of 50 kW. The installation ensures both a disinfection of sewage and partial decomposition of delete-

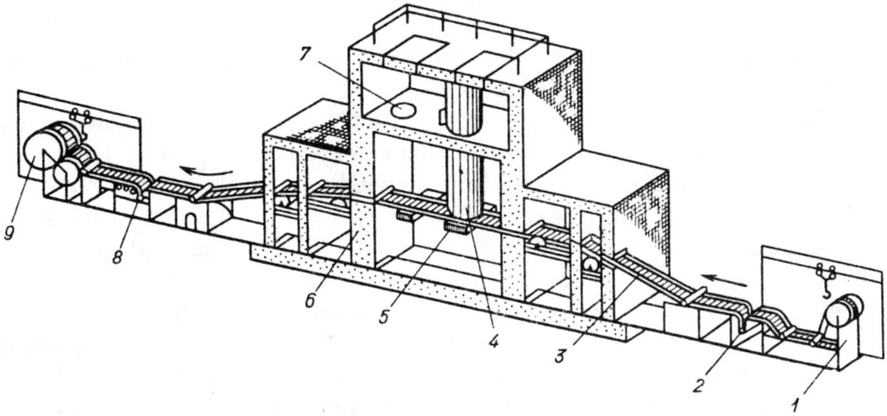


**Figure 9.20** Diagram of continuous line for irradiation of calendered blanks of tire components. *1*—Unwinding device; *2* and *8*—compensators; *3*—conveyer; *4*—electron accelerator; *5*—lead shutter; *6*—radiation shielding; *7*—location of standby electron accelerator; *9*—winder.

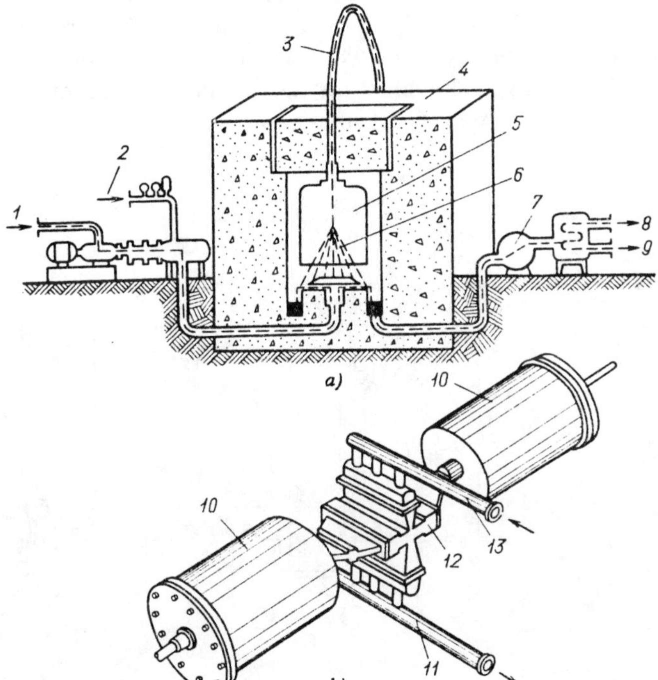


**Figure 9.21** *a)* Diagram of unit used for radiation treatment of sewage. *b)* Version of radiation zone. *1*—Sewage; *2*—compressed gas or oxygen; *3*—high-voltage cable; *4*—concrete shielding walls; *5*—electron source; *6*—electron beam; *7*—pump; *8*—dehydrated refuse; *9*—water; *10*—accelerator; *11* and *13*—inlet and outlet of seawage flow under radiation treatment; *12*—appliance for shaping electron beams.

rious organic substances. The productive capacity of the installation is 400 $m^3$ of thick silt (sewage) (2 to 8 percent of solid matter) per day, with a radiation dose of 4 kGy. The sewage flow may be up to 1.5 m wide. The installation under consideration is intended to find the best radiation conditions depending on depth and rate of flow. Provisions have been made for several different variants of flow formation, viz.: (1) high-speed rotary drum entrailing layers of silt; (2) endless band; (3) chute; (4) water flow falling by gravity. The probable version of construction of a radiation zone is shown in Fig. 9.21*b* (patent of the firm HVEC). It was anticipated that the installation should be equipped with an accelerator of 1.5 MeV and 100 kW, and that its productive capacity should exceed 550 $m^3$/day. The cost of the sewage treatment at a dose of 4 kGy is estimated to be less than 50 cents/$m^3$.

## 9.7.5 Purification of Exhaust Gases

Figure 9.22 illustrates the diagram of an experimental installation manufactured by Nippon Steel Corp. (Japan) for purification of exhaust gases. Research done by the use of this installation was carried out in 1977–1978 [23]. The electron beam energy amounts to 0.6 MeV and the productive capacity of the equipment is 3000 $m^3$/h

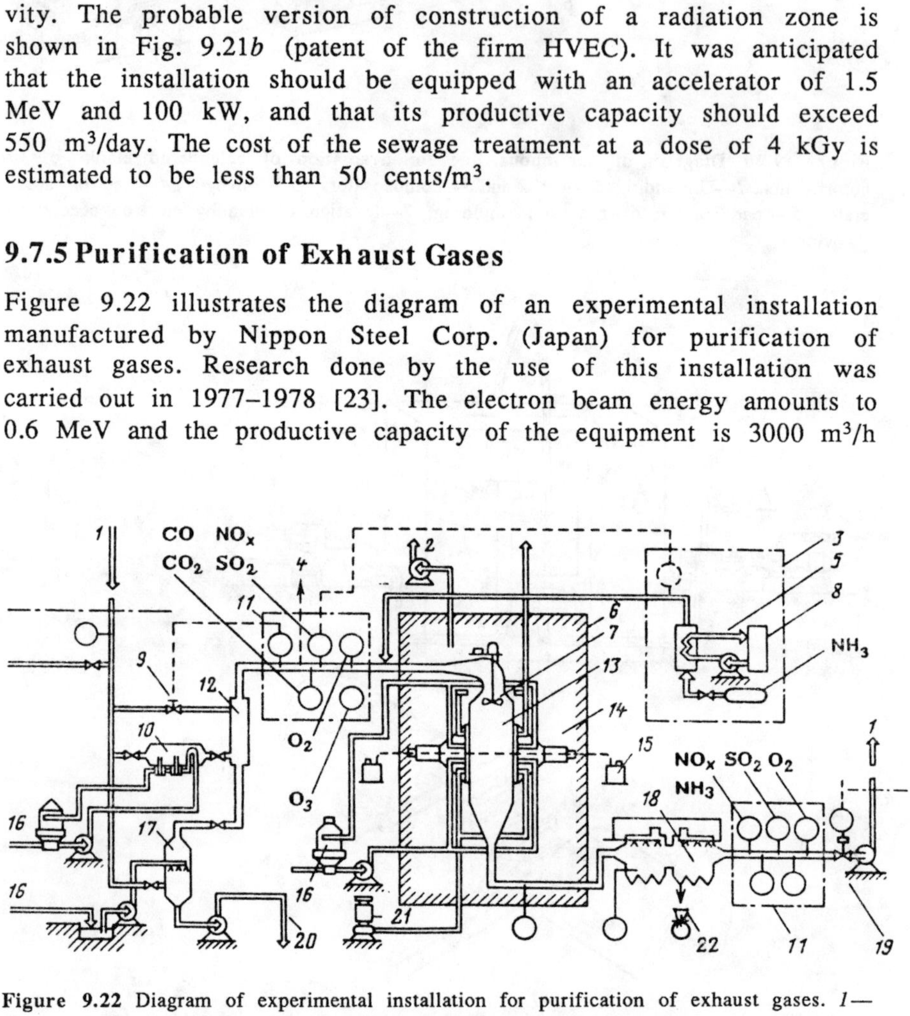


**Figure 9.22** Diagram of experimental installation for purification of exhaust gases. *1*—Trunk pipeline; *2*—ventilator; *3*—ammonia supply chamber; *4*—to chamber *3*; *5*—evaporator; *6*—reaction chamber; *7*—impeller; *8*—tank with hot water; *9*—bypass valve; *10*—heat exchanger; *11*—gas analysis room; *12*—mixing chamber; *13*—radiation zone; *14*—accelerating device; *15*—high-voltage source; *16*—water; *17*—cooling tower; *18*—electric filter; *19*—fan; *20*—removing system; *21*—air blower for cooling of reactor windows; *22*—collected product.

(maximum 10,000 $m^3/h$). The characteristic dose of radiation is 15 kGy; the optimum gas temperature is ~70°C. The gas to be purified is first cooled in the heat exchanger (10) and then is delivered to a gas analysis chamber (11) where the contents of $NO_x$, $SO_x$ and other components are measured. The ammonia ($NH_3$) is supplied to the gas under treatment from a chamber (3). The obtained mixture is forced into the radiation zone (13). At the inlet of the radiation zone, the gas is caused to swirl. As a result of the reaction between $NO_x$, $SO_x$ $NH_3$, and moisture, an aerosol is produced which is removed from the gas through the electric filter (18). In the course of continuous operation during 1 month the installation ensured that 80 percent of $NO_x$ and more than 95 percent of $SO_x$ were removed from the exhaust gas under treatment.*

## 9.8 PROSPECTS OF USING THE RELATIVISTIC ELECTRON BEAMS IN NEW FIELDS OF APPLICATION

Apart from the discussed applications of electron beams in technology and in medicine, as well as use in experimental facilities, new fields of application of relativistic electrons and associated radiations are presently under study.

Some promising fields of applications are unique in that alternatives are virtually nonexistent. Introduction of some processes into new fields of application depends mainly on the time required for development and practical use of novel accelerating facilities and technological equipment. Prior to starting industrial application of other new processes, it is necessary to determine their economical efficiency and the ability to successfully compete with the existing technological methods.

Among the first kinds of new applications are for synchrotron radiation (Section 2.9). High spectral-density of synchrotron radiation (several orders higher than in bremsstrahlung) makes it possible to obtain an intense photon flux with a required wavelength and to use it for diagnosis and therapy (with a considerably less dose of radiation applied to patients as compared to the commonly practiced x-ray fluorograph), to reveal distribution of heavy and light chemical elements in various human organs and to determine the change of their concentration for different diseases. The beams of x-ray synchroton radiation of 0.1 and 1 μm in diameter may be used for molecular microsurgery when solving some problems dealt with genetics. In x-ray lithography the synchrotron radiation ensures a resolution of less than 1 μm, that is pronouncedly better than in the existing units. It is conceivable that

* See new material added in proof starting on page 259.

these factors will give rise to radically new semiconductor devices. Future industrial sources of synchrotron radiation are likely to be installations resembling the modern accumulators of electrons, rated at $10^3$ to $2 \times 10^4$ MeV with superconductive magnets and powerful high frequency supply systems (see, for example, [10, 11]). Although synchrotron radiation generators are considered to be unique and rather expensive in use and require proficient attending personnel, they surely will find expanding applications in industry and medicine as they improve (primarily in intensity and efficiency).

A promising field of application in which the electron beam may display a radically new capability is distant transmission of energy [157, 159]. Building power transmission lines rated at a power of 25 to 100 GW (as compared with a few gigawatts in the modern lines) coupled with line length of thousand of kilometers should allow the power supply system to be substantially modified. For example, with the power transmission lines arranged latitudinally, power plants can be connected in succession to different industrial regions. This is in the arrangement used in many new power stations. The circuit diagram of a version of such a transmission line is shown in Fig. 9.23. The electric power generated by the power plant (hydroelectric power station, nuclear power station, etc.) at the standard voltage and frequency is applied to a high-voltage generator where the voltage is stepped up and rectified. Under the action of the applied voltage, the electric power in the accelerating device is converted to kinetic energy of particles. The line for conveying the electron beam can be positioned underground, in which case there will be no need to provide the right of way, and proper radiation shielding will be ensured in emergencies. The electron kinetic energy then will be transformed in the decelerating device again into ordinary electric energy. The main advantages of such transmission line are the practicability of conveying electric power of tens of gigawatts and absence of high voltages over the line. The main difficulty in realizing this project is

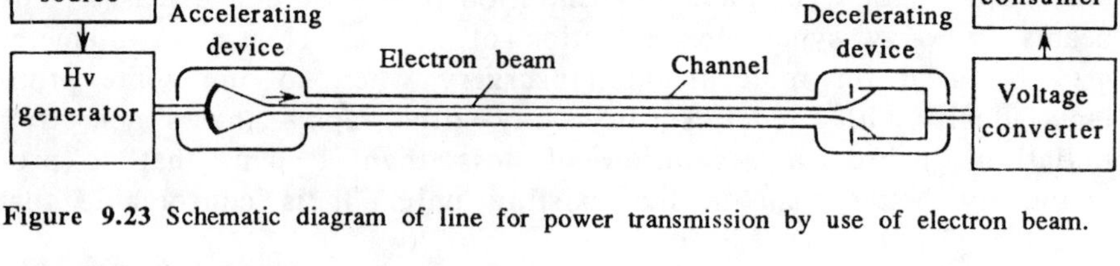


**Figure 9.23** Schematic diagram of line for power transmission by use of electron beam.

the necessity of developing economical systems for holding the electron beam in the channel, as well as pumping systems, and a system for maintaining vacuum throughout a tremendous mileage of electron transmission duct.

Stability of distribution of the beam in the power transmission line increases with the rise of the electron energy. For a line with a power of 25 to 100 GW, a most suitable combination of parameters is voltage of 5 to 10 MV and current of 5 to 10 kA [2].

As pointed out in Chapter 5, the use of a concentrated beam extracted into the atmosphere may also become profitable in the future for heat treatment of metals and particulary for crushing rocks in mining industry. Figure 9.24 illustrates a conceptual project of an installation with an electron beam intended for tunneling [160]. Here a linear electronic accelerator is used as an electron generator. The beam parameters are: energy, 5 MeV; current, 5 kA; pulse length, 1 μs and repetition frequency, 360 Hz. The energy content of each beam pulse is 25 kJ, with the average power being 9 MW. The accelerator comprises 64 modules (Fig. 9.24*c*) in each of which the electron energy increases by 80 keV. The construction of the linear electronic accelerator complies with the diagram depicted in Fig. 7.2*a*. It is expected that the efficiency of conversion of the energy applied to the accelerator into the beam energy may reach 50 percent. The installation incorporates devices for extraction and scanning of the beam. The rate of tunneling operation at the tunnel diameter of 6.4 m is 3.2 m/h. This is approximately one order higher than the speed provided by the other method. The possibility of obtaining a high-power beam in the atmosphere and its use in mining is, to a high degree, determined by solution of a number of physical and technological problems arising in the process of development of such equipment. A significant advance has been achieved lately in designing linear induction accelerators of electrons, which enables us to hope for their speedy application in various fields [161].

Consideration is also being given to applicability of concentrated electron beams for heating substances in the installations intended for performance of controllable reactions of synthesis of deuterium and tritium [162]. With development of the collective method of acceleration [163], electron accelerators of the megavolt range may become the basis of the installations generating intense ion fluxes for obtaining fissionable isotopes and for other purposes.

The other probable field of application of the intense accelerators of the electrons with an energy of a few MeV is apparently gen-

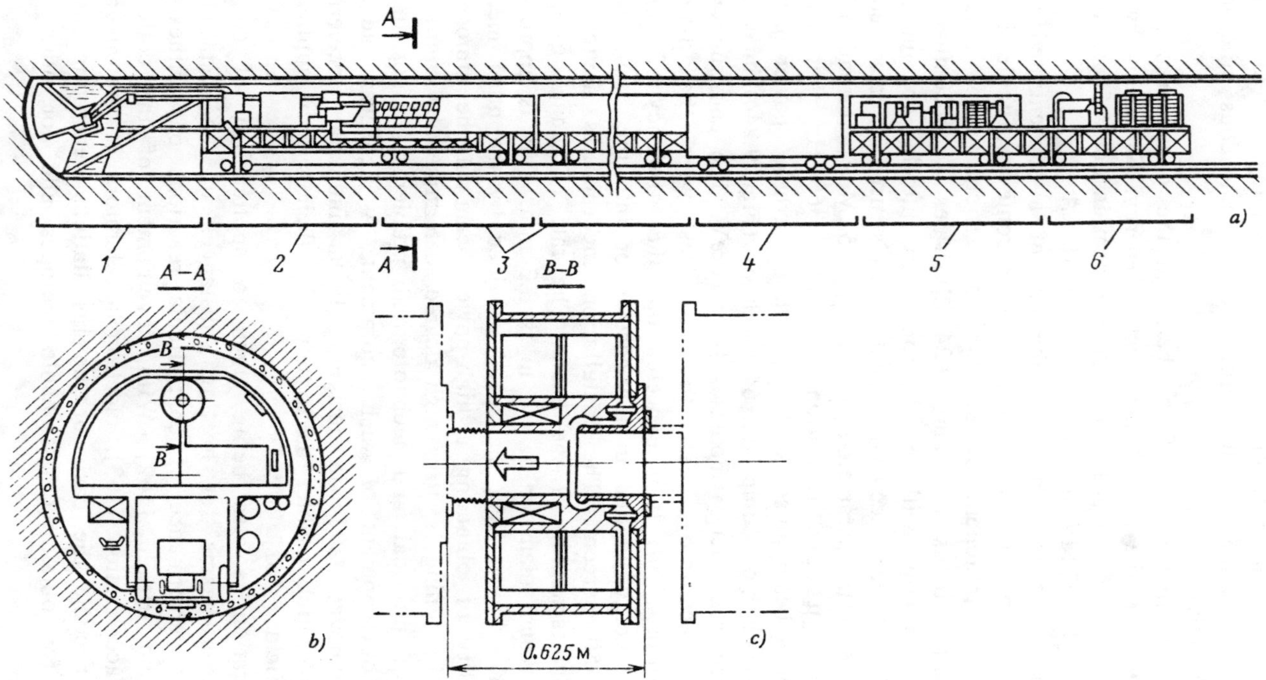


**Figure 9.24** Conceptual project of installation for tunneling by use of electron beam. *a)* General view. *b)* Cutaway view. *1*—Beam scanning unit; *2*—beam duct; *3*—LUE accelerator; *4*—radiation shielding; *5*—power supply and control system; *6*—heat exchanger

eration of high-power super-high-frequency radiation. The efficiency of conversion of energy of relativistic electron beams into SHF radiation in the existing generators reaches 30 percent [164].

The installations designed on the basis of electron accelerators permit generation of high-power SHF radiation with variable frequency.

## MATERIAL ADDED IN PROOF

A sewage sludge treatment plant with a capacity of up to 200 tons of sludge per day is considered in [184]. It is noted that the total amount of sludge in Japan in 1987 was 2.2 million tons (water content 65–80%) Fig. 9.21*c* shows the diagram of the supposed plant. At first, sludge cake is disinfected by electron beam irradiation and mixed with a bulking agent to make the sludge aerobic. Seed compost is also mixed to start fermentation. Fermentation temperature is controlled from 40–50°C. After 2 or 3 days, compost is obtained. In irradiation composting, disinfection and composting are separated. Optimum composting conditions are available in this case.

It takes less than 0.2 second for coliform not to be detected at a dose rate of 10 kGy/sec, but more than 20 days are required for the heat disinfection of conventional composting.

In the plant, provision is made for an accelerator with an electron energy of 1.5 MeV and power 15 kW. Thickness of the sludge to be treated is about 5 mm. Use is made of two fermentors of 12 m in diameter and 1.5 m in effective height (Fig. 9.21*d*). Four fermentors 16 m in diameter are necessary in conventional composting.

Irradiated sludge mixture comes from the outside into the fermentor, ferments as it moves toward the center, and is taken out as compost. The general view of the plant is shown in Fig. 9.21*e*.

A diagram of the stack gas diagnostic plant acting at the Badenwerk power plant (FRG) and using two electron accelerators, produced by Energy Sciences Inc. (USA), with an energy of 300 keV and a current of 0.3 A each is given in [185]. The work began in 1984 and is primarily aimed at reducing the emissions of nitrogen oxide from power plants to 200 mg/m$^3$. Ammonia was added to a $2 \cdot 10^4$ Nm$^3$/hr gas flow to bind $NO_x$. The dose rate was up to 20 kGy. A simultaneous decrease of $NO_x$ and $SO_x$ was registered. The general layout of the plant and the irradiation zone design are shown in Fig. 9.22*b, c*.

The total power of electron beams used to clean stack gases of the power plant should account for 2–4% of the total generated power. So to clean up stack gases of the 550 MW power plant more than fifty electrocurtain accelerators with energy 300 keV and current 0.9 A will be required. The most reliable and economical layout of such accelerators are shown in Fig. 9.22*d, e*. The lifetime of injector filaments in such accelerators is 10–12 thousands of hours. The lifetime of the electron beam window (its length is about 2 m) is a significant problem. Presently it is only several hundred hours due to the gas corrosive effects and the presence of particles in it, but in an inert atmosphere this lifetime reaches 1–2 thousands of hours. The other problems are to increase the power in a module up to 1 MW and to possibly reduce the dose to 10 kGy.

At the meeting of specialists from 17 countries devoted to the problem of stack gases cleanup and held in Karlsruhe in October, 1986, it was noted that the governments of a number of countries set new limits for the $NO_x$ and $SO_x$ content in stack gases. For example, in FRG these limits will come into effect in 1988–1993. Apparently the electron beam technology providing the simultaneous removal of $NO_x$ and $SO_x$ could turn out to be the most competitive technology and this should facilitate the concentration of efforts in solving the above-identified problems.

One of new approaches in designing powerful and long-lived plants to clean stack gases could turn out to be the use of a scheme shown in Fig. 8.14, that permits having several MW power of the beam in a single module with the energy of the order of 1 MeV.

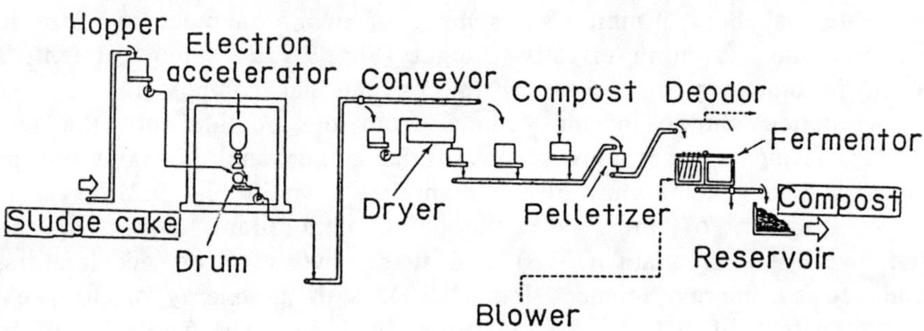


Fig. 9.21*c* Scheme of the sewage sludge treatment plant.

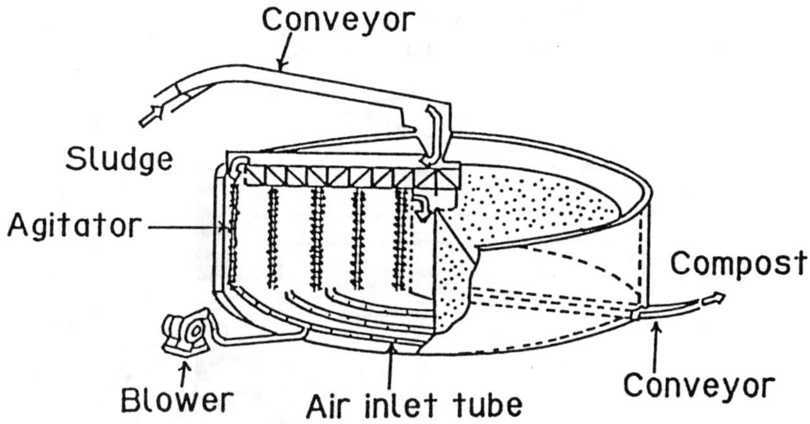


**Fig. 9.21*d*** An example of fermentor.

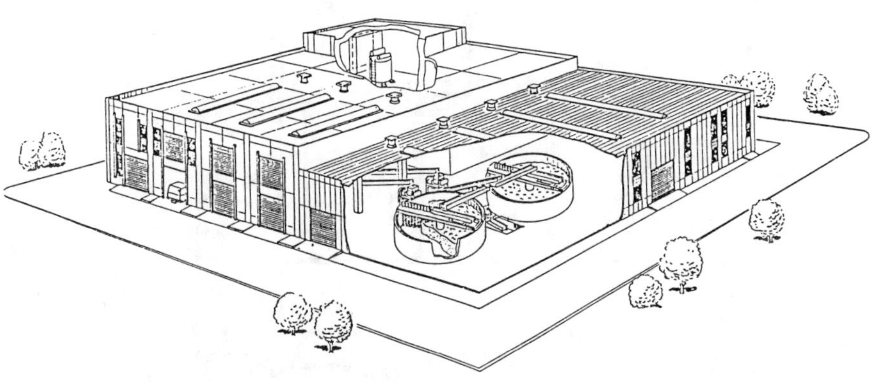

**Fig. 9.21*e*** Flow sheet and bird's vew of the sewage sludge treatment plant.

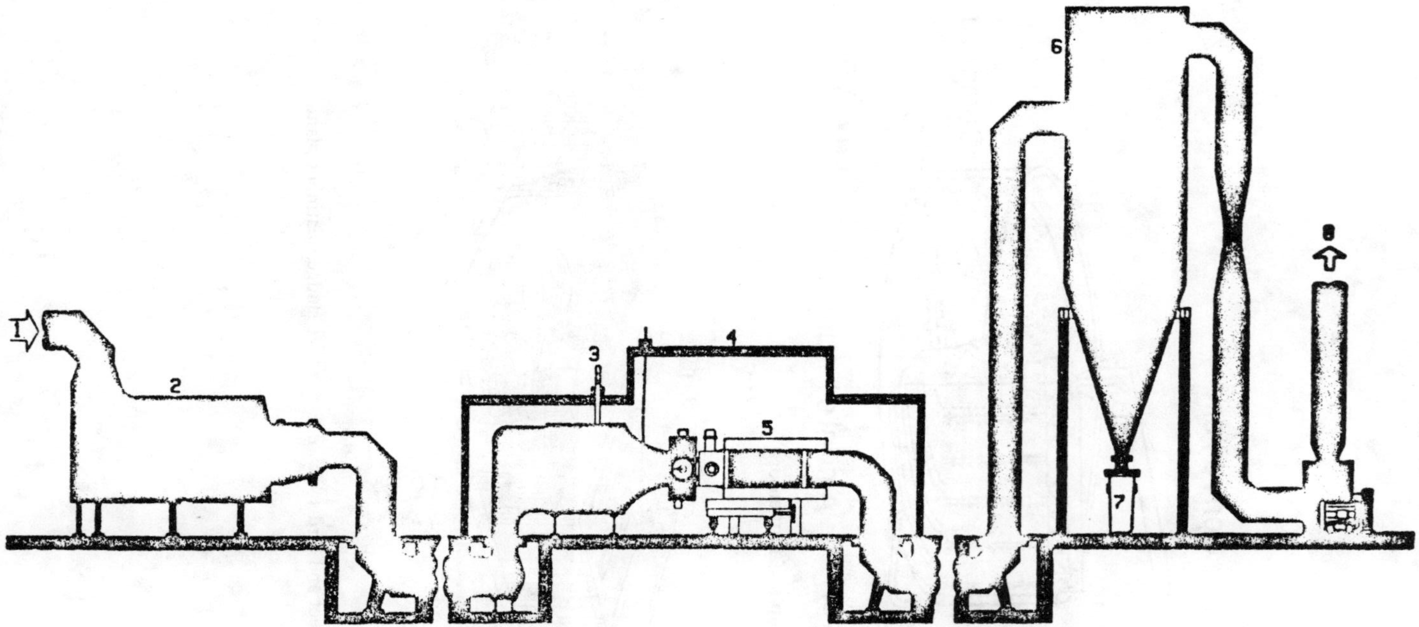


Fig. 9.22*b* Pilot system for electron beam treatment of stack gas. *1*—inflow of gas; *2*— graphite heater; *3*—ammonia addition and air mixer; *4*—x-ray shield for personnel; *5*—electron processors; *6*—impact filter; *7*—product bagging; *8*—outflow to stack.

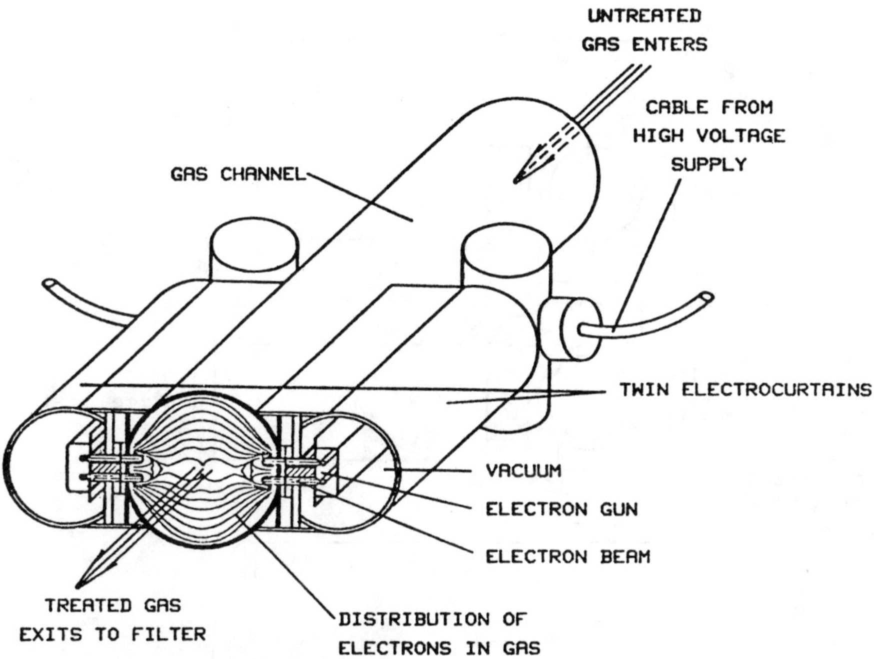


**Fig. 9.22***c* Electron beam stack gas treatment module with twin ESI Electrocurtains®.

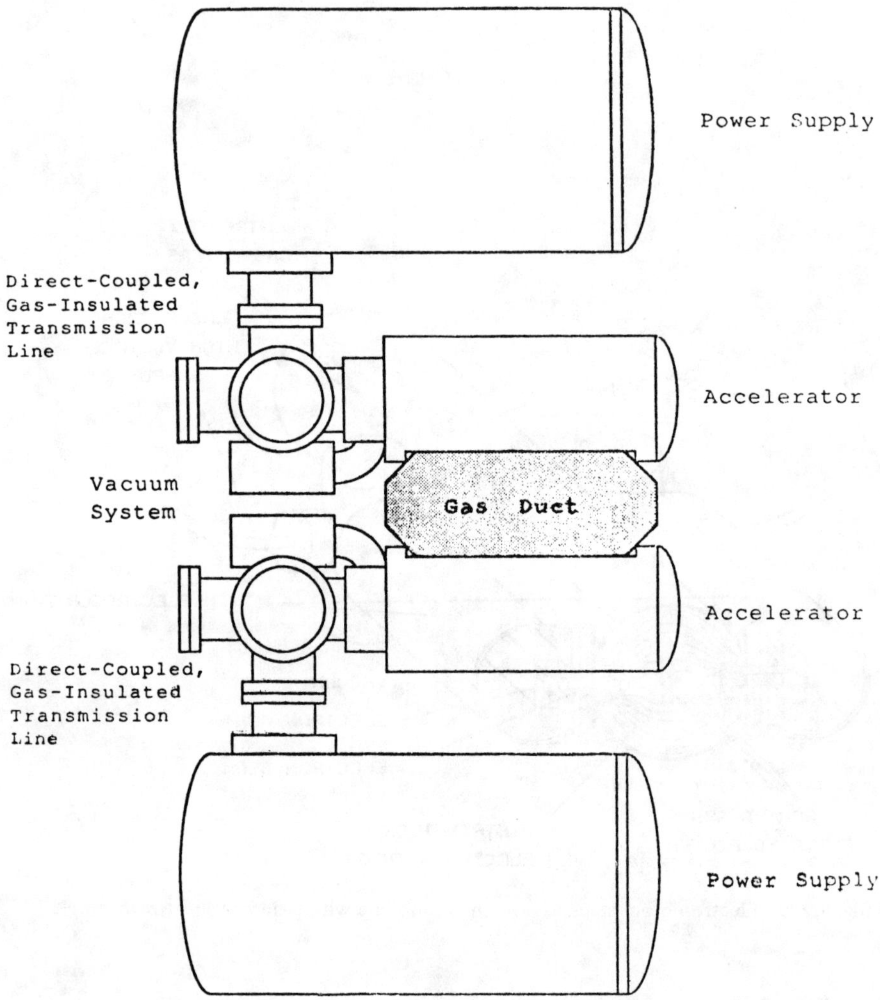


Fig. 9.22*d*, *e* Arrangement layouts of accelerators.

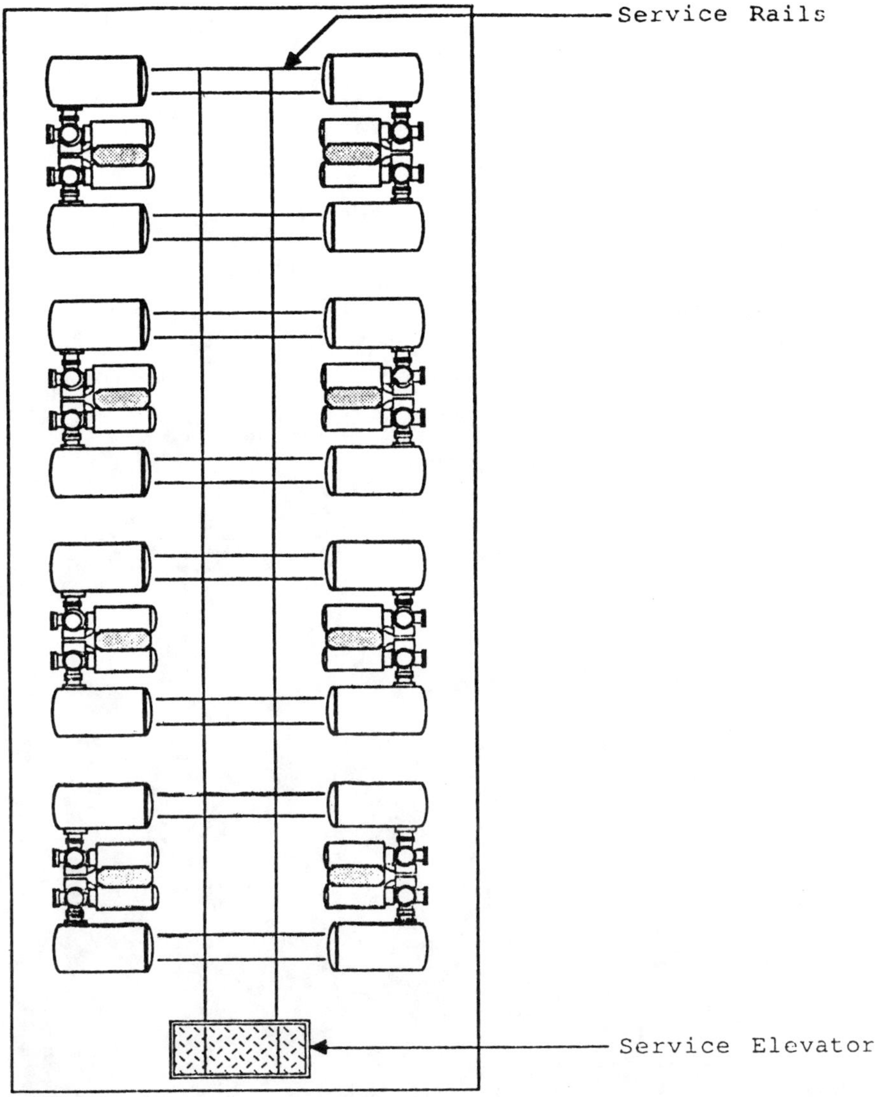


Fig. 9.22*e*

# APPENDIXES

## PARAMETERS OF ELECTRON ACCELERATORS USED FOR APPLIED PURPOSES

**Table A.1 Specifications of some basic versions of Soviet industrial electron accelerators**

| Type of accelerator (or HV generator) | Model | Characteristics: Energy, MeV | Characteristics: Average power of electron beam, kW | Characteristics: Operating conditions | Designing agency, manufacturer | Year of development | Reference |
|---|---|---|---|---|---|---|---|
| Electron transformer | ELT-1.5 | 1.5 | 25 | Pulses $5 \times 10^{-3}$ s, 50 Hz | Nuclear physics Institute, Novosibirsk, (NPI) | 1964 | [65] |
| Resonance shock transformer | ELIT-1 | 1 | 5 | Pulses $5 \times 10^{-8}$ s, to $2.5 \times 10^{-6}$ s, up to 100 Hz | NPI | 1965 | [58] |
| Electronic transformer-rectifier | ELV-1 | 1 | 20 | Continuous | NPI | 1973 | [80] |
| High-frequency resonator | ILU-6 | 1.8 | 20 | Pulses $4 \times 10^{-4}$ s, 50 Hz | NPI | 1975 | [80] |
| Resonance transformer | RTE-1 | 1 | 3 | Pulses 400 Hz | NIIEFA, Leningrad | 1968 | [165] |
| Transformer with bridge rectifier | Elektron-3 | 0.7 | 7 | Continuous | Same | 1970 | [79, 165] |
| Three-phase transformer with bridge rectifier | Elektron-4 | 0.5 | 10 | Continuous | NIIEFA | 1973 | [179, 165] |
| Same | Aurora-2 | 0.5 | 25 | Continuous | NIIEFA | 1974 | [79, 165] |
| Cascade generator | KGE-2.5 | 2.5 | 2.5 | Continuous | NIIEFA | 1968 | [172] |
| Three-phase electron accelerator | Teus-3 | 0.4 | 75 | Continuous | High-Temperature Institute, Acad. Sci. USSR, Moscow | 1974 | [81] |
| Three-phase transformer with bridge rectifier | EOL-400 | 0.4 | 14 | Continuous | Radioengineering Institute, Moscow | 1975 | [166] |

| | | | | | | | |
|---|---|---|---|---|---|---|---|
| Cascade transformer | RUP-400 | 0.4 | 2 | Pulses $10^{-2}$ s, 50 Hz | Mosrentgen, Moscow Actyubrentgen, Actjubinsk | | [167] |
| Resonance shock transformer | Rius-5 | 4 | Pulse power ~$10^8$ | Single pulses $4 \times 10^{-8}$ s | NPI | 1969 | [82] |
| High-frequency resonator | Raps-1 | 0.3 | 0.6 | Pulses $3 \times 10^{-7}$ s (800 kHz) trains of $5 \times 10^{-5}$, 300 Hz | Opticophysical measurements Institute, Moscow | 1979 | [85] |
| Linear accelerator | U-10 | 3 | Pulse current 0.3 A | | Physics Engineering Institute, Moscow | 1960 | [168] |
| Linear accelerator | ELU-5-1-5 | 5 | 5 A | Pulses $4.5 \times 10^{-6}$ s, repetition rate 200 Hz | Electronics Institute, Moscow | 1970 | [169] |
| Linear accelerator | LUE-1-8-5- | 8 | 5A | | Electrophysical Apparatus Institute, Leningrad | 1968 | [170] |
| Linear accelerator | LUE-15-10 | 13 | 9A | Pulse length 0.5, 2.5, 5.5 μs; repetition rate 150, 300, 900 Hz | Electrophysical Apparatus Institute, Leningrad | 1968 | [170] |
| Betatron | PMB-6 | 6 | | 50 Hz | Nuclear Physics Institute, Tomsk | 1968 | [171] |
| X-ray apparatus | IRA-1D | 0.25 | | Pulse $10^{-7}$ s, 0.1 Hz (reserve $10^4$ pulses) | NPO Burevestnik, Leningrad | 1963 | [172] |

*Note: The table indicates energies and powers obtained in tests of accelerators. Long-period operation of some installations was performed at lower parameters.

**Table A.2 Characteristics of four industrial accelerators produced by the Nuclear Physics Institute (Novosibirsk) [173]**

| Characteristic | ELV-2 | ELV-3 | ELV-4 | ILU-6 |
|---|---|---|---|---|
| Energy of electrons, MeV | 0.8–1.5 | 0.5–0.7 | 1–1.5 | 0.5–2 |
| Average current of electron beam, mA | 25 | 100 | 50 | 10 |
| Average power of electron beam, kW | 20 | 50 | 50 | 20 |
| Ripple of accelerating voltage, % | ±2.5 | ±2.5 | ±2.5 | 10 |
| Pulse length, μs | Continuous | Continuous | Continuous | Up to 700 |
| Pulse repetition frequency, Hz | – | – | – | Up to 300 |
| Power consumption, kW | 27 | 60 | 60 | 120 |
| Efficiency, % | 75 | 80 | 80 | 16.5 |

**Table A.3 Characteristics of Aurora industrial accelerators manufactured by the Electrophysical Apparatus Institute (Leningrad) [174]**

| Characteristic | Modifications 1 | | | 2 | | | 3 | | |
|---|---|---|---|---|---|---|---|---|---|
| Accelerating voltage, kV | 200–300 | | | 300–500 | | | 500–700 | | |
| Number of radiators | 1–3 | | | 1–3 | | | 1–3 | | |
| Maximum current of radiator beam, mA | 50 | 75 | 100 | 40 | 60 | 80 | 30 | 45 | 60 |
| Sweep width, cm | 100 | 150 | 200 | 100 | 150 | 200 | 100 | 150 | 200 |
| Total power of accelerated electron beams, kW | 50 | | | 75 | | | 100 | | |
| Scanning frequency of electron beam, Hz | | | | | | | | | |
| Along extraction window | 30 | | | 30 | | | 30 | | |
| Across extraction window | 300 | | | 300 | | | 300 | | |
| Heterogeneity of linear density of electron current, % | ±5 | | | ±5 | | | ±5 | | |
| Instability of electron current, % | ±2 | | | ±2 | | | ±2 | | |

*Note: Accelerating devices in all Aurora accelerators are arranged separately from generator and are coupled to generator through cables with solid insulation.

**Table A.4 Characteristics of Dynamitron industrial accelerators produced by Radiation Dynamics (New York)**

| Accelerating voltage, kV | Beam current, mA | Beam power, kW | Length of extraction window, mm | Method of connection of generator with accelerating tube |
|---|---|---|---|---|
| 400 | 100 | 40 | 1830 | Cable connection |
| 550 | 50 | 27.5 | 915 | |
| | 100 | 55 | 1830 | Gas-filled |
| 800 | 50 | 40 | 915 | coaxial |
| | 70 | 56 | 1220 | cable |
| | 100 | 80 | 1830 | |
| 1000 | 35 | 35 | 610 | Tube is combined with generator |
| | 35 | 35 | 610 | Gas-filled coaxial cable |
| | 70 | 70 | 1220 | Tube is combined with generator |
| | 70 | 70 | 1220 | Gas-filled |
| | 100 | 100 | 1830 | coaxial cable |
| 1500 | 25 | 37.5 | 610 | Combination of tube and generator or gas-filled coaxial cable |
| | 100 | 150 | 1830 | Gas-filled coaxial cable |
| 2000 | 30 | 60 | 610 | |
| | 50 | 100 | 915 | |
| 3000 | 20 | 60 | 610 | |
| | 40 | 120 | 915 | Tube is combined |
| | 50 | 150 | 915 | with |
| 4000 | 25 | 100 | 610 | generator |
| | 50 | 200 | 915 | |
| 4500 | 34 | 150 | 610 | |
| 5000 | 30 | 150 | 610 | |

Table A.5 Characteristics of industrial accelerators produced by High Voltage Engineering Corp. (HVEC) (Boston) in cooperation with the Nissin Voltage Co. Ltd. and Nissin Electric Co. Ltd. (Japan)

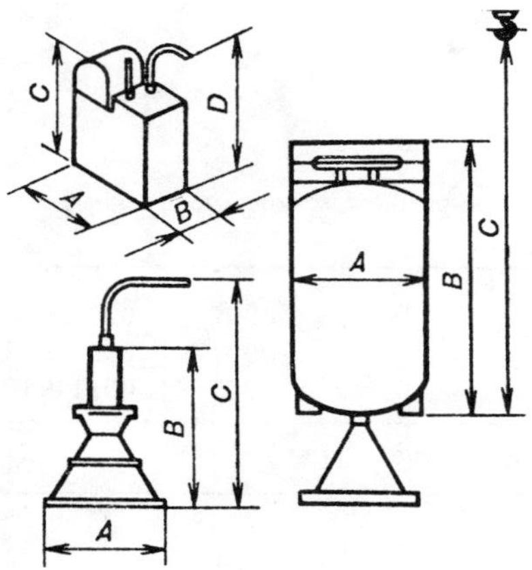


Accelerators with generator and accelerating tube connected by cable

| Type | Accelerating voltage, kV | Beam current, mA | Length of extraction window, cm | Generator | | Accelerating tube and extractor | |
|---|---|---|---|---|---|---|---|
| | | | | Size: A, B, C, D, m | Mass, t | Size: A, B, C, D, m | Mass, t |
| EPS-300 | 300 | 25 | 45 | 2; 1.4; 2.6; 4 | 4.5 | 0.7; 1.7; 3.7 | 0.2 |
| | | 65 | 120 | 2; 1.4; 2.6; 4 | 4.5 | 1.5; 2.3; 4.3 | 0.4 |
| | | 100 | 180 | 2; 1.4; 2.6; 4 | 4.5 | 2.2; 2.9; 4.9 | 0.9 |
| | | 200 | 180 × 2* | 2; 1.4; 2.6; 4 | 4.5 | | |

| | | | | | | | |
|---|---|---|---|---|---|---|---|
| EPS-500 | 500 | 25 | 45 | 2.6; 1.9; 3.3; 4.8 | 9 | 0.7; 1.7; 3.7 | 0.2 |
| | | 60 | 120 | 2.6; 1.9; 3.3; 4.8 | 9 | 1.5; 2.3; 4.3 | 0.4 |
| | | 100 | 180 | 2.6; 1.9; 3.3; 4.8 | 9 | 2.2; 2.9; 4.9 | 0.9 |
| | | 200 | 180 × 2* | 2.6; 1.9; 3.3; 4.8 | 9 | | |
| EPS-750 | 750 | 25 | 45 | 3.3; 2.5; 4.4; 6.3 | 20 | 0.7; 2.4; 5.4 | 1 |
| | | 65 | 120 | 3.3; 2.5; 4.4; 6.3 | 20 | 1.5; 3.6 | 1.2 |
| | | 100 | 180 | 3.3; 2.5; 4.4; 6.3 | 20 | 2.2; 3.6; 6.6 | 1.7 |
| EPS-1000 | 1000 | 25 | 45 | 3.3; 2.5; 4.5; 6.4 | 21 | 0.7; 2.6; 5.6 | 1 |
| | 1000 | 65 | 120 | 3.3; 2.5; 4.5; 6.4 | 21 | 1.5; 3.2; 6.2 | 1.2 |
| | 1000 | 100 | 180 | 3.3; 2.5; 4.5; 6.4 | 21 | 2.2; 3.8; 6.8 | 1.7 |

Accelerators with accelerating tube built in generator

| Type | Accelerating voltage, kV | Beam current, mA | Width of extraction window, cm | Size: A, B, C, D, m | Mass, t |
|---|---|---|---|---|---|
| EPS-1500 | 1500 | 65 | 120 | 2.5; 3.5; 5 | 12 |
| EPS-2000 | 2000 | 50 | 120 | 2.8; 4.5; 6.5 | 15 |
| EPS-3000 | 3000 | 30 | 120 | 2.9; 5.8; 6.8 | 23 |

*Note: Accelerator is provided with two accelerating devices.

## Table A.6 Characteristics of industrial accelerators produced by Energy Science Inc. (Boston)

| Type | Accelerating, voltage, kV | Beam current, mA | Length of extraction window, cm | Power of electron beam in front of extraction window, kW |
|---|---|---|---|---|
| Electrocurtain (with extended thermionic cathode and sheet beam of electrons) | | | | |
| CB 175/30/5 | 175 | 45 | 30 | 7.9 |
| CB 175/50/5 | 175 | 75 | 50 | 13 |
| CB 225/50/15 | 225 | 30 | 50 | 67 |
| CB 225/100/10 | 225 | 400 | 100 | 90 |
| CB 275/50/5 | 275 | 125 | 50 | 34 |
| CB 275/100/5 | 275 | 250 | 100 | 69 |
| Pilot 200 | 200 | | 45 | |
| Electrocure (with accelerating tube and scanning beam) | | | | |
| SB 300/30/20 | 300 | 25 | 30 | 7.5 |
| SB 300/50/30 | 300 | 37.5 | 50 | 11 |
| SB 300/65/40 | 300 | 50 | 65 | 15 |
| SB 300/80/50 | 300 | 62.5 | 80 | 19 |
| SB 300/95/60 | 300 | 75 | 95 | 22.5 |
| SB 300/110/70 | 300 | 87.5 | 110 | 26 |
| SB 300/125/80 | 300 | 100 | 125 | 30 |
| SB 300/140/90 | 300 | 112.5 | 140 | 34 |
| SB 300/155/110 | 300 | 125 | 155 | 37.5 |
| Electropulse | | | | |
| Electropulse® 208 | 200 | 2.5 | 15 | 0.5 |

*Similar units are manufactured by Charmilles-Energy Science International (Geneva, Switzerland).

## Table A.7 Characteristics of industrial accelerators, some high-voltage generators, and x-ray apparatuses

| Description and type | Maximum voltage, MV | Average power, kW | Manufacturer |
|---|---|---|---|
| Rotary generator | | | |
| KP-300-10 | 0.3 | 3 | Tunzini-Sames (Grenoble, France) |
| KP-750-3 | 0.75 | 2.2 | Tunzini-Sames |
| Cascade generator | 0.6 | 60 | E. Haefely (Basel, Switzerland) |
| | 1 | 100 | E. Haefely |
| | 0.6 | 6 | Brown Boveri (F.R. Germany) |
| | 3 | 18 | Hitachi (Japan) |
| Dayatron DP-1000 | 1 | 50 | Mitsubishi (Japan) |
| Resonance transformer | 2 | 12 | General Electric (U.S.) |
| | 0.18 | 18 | Polymer Physik GMBH (F.R. Germany) |
| 9802 | 0.3 | 30 | Texas Nuclear (U.S.) |

**Table A.8 Characteristics of x-ray pulsed apparatuses, series MIRA, produced by Objedinenije Burevestnik (Leningrad) [61]**

| Characteristics | MIRA-1D | MIRA-2D | MIRA-3D | MIRA-4D | MIRA-5D |
|---|---|---|---|---|---|
| Amplitude of voltage on x-ray tube, kV | 100 | 150 | 200 | 350 | 500 |
| Exposure dose of pulse radiation at distance of 0.5 m from anode, μC/kg | 0.05 | 0.2 | 0.5 | 1 | 2 |
| Pulse frequency, Hz | 20–25 | 10–15 | 4–5 | 2–3 | 1.5–2 |
| Duration of x-ray pulse at half-cycle, ns | 10 | 15 | 20 | 20 | 20 |
| Diameter of focal spot, mm | 2 | 3 | 3 | 4 | 4 |
| Cone angle of radiation beam, degree | 30 | 150 | 150 | 150 | 150 |
| Service capacity of x-ray apparatus, pulse | $5.10^6$ | $5.10^6$ | $10^6$ | $5.10^4$ | $5.10^4$ |
| Maximum thickness of steel accessible for radiography, mm | 10 | 20 | 40 | 60 | 100 |
| Power consumption, W | 300 | 400 | 600 | 800 | 1000 |
| Mass of x-ray unit, kg | 2 | 4 | 10 | 25 | 40 |
| Type of x-ray tube in use | IMA-6 | IMA2-150D | IMA5-320D | IA-6 | IA-6 |

**Table A.9 Characteristics of x-ray apparatuses RINA-3B/6, RINA-4B/4, MIRA-5B/1 (HPO Burevestnik) [61]**

| Characteristics | RINA-3B/6 | RINA-4B/4 | MIRA-5B/1 |
|---|---|---|---|
| Number of radiators | 6 | 4 | 1 |
| Amplitude of voltage on x-ray tube, kV | 400 | 500 | 600 |
| Exposure dose of pulse at distance of 1 m from x-ray tube, μC/kg | 0.5 | 2 | 1 |
| Diameter of focal spot, mm | 3 | 4 | 4 |
| Half-cycle pulse length, ns | 20 | 30 | 20 |
| Overall dimensions of x-ray unit, mm | 1110 × 270 × 270 | 840 × 340 × 240 | 870 × 300 × 350 |
| Mass of x-ray unit, kg | 50 | 80 | 40 |

## Table A.10 Characteristics of some high-voltage bremsstrahlung generators used for flaw detection (developed in United States)

| Description, | Type | Energy, MeV | Pulse current, A | Pulse length, s | Firm | Year when developed (approx.) |
|---|---|---|---|---|---|---|
| Resonance transformer | | 1 | | | General Electric (G.E.) | 1939 |
| | | 2 | $5 \times 10^{-3}$ | $\sim 10^{-3}$ (180 Hz) | G.E. | 1944 |
| Electrostatic generator with gas-filled coaxial line | Febetron-705 | 2 | $5 \times 10^{3}$ | Single pulses $30 \times 10^{-3}$ | Field Emission Corp. | 1965 |
| Electrostatic generator with gas-filled coaxial line | FX45 | 4 | $20 \times 10^{3}$ | Single pulses $2.5\ 10^{-3}$ | Ion Physics Corp. | 1965 |

*Although this table presents specifications of bremsstrahlung high-voltage generators used for flaw detection, some physics laboratories have developed more powerful generators for research with current of $10^5$ and $10^6$ A.

## Table A.11 Characteristics of x-ray pulsed apparatuses produced by Hewlett Packard (U.S.) [61]

| Model | Voltage amplitude, MV | Pulse length, ns | Exposure dose of pulse radiation, μC/kg (at distance from focal spot, cm) | Diameter of focal spot, mm |
|---|---|---|---|---|
| 43730A | 0.105 | 30, 70, 100 | 5–16 (12) | 2.6 |
| 43731A | 0.15 | 70 | 10 (20) | 3 |
| 43632A | 0.18 | 30 | 19 (10) | 18 |
| 43733A | 0.3 | 30 | 7–16 (38) | 4–5 |
| 43704A* | 0.3 | 25 | – | – |
| 43704A | 0.3 | 15 | 12 (30) | – |
| 43701A* | 0.6 | 30 | – | – |
| 43701A | 0.6 | 20 | – | – |
| 43706A | 0.6 | 3 | – | – |
| 43736A | 0.6 | 30 | 10 (100) | 5 |
| 43708A | 0.8 | 5 | – | – |
| 43710A | 1 | 25 | – | – |
| 43703A | 2 | 40 | 19 (200) | 5 |
| 43705* | 2.3 | 50 | – | – |
| 43705 | 2.3 | 20 | – | – |

*Models marked with asterisks electron beams are subject to extraction and direct use. All units listed in this table incorporate Marx generators and accelerating devices with multipoint autoemission cathodes. As a rule, they are assembled in a common tank filled with compressed gas. Model 43730A is also manufactured with an accelerating device accommodated in an individual tank connected to the generator by the use of a cable of up to 30 m in length.

## Table A.12 Characteristics of x-ray apparatuses produced by Scanditronics

| Characteristics | Scandiflash 300 | Scandiflash 600 | Scandiflash 1200 |
|---|---|---|---|
| Amplitude of voltage on x-ray tube, MVs | 0.1–0.3 | 0.250–0.6 | 0.5–1.2 |
| Pulse current, kA | 10 | 10 | 10 |
| Exposure dose of pulse radiation 0.5 m from focal spot of tube, μC/kg | 9 | 20 | 50 |
| Half-cycle pulse length, ns | 20 | 15 | 10 |
| Dimensions of high-voltage unit, mm | | | |
| Diameter | 700 | 800 | 800 |
| Length | 600 | 885 | 1205 |
| Mass of high-voltage unit, kg | 150 | 300 | 450 |

## Table A.13 Parameters of Marx generators used for supply of pulse accelerators (Maxwell, U.S.)

| Voltage, kV | Capacitance (at discharge moment), nF* | | | Stored energy, kJ | | | Number of cascades |
|---|---|---|---|---|---|---|---|
| | A | B | C | A | B | C | |
| 300 | 13.0 | 26.0 | 60.0 | 0.6 | 1.2 | 2.7 | 3 |
| 400 | 10.0 | 20.0 | 45.0 | 0.8 | 1.6 | 3.6 | 4 |
| 500 | 8.0 | 16.0 | 36.0 | 1.0 | 2.0 | 4.5 | 5 |
| 600 | 6.7 | 13.4 | 30.0 | 1.2 | 2.4 | 5.4 | 6 |
| 700 | 5.7 | 11.4 | 25.7 | 1.4 | 2.8 | 6.3 | 7 |
| 800 | 5.0 | 10.0 | 22.5 | 1.6 | 3.2 | 7.2 | 8 |
| 900 | 4.4 | 8.8 | 20.0 | 1.8 | 3.6 | 8.1 | 9 |
| 1000 | 4.0 | 8.0 | 18.0 | 2.0 | 4.0 | 9.0 | 10 |
| 1100 | 3.6 | 7.3 | 16.4 | 2.2 | 4.4 | 9.9 | 11 |
| 1200 | 3.3 | 6.7 | 15.0 | 2.4 | 4.8 | 10.8 | 12 |

*Note: Capacitance of cascade in versions A, B, and C are, respectively, 40, 80, and 180 nF.

## Table A.14 Betatrons [175, 176]

| Model | Max. energy of electrons, MeV | Power con sumption, kVA | Mass of radiator, t | Country or company of manufacture |
|---|---|---|---|---|
| B5-25M | 25 | 20 | | Kuibyshev electroengineering plant (Moscow, USSR) |
| PMB-6M | 6 | 2 | 0.1 | Nuclear Physics Institute (Tomsk, USSR) (NPI) |
| MIB-4 | 4 | 2 | 0.045 | NPI |
| B-18-30 | 18 | 18 | 0.7 | NPI |
| B-25-10 | 25 | 12 | 2.5 | NPI |
| B-35/8 | 35 | 30 | 4 | NPI |
| BC-8-25 | 25 | 30 | 4.2 | NPI |
| BC-9-30 | 30 | 30 | 4.2 | Radioscopy Institute (Tomsk, USSR), (RI) |
| B-32/6 | 32 | 40 | 5 | RI |
| | 18 | 40 | 1.9 | F. R. Germany |
| 25RTM | 25 | | 5.2 | United States |
| | 31 | 16 | 5.5 | Brown Boveri, (Switzerland) |
| BT-25 | 25 | 102 | 4.5 | Japan |
| | 19 | 12 | 0.6 | CzSSR |

## Table A.15 Characteristics of linear electron accelerators used for radiography [177]

| Model | Boundary quantum energy of bremsstrahlung, MeV | Diameter of focal spot, mm | Exposure dose power, $m^2/min$ | Type of electromagnetic wave* | Length of accelerating device, mm | Type of SHF generator | Input SHF power, MW | Max. power consumption, kVA | Overall dimensions of radiator unit, m | | | Mass of radiator | Firm designer or country |
|---|---|---|---|---|---|---|---|---|---|---|---|---|---|
| | | | | | | | | | Length | Width | Height | | |
| LUEV-2-200D | 2 | 2 | 2 | TW | 0.4 | K | ~2.25 | 15 | 1.8 | 0.83 | 0.92 | 850 | EFAI (USSR) |
| LUE-5-500D | 5 | 2 | 5 | TW | 1.1 | M | 3 | 40 | 2.88 | 0.83 | 1.16 | 1900 | EFAI |
| LUEF-5-1500D | 5 | 2 | 13.7 | TW | 1.2 | K | ~2.25 | 20 | 2.12 | 0.83 | 0.92 | 900 | EFAI |
| LUEV-10-5000D | 8 | 3 | 41 | TW | 1.2 | K | 4.5 | 25 | 2.12 | 0.83 | 0.92 | 900 | EFAI |
| LUE-8-2000D | 8 | 3 | 20 | TW | 1.1 | M | 5 | 50 | 2.88 | 0.83 | 1.16 | 1900 | EFAI |
| LUE-15-15000D | 18.5 | 3 | 150 | TW | 2.56 | M | 6–7 | 70 | 4.3 | 0.9 | 11.15 | 2450 | EFAI |
| Linatron 200 | 2 | 2 | 2 | SW | 0.1 | M | ~1.6 | 10 | 0.76 | 1.03 | 0.84 | 816 | Varian (U.S.) |
| Linatron 1000 | 6 | 2 | 10 | SW | – | M | ~1.6 | 10 | 1.52 | 0.71 | 0.74 | 900 | Varian |
| Linatron 6000 | 16 | 3 | 60 | SW | 1.4 | K | ~5 | 45 | 2.52 | 1.32 | 1.52 | 3900 | Varian |
| Super XX6000 | 8 | 3 | 60 | TW | 2.3 | K | ~5 | 65 | – | – | – | 3000 | Brown-Boveri (Switzerland) |

*SW = standing wave; TW = traveling wave; K = klystron; M = Magnetron.

**Table A.16 Characteristics of linear electron accelerators intended for medical applications* [177, 178]**

| Model | Electron energy, MeV | Type of accel-erating wave | SHF generator: Wave length, m | SHF generator: Type | SHF generator: Power, MW | Turn angle of acceler-ated elec-tron beam, degree | Size of radia-tion zone at distance of 1 m from target, cm | Firm (designer, country) |
|---|---|---|---|---|---|---|---|---|
| LUE-25 | 10–25 | TW | 6 | K | 20 | 90 | 18 × 18 | NIIEFA |
| Lue-15 | 10–20 | TW | 2.6 | M | 9 | 270 | 30 × 30 | (USSR) |
| Clinac 4 | – | SW | 0.3 | M | 2 | – | 30 × 30 | |
| Clinac 6X | – | SW | 0.3 | M | 2 | – | 30 × 30 | |
| Clinac 12 | 6–12 | SW | 1 | M | 2 | 270 | 35 × 35 | Varian |
| Clinac 18 | 6–18 | SW | 1.4 | K | 5 | 270 | 35 × 35 | (U.S) |
| Clinac 35 | 7–28 | TW | 2.25 | K | 20 | 57 or 90 | 35 × 35 | |
| Clinac 2500 | 6–28 | SW | 1.9 | K | 5.5 | 270 | 40 × 40 | |
| SL 75-5 | – | TW | 1.25 | M | 2 | 90 | 40 × 40 | Philips/MEL |
| SL 75-10 | 4–10 | TW | 2.25 | M | 2 | 90 | 30 × 30 | (U.K.) |
| SL 75-20 | 5–20 | TW | 2.5 | M | 5 | 90 | 30 × 30 | |
| ML-4M | – | SW | 0.21 | M | 2 | – | 30 × 30 | Mitsubishi |
| ML-15M11B | 8–15 | TW | 1.7 | K | 5 | 110 | 30 × 30 | (Japan) |
| LMR-4 | – | SW | 0.31 | M | 2 | – | 40 × 40 | |
| LMR-13 | 8–13 | TW | 1.6 | M | 4.8 | 105 | 30 × 30 | Toshiba (Japan) |
| LMR-15 | 8–16 | TW | 1,73 | M | 4.8 | 105 | 30 × 30 | |
| Saturne-10 | 6–10 | SW | 1.2 | M | 2 | 262 | 40 × 40 | CGR-MeV (France) |
| Saturne-25 | 7–25 | TW | 2.3 | K | | | 40 × 40 | AECL (Canada) |
| Mevatron (XII) 74 | 5–11 | SW | 1.3 | M | 2 | 270 | 40 × 40 | ARCO |
| Mevatron (XX) 77 | 3–18 | SW | 1.3 | M | 7 | 270 | 40 × 40 | (F.R. Germany) |
| Dynaray 4 | – | TW | 0.75 | M | 2 | 90 | 30 × 30 | Radiation |
| Dynaray 10 | 4–10 | TW | 2.25 | M | 2 | 266 | 35× 35 | Dynamics |
| Dynaray 18 | 6–18 | TW | 2 | K | 5 | 266 | 30 × 30 | (US) |
| Therapi 4 | – | SW | 0.3 | M | 2 | – | 40 × 40 | SHIN/EMI |
| Therapi 400 | – | SW | 0.3 | M | 2 | – | 40 × 40 | (U.K.) |

*Note: Designations are as for Table A.15. Tables A.15 and A.16 give specifications of specialized installations based on linear electron accelerators. Some such accelerators find application especially in radiation chemical facilities, for sterilization, etc. Basic performance characteristics of linear electron accelerators ELU-4 and ELU-6 are given in Section 7.4.

## Table A.17 Characteristics of some microtrons

| Energy, MeV | Pulse current, MA | Number of orbits | Diameter of poles, cm | Mass of magnet, t | User (Place of installation) | Reference |
|---|---|---|---|---|---|---|
| 15 | 35 | 17 | 75 | 0.9 | Institute of Physical Problems, Acad. Sci. USSR (Moscow) | [111] |
| 32 | 50 | 30 | 110 | 5 | | [111] |
| 13.6 | 20 | 21 | 90 | 1.6 | Saratov regional oncology dispensary (Engelstown, USSR) | |

# REFERENCES


1. J. D. Lawson, *The Physics of Charged Particle Beams*, Clarendon Press, Oxford 1977.
2. E. A. Abramyan, B. A. Alterkop, G. D. Kuleshov, *Intense Electron Beams*, Energo-atomizdat, Moscow, 1984.
3. R. B. Miller, *An Introduction to the Physics of Intense Charged Particle Beams*, Plenum Press, New York, 1982.
4. Y. S. Ryabukhin, A. V. Shalnov, *Accelerated Beams and Their Application*, Atomizdat, Moscow, 1980.
5. M. Ardenne, *Tabellen zur Angewandten Physik*, VEB, Deut. Verlag. d. Wiss., Berlin, 1962.
6. S. M. Settzer, M. J. Berger, Transmission and Reflection of Electrons by Foils, *Nucl. Instrum. Methods*, vol. 119, pp. 157–159, 1974.
7. A. Charlesby, *Atomic Radiation and Polymers*, Pergamon Press, Oxford, 1960.
8. I. K. Kikoin, ed., *Tables of Physical Data, Handbook*, Atomizdat, Moscow, 1976.
9. V. V. Gromov, *Electric Charge in Materials under Radiation*, Energoizdat, Moscow, 1982.
10. C. Kuntz, *Synchrotron Radiation: Techniques and Applications*, Springer-Verlag, Berlin, 1979.
11. G. N. Kulipanov, A. N. Skrinskiy, Synchrotron Radiation and Prospects of Their Application, *Vestn. Acad. Sci., USSR*, no. 8, pp. 46-61, 1978.
12. A. K. Pikaev, *Modern Radiation Chemistry*, Nauka, Moscow, 1985.
13. E. A. Borisov, F. Z. Raichuk, G. V. Shiryaeva, Technology of Radiochemical Production of Concrete-Polymer and Wood-Plastic Materials, *Moscow, Energoizdat, Ser. Radiat. Khim. Tech.* Vol. 12, 1982.
14. E. E. Finkel, R. P. Braginskiy, *Heat Stability Cables and Wires with Radiation-Modified Insulation*, Energiya, Moscow, 1976.
15. E. E. Finkel, R. P.Braginskiy, S. S. Leshcenko, *Stabilization of Radiation Modified Polymers*, Khimiya, Moscow, 1976.
16. N. Sagane, H. Narayana, Plastic Foam, *3rd Int. Meet. Radiat. Proc.*, Tokyo, 1980.
17. E. E. Finkel, V. L. Karpov, S. M. Berlyant, Technology of Radiation Modification of Polymers, *Moscow, Energoatomizdat Ser. Radiat. Khim. Tech.* vol. 15, 1983.
18. K. Veno, K. Kojima, N. Clear, Flame Retardant, Radiation Cross-linked Polyolefine Heat-Shrinkable Tubing, *Radiat. Phys. Chem.*, vol. 25, no. 4–6, pp. 867–877, 1985.

19. V. N. Shubin, S. A. Brusentsova, G. K. Nikonorova, *Radiation-Polymerization Treatment of Industrial Sewage,* Atomizdat, Moscow, 1979.
20. E. B. Juravskaiy, G. G. Degtyareva, et.al., Two-stage Method of Vulcanization of Rubber Technical Articles with Electron Accelerator, *Kauch. Rezina,* no. 2, pp. 24–27, 1977.
21. T. F. O'Brien, Electron Beam Curing of Intaglio Inks, Radiat. Phys. Chem. vol. 25, no. 4–6, pp. 609–616, 1985.
22. G. N. Piyankov, V. V. Chlapatskaya, Y. F. Gologniy, et. al., Technology Parameters and Strength Properties of Wound, Fibre-Glass Materials Cured by Radiation Method, *Tekhnol. Sudostroen.*, no. 1, pp. 56–58, 1977.
23. K. Kawamura, T. Katayama, K. Kawamura, The Pilot Plant Experiment of Electron Beam Irradiation Process for Removal of $NO_x$ and $SO_x$ from Sinter Plant Exhaust Gas in the Iron and Steel Industry, 3rd. Int. Meet. Radiat. Proc., Tokyo, 1980.
24. R. V. Jagatspanyan, V. I. Kosorotov, M. T. Filippov, *Introduction in Radiation—Chemistry Technology*, Atomizdat, Moscow, 1979.
25. P. Harteck, Radiation Chemistry of the Fixation of Nitrogen, *Sciences,* vol. 146, pp. 30–35, 1964.
26. V. V. Rachinsky, *Course of Atomic Technic Principles for Agriculture,* Atomizdat, Moscow, 1978.
27. A. H. Aidus, *Physical-Chemical Principles of Radiation Biological Processes and Radiation Shielding,* Atomizdat, Moscow, 1979.
28. W. Bögl and L. Heide, Chemiluminescence Measurements as an Intensification Method for Gamma-Irradiated Foodstuffs, *Radiat. Phys. Chem.*, vol. 25, no. 1–3, pp. 173–186, 1985.
29. J. Masefield, Factors Influencing the Growth of Radiation Sterilization in North America, *Abstract Book 5th Int. Meet. on Radiation Processing*, San-Diego, Calif. p. 93, 1984.
30. E. S. Josephson, Food Irradiation and Sterilization, *3rd Int. Meet. Radiat. Proc.,* Tokyo, 1980.
31. H. Leboutet, J. Aucouturier, Theoretical Evaluation of Induced Radioactivity in Food Products by Electron or X-ray Beam Sterilization, *Trans. 5th Int. Meet. Rad. Proc.,* San Diego, 1984, Radiat. Phys. Chem., vol. 25, no. 1–3, pp. 233–242, 1985.
32. E. S. Pertsovskiy, E. V. Sakharov, V. A. Dolinin, *Application of Radionuclides and Radiation in Food Industry,* Atomizdat, Moscow, 1980.
33. N. M. Berezina, D. A. Kanshanskiy, *Presowing Irradiation of Agriculture Plants,* Atomizdat, Moscow, 1975.
34. K. Otsuhata, M. T. Razzak, R. L. Castanares, Y. Tabata, Effect of Surface Texture of Grafted Films on Antithrombogenicity, *Trans. 5th Int. Meet. Radiat. Proc.*, San Diego, Calif. 1984, Radiat. Phys. Chem., vol. 25, 4–6, pp. 537–548, 1985.
35. A. Suess, T. Lessel, Radiation Treatment of Sewage Sludge—Experience with an Operating Pilot Plant, *Radiat. Phys. Chem.*, vol. 9, no. 3–6, pp. 353–365, 1977.
36. N. N. Rykalin, J. V. Zuev, A. A. Uglov, *Principles of Electron-Beam Treatment of Metals*, Mashinostroeniye, Moscow, 1978.
37. H. Schvarts, *Mechanism of Penetration in Metal of High Density Electron Beams, Electronno i Ionno-Luchevaya Tekhnologiya,* Metallurgia, Moscow, 1968.
38. S. Schiller, V. Heisig, S. Panzer, *Electronens-trahltechnologie,* VEB Verlag Technik, 1976.
39. B. E. Paton, O. K. Nazarenko, V. I. Shalov, Peculiarities of Apparatus and Processes of Electron-Beam Welding and Cutting in Space, Avtomat. Svarka, no. 3, pp. 3–8, 1971.
40. A. I. Chvertko, O. K. Nazarenko, A. M. Svyatskiy, A. I. Nekrasov, *Equipment for Electron-Beam Welding,* Naukova Dumka, Kiev, 1973.
41. S. Schiller, U. Heisig, G. Jasch. Verhalten von Electronenstralen in vorvakuum, Schweisstechnik, no. 21, S. 11–16, 1971.

42. J. F. Lowry, B. W. Schumacher, Extended Working Range for Electron Beams in the Atmosphere, *Nucl. Instrum. Methods,* no. 130, pp. 577–596, 1975.
43. V. A. Gorbunov, N. K. Kuksanov, R. A. Salimov, Peculiarities of Heating of Metallic Plate by Concentrated Electron Beam with Energy of 800–1500 keV and Power up to 70 kW, *Svar. Proizvod.* vol. 1, pp. 10–12, 1979.
44. A. Visser, Strahltechnik 2, DVS-Berichts 4. Deut. Verlag Schweisstechnik GmB H, Dusseldorf, 1968, S. 173–216.
45. A. N. Skrinsky, V. G. Mizin, L. P. Fominskiy, High Productive Overlaying and Fusion of Power Coatings by Relativistic Electron Beam, *Dokl. Acad. Sci., USSR,* vol. 283, no. 4, pp. 865–869, 1985.
46. B. W. Schumacher, G. R. Taylor, Rock Breakage by Means of Electron Beam Piercing (Laboratory Tests), *Rep. 10th Symp. on Electron, Ion and Laser Beam Technology,* San Francisco Press, pp. 271–284, 1969.
47. F. Winterberg, The Potential of Electric Cloud Modification by Intense Relativistic Electron Beams, Z. Meteorol., vol. 25, H 3, S. 180–191, 1975.
48. B. V. Alekseev, *Mathematic Kinetics of Reacted Gases,* Nauka, Moscow, 1982.
49. V. M. Ievlev, A. S. Koroteev, Extraction in Atmosphere and Investigation of High Power Stationary Electron Beams, *Izv. Acad. Sci. USSR, Ser. Energ. Transp.* no. 3, pp. 3–13, 1981.
50. E. P. Velikhov, V. S. Letohov, Laser Atomic-Molecular Technology, Nauka i Chelovechestvo, Megdunorodniy ejegodnic, Znanie, Moscow, pp. 281–297, 1982.
51. S. A. Pletniov, *Lasers in Clinical Medicine,* Medicina, Moscow, 1981.
52. E. A. Abramyan, *Industrial Electron Accelerators,* Moscow, preprint IVTAN, 1984, N 6–137.
53. A. K. Valter, *Electrostatic Charged Particle Accelerators,* Gasatomizdat, Moscow, 1963.
54. R. G. Herb, Pelletron Accelerators for Very High Voltage, *Nucl. Instrum. Methods,* vol. 122, pp. 267–276, 1974.
55. W. Herchenbach, Hochspannung-serzeugung durch Ladungstransport auf rotierenden Isolatorflachen., *Angew. Phys.,* vol. 7, no. 1–6, pp. 32–43, 1955.
56. E. E. Charlton, W. F. Westendorp, Industrial X-ray Unit, *Electronics,* vol. 16, no. 12, pp. 128–135, 1944.
57. B. I. Albertinsky, 1MV X-ray Flaw Detector, *Electrofizicheskaya Apparatura,* Atomizdat, Moscow, vol. 10, pp. 93–96, 1973.
58. E. A. Abramyan, On Possibilities of Transformer Type Accelerators, Nucl. Instrum. Methods, vol. 59, no. 1, pp. 22–28, 1968.
59. E. A. Abramyan, V. A. Gaponov, S. B. Vasserman, G. S. Krajnov, Charged-particle accelerator, SU patent no. 224712, Filed 27.07.1965, U.S. patent no. 3450996, DE patent no. 1564166.
60. G. J. Rohwein, I. A. Trace, Transformerchanged Electron Beam Generator, *IEEE Trans. Nucl. Sci.,* NS-22, no. 3, pp. 1013–1015, 1975.
61. G. A. Mesiats, S. A. Ivanov, N. I. Komijak, E. A. Peliks, *Heat Power Nanoseconde Bremsstrahlung Pulse,* Energoatomizdat, Moscow, 1983.
62. A. A. Egorov, V. S. Panasyuk, V. M. Radchenko, Supply and Control of Tesla Transformer Type Accelerator, *Prib. Tech. Eksp.,* no. 2, pp. 18–23, 1968.
63. Yu. V. Kiselev, V. P. Cherepanov, *Spark gaps,* Sovetskoje Radio (Ser. Elementy radioelectronnoj apparatury), Moscow, 1976.
64. E. A. Abramyan, V. A. Gaponov, A source of accelerated electrons of uniform energy, SU patent no. 208144, Filed 15.06, 1964. U.S. Patent no. 3390303, DE Patent, no. 1514036, JP Patent no. 501942.
65. E. A. Abramyan, V. A. Gaponov, High-Current Accelerator of the Base of the Transformator, *At. Energ.,* vol. 20, no. 5, pp. 385–392, 1966.
66. D. M. Margolin, L. A. Vasil'ev, Apparatus RUP-400-5 as a Radiation Source for Radia-

tion-chemical Research, *Probl. Fiz. Khim.*, vol. 3, Goskhimizdat, Moscow, pp. 134–139, 1963.

67. E. A. Abramyan, Pulse Transformer with Shock Excitation, SU Patent no. 949856, Filed 6.09.1978, US Patent no. 4409492.
68. Three High-Power Accelerators for Radiation Processing, *Nucleonics*, vol. 18, no. 8, pp. 54–55, 1960.
69. G. I. Budker, V. A. Gaponov, B. M. Korabel'nikov, Charge-Particle Accelerator, SU Patent no. 589698, Filed 18.05.1973, US Patent no. 4016499, GB Patent no. 1454485.
70. V. L. Auslender, R. A. Salimov, Electron Accelerators for National Economy Produced by Novosibirsk Nuclear Physics Institute, *At. Energ.*, vol. 44, no. 5, pp. 403–408, 1978.
71. V. V. Akulov, M. P. Svin'in, High Voltage transformer, SU Patent No. 52346 Filed 5.07.1969. Bulletin of Inventions, no. 28, (1976).
72. B. I. Albertinskiy, M. P. Svin'in, *Cascade Generators*, Atomizdat, Moscow, 1980.
73. Accelerators, Instrumentelle Hilfsmittel der Kernphysik, in *Handbuch der Physik, Herousgegeken von S. Flugge*, vol. 64, Springer-Verlag, 1959.
74. W. Heilpern, Kaskadengeneratoren zur Partikelbeschleuning auf 4 MeV, *Helv. Phys. Acta*, vol. 28, no. 5–6, pp. 485–491, 1955.
75. B. I. Albertinskiy, I. V. Kuritsina, O. F. Nikonov, O. B. Ovchinnikov, *Prib. Tekh. Eksp.*, no. 3, pp. 43–46, 1971.
76. M. Schenkel, Eine Neune Schaltung fur die Erzugung hoher Gleichspannungen, *Electrotech. Z.*, vol. 40, no. 28, pp. 333–334, 1919.
77. M. R. Cleland, K. M. Morganstern, Dynamitron-A High-Power Electron Accelerator, Nucleonics, vol. 18, 8, pp. 52–53, 1960.
78. M. R. Cleland, C. C. Thompson, *RDI TIS 79–1*, Melville, N.Y., 1979.
79. I. M. Dmitrieva, I. V. Gruzgal'skij, A. K. Evseev, et al., Problems of Standardizing High-Voltage Electron Accelerators, *Proc. 2nd All-Union Meeting on Use of Charged-Particle Accelerators in the National Economy (Leningrad, 1975)*, vol. 1, NIIEFA, pp. 142–157, 1976.
80. G. I. Budker, Charged Particle Accelerators for National Economy Produced by Novosibirsk Nuclear Physics Institute, *Proc. 2nd All-Union Meeting on Use of Charged Particle Accelerators in the National Economy (Leningrad, 1975)*, vol. 1, NIIEFA, pp. 48–75, 1976.
81. E. A. Abramyan, H. F. Barakaev, et al., 300 keV High Current Electron Accelerator, *Proc. 2nd All-Union Meeting on Charged-Particle Accelerators, Dubna, 1978*, vol. 2, pp. 120–124,
82. E. Marx (Brannschweig), Verfahren zur Schlagprufung von isolatoren und anderen elektrischen Vorrichtungen, Patent Germany no. 455933.
83. E. A. Abramyan, S. B. Vasserman, V. M. Dolgyshin, et al., Short Pulse Hard X-ray Generator of High Intensity, *Dokl. Acad. Sci.*, vol. 192, n. 1, pp. 76–77, 1979.
84. V. L. Auslender, A. A. Livshic, A. D. Panfilov et al., Industry ILU Type Accelerator, *Proc. 2nd All-Union Meeting on Use of Charged-Particle Accelerators in the National Economy (Leningrad, 1982)*, vol. 1, NIIEFA, pp. 25–31, 1982.
85. V. F. Romanovskiy, V. S. Panasyuk, B. M. Stepanov, X-ray Generator, SU Patent no. 748926, Filed 9.01.1978, US Patent no. 4120813.
86. A. M. Ovcharov, V. F. Romanovsky, B. M. Stepanov, et al., X-ray Unit RAPS-1 with HF Electron Accelerator, *Defectoskopiya*, no. 9, pp.94–97, 1980.
87. E. A. Abramyan, A. A. Kurkov, B. A. Shternin, Method of Manufacturing of Multilayer High Voltage Cone Welding, Patent no. 657471 USSR, Bulletin of Inventions no. 14, 1979, Filed 21.12.1978.
88. A. I. Poltev, Dosing and Calculation of High Voltage $SF_6$-Apparatus, Energiya, Leningrad, 1979.

89. L. V. Chepel, Some Problems of Using High-Voltage Alternating Transmission Lines in Industry Radiation Processes, *At. Energ.*, vol. 30, no. 6, p. 448, 1971.
90. E. A. Abramyan, V. V. Vecheslavov, V. A. Gaponov, A. N. Rezakov, High-Intensity Electron Beams in Accelerating Tubes with Strong Focusing, *At. Energ.*, vol. 25, no. 2, p. 135, 1968.
91. E. A. Abramyan, G. D. Kuleshov, Accelerating Tube, SU Patent no. 657471, Filed 11.03.1976, Bulletin of Inventions no. 28, 1981.
92. S. A. Ivanov, X-ray Tubes for Science and Industry, Obsory po elektronnoj tekhnike, Moscow, izd. CNII Elektronika, 1982, (Ser. 4, Elektrovakuumnye i gazorasriydnye pribory, vol. 1).
93. E. A. Abramyan, G. D. Kuleshov, Linear Induction Accelerator, Patent no. 865115, Bulletin of Inventions 1983, no. 25, Filed 12.05.1980.
94. E. A. Abramyan, G. I. Budker, G. B. Glagolev, A. A. Naumov, Betatron with Spiral Accumulation of Electrons, *Zh. Tech. Fiz.*, vol. 35, no. 4, pp. 605–611, 1965.
95. A. I. Pavlovskiy, G. D. Kuleshov, A. I. Gerasimov, et al., Investigation of Electron Beam Injection in Betatron, *Zh. Tech. Fiz.*, vol. 47, no. 2, pp. 370–372, 1977.
96. E. A. Abramyan, L. N. Bondarenco, V. I. Volosov, et al., Magnetic Screens Transparented for Induction Electric Filed, *Prib. Tekn. Exp.*, no 1, pp. 178–181, 1965.
97. Yu. P. Vakhrushin, A. I. Anatskiy, *Linear Induction Accelerators*, Atomizdat, Moscow, 1978.
98. N. C. Christofilos, High Current Linear Induction Accelerator for Electrons, *Rev. Sci. Instrum.*, vol. 34, no. 7, pp. 886–890 1964.
99. V. S. Bosamykin, A. I. Pavlovskiy, Linear Induction Accelerators, SU Patent no. 270913, Bulletin of Inventions 1971, no. 34, Filed 10.08.1968.
100. E. A. Abramyan, Charged-Particle Accelerator, SU Patent no. 273893, Bulletin of Inventions, 1976, no. 9, Filed 12.11.1968.
101. A. I. Pavlovskiy, V. S. Bosamykin, V. A. Savchenko, High-Power Electron Accelerator LIU-101, *Dokl. Acad. Sci. USSR*, vol. 25, no. 5, pp. 1118–1122, 1980.
102. Yu. P. Vakhrushin, Linear Electron Accelerators for Industry and Medicine, *At. Energ.*, vol. 55, no. 6, pp. 429–433, 1983.
103. M. C. Laguros-Solar, S. M. Matthews, Radioscope and Electric Accelerator Sources for Food Irradiation, *Trans. 5th Int. Meet. Radiat. Proc. San Diego, Calif.*, 1984.
104. E. A. Abramyan, G. I. Budker, A. A. Naumov, Method of Forming of Alternating Magnetic Field for Cyclic Charged-Particle Accelerator, SU Patent no. 168809, Bulletin of Inventions 1965, no. 5, Filed 8.04.1963.
105. A. L. Zvontsov, V. L. Chakhlov, A. A. Filimonov, Electro-Magnet of Betatron with Azimuth Variation of Control Magnetic Field, *Prib. Tech. Eksp.*, no. 2, pp. 40–43, 1975.
106. A. I. Pavlovskiy, G. D. Kuleshov, Pulse Ironless Betatrons, Electron Accelerators, *Proc. 7th Inter-High School Electron Accelerators Conference Tomsk 1968*, Atomizdat, Moscow, vol. 2, pp. 5–8, 1975.
107. A. N. Lebedev, A. V. Shal'nov, *Principles of Accelerator Physics and Technique*, vol. 2, Energoatomizdat, Moscow, 1983.
108. S. P. Kapitsa, V. N. Melekhin, *Microtron*, Nauka, Moscow, 1963.
109. Linear Accelerator "Elektronika", Advertisement.
110. H. Lebouter, J. Anconturier, An Electron Accelerator for Irradiation "Cassitron", *Trans. 4th Int. Meet. Radiat. Proc., Dubrovnik, Yugoslavia*, pp. 403–408, 1982.
111. A. N. Lebedev, A. V. Shal'nov, *Principles of Accelerator Physics and Technique*, vol. 3, Energoatomizdat, Moscow, 1983.
112. S. P. Bugaev, Yu. E. Kreindel, P. M. Schmanin, Big Cross-Section Electron Beam, Energoatomizdat, Moscow, 1984.
113. S. P. Dmitriev, A. S. Ivanov, M. N. Svin'in, Installation for Radiation Treatment by Electron Beam, Patent no. PCT (SU80/00065), no. WO 081/03104.

114. O. E. Gusev, S. P. Dmitriev, A. S. Ivanov, et al., Small-Size System for Forming of Accelerated Electron Radiation Field, *Proc. 4th All-Union Meeting on Use of Charged-Particle Accelerators in the National Economy (Leningrad, 1982)*, NIIEFA, vol. 2, pp. 89–93, 1982.
115. D. S. Val'man, V. P. Ovchinnikov, Yu. B. Rusak, Supply System for Scanning Magnets, *Proc. 4th All-Union Meeting on Use of Charged-Particle Accelerators in the National Economy (Leningrad, 1982)*, NIIEFA, vol. 2, pp. 94–97, 1982.
116. E. A. Abramyan, Y. S. Simakhov, I. L. Chertok, Installations for Grain Disinsectization, *Proc. All-Union Meeting on Use of Charged-Particle Accelerators in the National Economy, (Leningrad, 1971)*, p. 2-8.
117. V. L. Auslender, N. K. Kuksanov, V. A. Poliyakov et al., Electron Beam Extraction System with Circular Radiation Field, *Proc. 3rd All-Union Meeting on Use of Charged-Particle Accelerators in the National Economy (Leningrad, 1979)*, NIIEFA, vol. 1, pp. 126–130, 1979.
118. Yu. N. Timko, L. L. Akrachkova, E. E. Finkel et al., Electromagnetic System Application for Optimal Radiation Field Topography Formation for Cable Irradiation by Electron Acelerators, *Proc. 4th All-Union Meeting on Use of Charged-Particle Accelerators in the National Economy (Leningrad, 1982)*, NIIEFA, vol. 2, pp. 3–10, 1982.
119. E. A. Abramyan, I. L. Chertok, Extraction of an Accelerated Electron Beam from a Vacuum, *Prib. Tekh. Eksp.*, no. 6, pp. 164–165, 1969.
120. W. E. Pauli, Uber line neue Entladungsrohre, die den Austritt aller in der Rohre entstehenden Strahlen in die atmospharische Luft gestatter, *Zeitschrift*, vol. 21, no. 1, pp. 11–14, 1920.
121. S. T. Sinitsyn, Corpuscle Ray Extraction from Open Window Tube, Zl. Eksp. Teor. Fiz., vol. 33, no. 4, pp. 277–282, 1933.
122. V. A. Gorbunov, N. K. Kuksanov, R. A. Salimov, et al., Extraction in Atmosphere of 60 kW Concentrated Electron Beam of Accelerator ELV-4, *Proc. 3rd All-Union Meeting on Use of Charged-Particle Accelerators in the National Economy (Leningrad, 1979)*, NIIEFA, vol. 1, pp. 122–125, 1979.
123. B. W. Schumacher, J. F. Lowry, R. C. Smith, High Power Electron Beam in Atmosphere, *Proc. 4th Int. Electron Beam Proc. Symp.*, 1976.
124. E. A. Abramyan, G. D. Kuleshov, Method of Welding by Fusion in the Atmosphere, SU Patent no. 904942, Bulletin of Inventions, 1982, no. 6, Filed 17.01.1978, GB Patent no. 2886289, FR Patent no. 2492708.
125. E. A. Abramyan, I. V. Liyashenko, V. M. Radchenko, I. L. Chertok, Installation for Charged-Particle Beam Generation in Atmosphere, SU Patent no. 403370, Bulletin of Inventions, 1984, no. 28, Filed 1.06.1970.
126. Drehanoden-Röntgenrohren (Advertisement of Firm Rohrenwerk).
127. V. V. Klyuev, Bremsstrahlung Technology, vol. 1, Mashinostroeniye, Moscow, 1980.
128. J. P. Farell, High-Power Bremsstrahlung Sources for Radiation Sterilization, *Trans. 2nd Int. Meet. Radiat. Proc., Miami*, 1978.
129. A. K. Pikaev, *Dosimetry in Radiation Chemistry*, Nauka, Moscow, 1975.
130. V. V. Generalova, M. N. Gurskiy, *Dosimetry in Radiation Processing*, Izd. Standartov, Moscow, 1981.
131. W. L. McLaughlin, R. M. Uribe, A. Miller, Megagray Dosimetry, *Trans. 4th Int. Meet. Radiat. Proc., Dubrovnik, Yugoslavia*, 1982.
132. E. E. Finkel, R. P. Braginskiy, Radiation Technology in Cable Industry, Elektrotekh. Promyshlennost' Ser. Kabel'naya Tekh., vol. 9, no. 91, pp. 26–30, 1972.
133. E. A. Abramyan, E. E. Finkel, Electron Beams in Cable Engineering, *Trans. 2nd Int. Meet. Radiat. Proc., Miami*, 1978.
134. E. E. Finkel, G. I. Meshchanov, Status and Prospects of Electron Accelerator Uses in Cables and Wire Manufacturing, *Proc. 3rd All-Union Meeting on Use of Charged-Particle Accelerators in the National Economy (Leningrad, 1979)*, NIIEFA, vol. 2, pp. 29–44, 1979.

135. V. V. Krayushkin, Mathematic Simulation of Radiochemical Installation with Electron Accelerators for Modular Unagitated System Irradiation, *Proc. 3rd All-Union Meeting Use of Charged-Particle Accelerators in the National Economy (Leningrad, 1979)*, NIIEFA, vol. 2, pp. 219-225, 1979.
136. Advances in Radiation Processing, *2nd Intern. Meet. Radiat. Proc., Miami, 1978, Rad. Phys. and Chem.*, vol. 14, No. 1–6, 1979.
137. L. V. Chepel, *Application of Accelerators in Radiation Chemistry*, Moscow, Atomizdat, 1975.
138. S. Ota, Current Status of Irradiated Heat-Shrinkable Tubing in Japan, *Proc. Full Text of Invited Papers, 3rd Int. Radiat. Proc. Tokyo*, pp. 73–81, 1980.
139. S. V. Nablo, E. P. Tripp, *Low Energy Electron Process Applications*, Energy Sci., TR-54, 1976.
140. W. A. Frutiger, S. V. Nablo, Performance Characteristics of High Power, Multiple Gun, Electrocurtains, *Trans. 5th Int. Meet. Radiat. Proc. Dubrovnik, Yugoslavia, Oct. 1982, Radiat. Phys. Chem.*, vol. 22, no. 3–5, pp. 431–440, 1983.
141. W. A. Frutiger, S. V. Nablo, High Performance Electron Processing Systems, Energy Sci., TR-124, 1984.
142. V. E. Pospelov, V. I. Troenko, V. P. Koriyakin et al., Radiochemical Installation with Accelerator "Electron-3" for varnished curing, *Lakokrasoch. Mater. Ikh. Primen.* No. 4, pp. 62–63, 1974.
143. G. D. Shiryaeva, Yu. D. Koslov, *Electron Beam Curing Technology*, Atomizdat, Moscow, (Ser. Radiat Khim, Tekh.), 1980.
144. G. R. Berbeko, S. V. Nablo, Electron Beam Curing, *Paint Varnish Production*, no. 8, pp. 39–43, 1974.
145. A. Danno, Modification of Plastics by Cross-Linking and Grafting, At. Energ. Rev., vol. 9, no. 2, pp. 399–422, 1971.
146. V. K. Burov, B. M. Van'yushkin, A. E. Voskoboev et al., Pilot Installation for Radiochemical Treatment of Fabrics, *Proc. 2nd All-Union Meeting on Use of Charged-Particle Accelerators in the National Economy (Leningrad, 1975)*, NIIEFA, vol. 1, pp. 391–397, 1976.
147. M. P. Svin'in, K. D. Pismannik, Electron Accelerators for Radiation Modification of Textiles, *Proc. 4th All-Union Meeting on Use of Charged-Particle Accelerators in the National Economy (Leningrad, 1982)*, NIIEFA, vol. 1, pp. 16–24, 1982.
148. V. B. Osipov, S. V. Mamiconayn, G. D. Stepanov et al., Commercial Units for Radiation Sterilizing Medical Supplies, Sterilization by Ionizing Radiat. Proc. Int. Conf., Vienna, Austria, April 1974, Multiscience, Publication Ltd., Montreal, Canada, 1974, pp. 136–144.
149. S. V. Nablo, R. N. Cheever, Low Energy Electron for Pasteurization and Aseptic Practice, *Radiat. Phys. Chem.*, vol. 14, no. 3, pp. 267–276, 1979.
150. G. A. Zakladnoy, A. I. Men'shenin, E. S. Pertsovskiy et.al., Industry Application of Radiation Disinsectization of Grain, *At. Energ.*, vol. 52, no. 1, pp. 57–61, 1982.
151. I. A. Prudnikov, B. N. Gromov, V. N. Volosatov et al., Chamber for Flaw Inspection of Atomic Power Plant Equipment with Linear Accelerator, *Proc. 3rd All-Union Meeting on Use of Charged-Particle Accelerators in the National Economy (Leningrad, 1979)*, NIIEFA, vol. 4, pp. 169–173, 1979.
152. Das Brown Boveri 31 Mev Betatron fur zerstarungsfreie Werkstoffprufung,, "Brown Boveri".
153. The Asklepitron 35, a 35 MeV Betatron for Modern Radiology, *Brown Boveri Rev.* no. 5, pp. 20–22, 1962.
154. E. B. Bajanov, B. V. Zabrodin, A. P. Kozlov, Formation of Linear Accelerator LUE-25 Electron Beam by Dissipating Filter System, *Proc. 3rd All-Union Meeting on Use of Charged Particle Accelerators in the National Economy (Leningrad, 1979)*, NIIEFA, vol. 3, pp. 31–37, 1979.

155. A. S. Kuz'minskiy, T. S. Fedoseeva, M. Y. Kaplunov, *Technology of Radiation Vulcanization and Elastomer Modification*, Energoizdat, Moscow, 1982, (Ser. Radiat. Khim. Tech.).
156. J. I. Trump, K. A. Wright, A. J. Sinskey, D. Shah, S. Sommer, Prospects for High-Energy Electron Irradiation of Waste-Water Liquid Residuals, *Proc. Symp. Radiat. Clean Environment, Munich, March, 1975*. IAEA-SM-1941503, pp. 343–354.
157. N. N. Tikhodeev, *Electrical Energy Transport*, Energoatomizdat, Moscow-Leningrad, 1984.
158. N. Christoffilos, Focusing System for Ions and Electrons, US Patent no. 2736799, Feb. 28, 1956.
159. E. A. Abramyan, B. A. Alterkop, G. D. Kuleshov, Electron Beam Energy Transport, *Elektrichestvo*, no. 7, pp. 1–6, 1983.
160. R. T. Avery, D. Keefe, T. L. Brekke, I. Finnie, Hard-Rock Tunnelling Using Pulsed Electron Beams, *IEEE Trans Nucl. Sci.*, vol. NS-22, no. 3, pp. 1798–1801, 1975.
161. R. I. Briggs, High Current Electron Linacs (Advanced Test Accelerator/Experiment Test Accelerator/, *Rept. on LINAC, 84 Conf., Darmstadt, June 1984*, pp. 507–509.
162. J. Yonas, Fusion Power and Charged-Particle Beams, *Usp. Fiz. Nauk*, vol. 133, No. 1, pp. 159–180, 1981.
163. V. P. Sarantsev, E. A. Perel'shtein, Collective Heavy Ion Accelerator in Dubna, *Fiz. Plasm.*, vol. 3, no. 3, pp. 449–464, 1977.
164. A. V. Gaponov-Grekhov, M. I. Petelin, Relativistic HF Electronics, Vestn. Acad. Sci., USSR, no. 4, pp. 11-13, 1979.
165. V. A. Glukhikh, M. P. Svin'in, Status and Prospects of High Voltage Accelerator Progress, *Proc. 2nd All-Union Meeting on Use of Charged-Particle Accelerators in the National Economy (Leningrad, 1975)*, NIIEFA, vol. 1, pp. 76–86, 1976.
166. B. P. Murin, G. I. Batskikh, A. A. Kuz'min et al., Industry Electron Accelerator, *Proc. 2nd All-Union Meeting on Use of Charged-Particle Accelerators in the National Economy (Leningrad, 1975)*, NIIEFA, vol. 1, pp. 265–270, 1976.
167. S. V. Rumyantsev, Radiation Flaw Detection, Atomizdat, Moscow, 1968.
168. O. A. Val'dner, Development in Physics Engineering Institute of Serial Linear Electron Accelerators, in G. A. Tiyagunov, ed., *Accelerator*, no. 3, Gosatomizdat, Moscow, 1962
169. A. P. Bykov, V. P. Marvin, L. V. Chepel, F. V. Chemarov, Linear Electron Accelerator for Industry and Science, *All-Union Meeting on Use of Accelerator in National Economy and Medicine (Leningrad, 1971)*.
170. E. G. Komar, Charged-Particle Accelerator Application in National Economy and Medicine, *All-Union Meeting Use of Accelerator in National Economy and Medicine (Leningrad, 1971)*.
171. V. I. Gorbunov, L. G. Krasin, G. V. Titov, Experience and Application of Small-Size Betatron PMB-6 for Non-Destructive Testing of Joint Welds, *All-Union Meeting on Use of Accelerator in National Economy and Medicine (Leningrad, 1971)*.
172. V. A. Tsukerman, Progress of Pulse Bremsstrahlung Technology in USSR, Apparatura i Metody Rentg. Analis., Leningrad, *Machinostroeniye*, vol. 24, pp. 22–23, 1980.
173. V .L. Auslender, R. A. Salimov, A. N. Skrinskiy, G. A. Spiridonov, High-Power Electron Accelerators and Radiation Technology, *Vestn. Acad. of Sci. USSR*, no. 6, pp. 48–57, 1981.
174. M. P. Svin'in, Accelerators for Radiation Technology—Status and Ways of Improvement, *Proc. 4th All-Union Meeting on Use of Charged-Particle Accelerators in National Economy (Leningrad, 1982)*, NIIEFA, vol. 1, pp. 5–15, 1982.
175. B. M. Yakovlev, Activation Analyses of Element Matter Composition Using Bremsstrahlung and Photoneutrons Generated by Electron Accelerator, *Proc. 4th All-Union Meeting on Use of Charged-Particle Accelerators in National Economy (Leningrad, 1982)*, NIIEFA, vol. 4, pp. 204–211, 1982.

176. O. F. Bulaev, V. A. Moscalev, V. L. Chakhlov, Welded Joints—Quality Testing with Betatrons, *10th World Conf. on Nondestructive Testing: Contributed Papers, Moscow, 1982,* pp. 131–138.
177. Y. P. Vakhruchin, I. A. Prudnikov, B. N. Tronov, et al., Flaw Inspection of Atomic Power Plant Elements Using Linear Electron Accelerators, Vop. At. Nauki. Tekh., Ser. Tekh. Fiz. Eksp., vol. 2, no. 14, pp. 39–42, 1983.
178. Y. Y. Kirillin, V. M. Nikolaev, I. A. Prudnikov, Foreign Linear Accelerators for Radiotherapy, Obzor, OV-60, Leningrad, NIIEFA, 1983.
179. V. S. Kozintsev, O. P. Popov, P. E. Smirnov, et al., Microtron for Radiotherapy, *Proc. 2nd All-Union Meeting on Use of Charged-Particle Accelerators in National Economy (Leningrad, 1975), NIIEFA,* vol. 2, pp. 48–54, 1976.
180. Radical-Kur, *Energie,* vol. 37, no. 1/2, pp. 26–32., 1985.
181. D. J. Heldritsh, P. L. Feldman, Flue gas $SO_2/NO_x$ control by combination of dry scrubber and electron beam, *Radiat. Phys. Chem.*, vol. 24, no. 1, pp. 129-143, 1984.
182. K. Yotsumoto, H. Sunaga, S. Tanaka et al. High Power Bremsstrahlung X-Ray Source For Radiation Processing, Report on 6th Intern. Meet. Rad. Proc., Ottawa, Canada, June 1987. To be published in Radiation Physics and Chemistry.
183. J. R. Bayless, R. J. Adler. Linear Induction Accelerators for Radiation Processing, ibid.
184. S. Hashimoto, K. Nishimura, S. Machi. Economic Feasibility of Irradiation-composting Plant of Sewage Sludge, ibid.
185. K. Williams, W. A. Frutiger, J. Hiley, S. V. Nablo. Requirements for Very High Power Electron Beam Systems for Utility Stack Gas Treatment, ibid.

# INDEX